An American National Standard

IEEE Guide for Safety in AC Substation Grounding

Published by
The Institute of Electrical and Electronics Engineers, Inc

ANSI/IEEE
Std 80-1986
(Revision of IEEE
Std 80-1976)

An American National Standard

IEEE Guide for Safety in AC Substation Grounding

Sponsor

The Substation Committee
of the
IEEE Power Engineering Society

Approved March 21, 1985

IEEE Standards Board

Approved July 26, 1985

American National Standards Institute

IEEE Power Engineering Seminars

The IEEE sponsors seminars on the Color Books and other power engineering standards throughout the year.
Our seminars include:

- Protection and Co-Generation Plants Paralleled with Utility Transmission Systems
- Health Care Facilities Power Systems
- Planning, Design, Protection, Maintenance, and Operation of Industrial and Commercial Power Systems
- Electric Power Supply Systems for Nuclear Power Generating Stations
- Large Storage Batteries — Nickel-Cadmium and Lead

IEEE-sponsored seminars and training programs may also be brought to your plant. For details, write to the IEEE Standards Seminar Manager, 445 Hoes Lane, PO Box 1331, Piscataway, NJ 08855-1331 USA. In the US and Canada call us Toll Free at 1-800-678-IEEE and ask for Standards Seminars and Training Programs. Our fax number is 201-562-1571.

IEEE Training Programs on the NESC

The new 1990 National Electrical Safety Code (NESC) was published by the IEEE ... and if you're an utility engineer, manager, or consultant, you'll need to know what the new Code covers. We offer two intensive training programs to help you:

- A thoroughly updated 3-day training program that covers the development and application of the Code.
- A new 4-day training program that covers the same material as our 3-day program ... plus a special section on work rules.

Special team discounts are available. IEEE-sponsored seminars and training programs may also be brought to your plant. For details, write to the IEEE Standards Seminar Manager, 445 Hoes Lane, PO Box 1331, Piscataway, NJ 08855-1331 USA. In the US and Canada call us Toll Free at 1-800-678-IEEE and ask for Standards Seminars and Training Programs. Our fax number is 201-562-1571.

Corrected Edition

Second Printing
February 1990

ISBN 471-85393-3

Library of Congress Catalog Number 86-081053

The Institute of Electrical and Electronics Engineers, Inc
345 East 47th Street, New York, NY 10017 USA

August 8, 1986

SH10579

Foreword

(This Foreword is not a part of ANSI/IEEE Std 80-1986, IEEE Guide for Safety in AC Substation Grounding.)

This third edition represents the first major revision of this guide since its first issue in 1961. Major modifications include the redefinition of simplified equations for calculating touch and step voltages, changes in safety criteria, and expansion of examples illustrating the use of this guide. Other changes and additions concern a section on gas-insulated substations, introduction of a derating factor for crushed stone surfacing, the effects of ground rods, equations for calculation of grid resistance, current division among available ground paths, sizing of conductors of various materials, and discussion of multilayer soil models. Although the scope of this document is considerably larger than those of the 1961 and 1976 editions, there is a continuity of principles and in the general approach to safe grounding practices. The third edition thus continues to build upon the foundations laid by two earlier working groups, AIEE Working Group 56.1 and IEEE Working Group 69.1.

The work of preparing this standard was done by Working Group 78.1 of the Distribution Substation Subcommittee and was sponsored by the Substation Committee of the IEEE Power Engineering Society. The membership of this working group was as follows:

Jiri G. Sverak, *Chairman* **Donald N. Laird,** *Vice Chairman*

B. Yamin Afshar, *Secretary*

K. A. Alexander
S. J. Arnot
R. Benson
C. J. Blattner
J. B. Cannon
J. W. Cheever
E. F. Counsel
C. F. Dalziel
F. Dawalibi
F. A. Denbrock
W. K. Dick
T. H. Dodds
E. Eldridge
L. N. Ferguson
H. H. Ferris, Jr
D. L. Garrett
R. J. Heppe
J. E. Idzkowski
D. T. Jones
E. F. Jones
W. R. Jones
R. P. Keil
F. W. Knorr
A. E. Kollar
K. P. Lau
R. H. Lee
W. C. McCart
T. S. McKlenahan
A. P. Meliopoulos
J. L. Mixon, Jr
D. Mukhedkar
J. M. Nahman
J. T. Orrell
S. G. Patel
G. J. Paul
A. B. Purdy
E. S. Raila
W. B. Raley
J. Ratkowski
G. E. Smith
W. K. Switzer
R. Verma
L. F. Volf, Jr
R. P. Webb
J. Wehling
B. W. Wray
L. G. Zukerman*

*Deceased

Those individuals who made the bulk of technical and editorial contributions to this revision were:

B. Y. Afshar
J. B. Cannon
F. Dawalibi
W. K. Dick
T. H. Dodds
H. H. Ferris
D. L. Garrett
R. J. Heppe
J. E. Idzkowski
T. S. McLenahan
R. P. Keil
A. E. Kollar
A. P. Meliopoulos
S. G. Patel
J. G. Sverak
W. K. Switzer
R. Verma

The following persons were on the balloting committee that approved this document for submission to the IEEE Standards Board:

W. J. Ackerman
B. Y. Afshar
J. D. Betz
C. J. Blattner
N. A. Bleshman
W. R. Block
R. H. Bower
G. T. Brecht, Jr
R. A. Byron
J. B. Cannon
J. W. Cheever
S. Cluts
F. A. Denbrock
W. K. Dick
D. Dobbs
P. R. Dolan
L. N. Ferguson
D. J. Gaushell
L. M. Gordon
A. Haban
L. J. Haycock
G. E. Heidenreich
N. G. Hingorani
J. E. Holladay
M. L. Holm
D. E. Johannson
A. P. Johnson
D. C. Johnson
E. F. Jones
G. Karady
J. J. Keane
R. P. Keil
H. C. Koehler
D. F. Koenig
T. J. Kolenda
L. W. Kurtz
D. L. Laird
K. P. Lau
M. D. Limerick
P. Lolich
J. A. Maneatis
A. Matthey
W. C. McCart
C. Minshew
P. R. Nannery
E. A. Neal, Jr
J. T. Orrell
J. Paolozzi
S. G. Patel
G. Paul
R. J. Perina
T. A. Pinkham III
E. S. Raila
W. E. Richard
B. D. Russell
J. Sabath
D. R. Schafer
K. Sommer
R. C. St. Clair
J. G. Sverak
W. K. Switzer
E. R. Taylor
J. Thorson
R. J. Wehling
W. J. Weston
R. M. Youngs
C. E. Zanzie

When the IEEE Standards Board approved this standard on March 21, 1985, it had the following membership:

John E. May, *Chairman* **John P. Riganati,** *Vice Chairman*

Sava I. Sherr, *Secretary*

James H. Beall
Fletcher J. Buckley
Rene Castenschiold
Edward Chelotti
Edward J. Cohen
Paul G. Cummings
Donald C. Fleckenstein
Jay Forster
Daniel L. Goldberg
Kenneth D. Hendrix
Irvin N. Howell
Jack Kinn
Joseph L. Koepfinger*
Irving Kolodny
R. F. Lawrence
Lawrence V. McCall
Donald T. Michael*
Frank L. Rose
Clifford O. Swanson
J. Richard Weger
W. B. Wilkens
Charles J. Wylie

*Member emeritus

Contents

FIGURES

FIGURES PAGE

TABLES

APPENDIX FIGURES

APPENDIX FIGURES PAGE

APPENDIX TABLES PAGE

An American National Standard

IEEE Guide for Safety in AC Substation Grounding

1. Introduction

1.1 Purpose and Scope. The intent of this guide is to provide guidance and information pertinent to safe grounding practices in ac substation design. This guide is primarily concerned with outdoor substations, either conventional or gas-insulated. These include distribution, transmission, and generating plant substations. With proper caution, the methods described herein are also applicable to indoor portions of such substations, or to substations that are wholly indoors.[1]

The specific purposes of this guide are:

(1) To establish, as a basis for design, the safe limits of potential differences that can exist in a substation under fault conditions between points that can be contacted by the human body

(2) To review substation grounding practices with special reference to safety, and develop criteria for a safe design

(3) To provide a procedure for the design of practical grounding systems, based on these criteria

(4) To develop analytical methods as an aid in the understanding and solution of typical gradient problems

(5) To provide a bibliography of pertinent literature on grounding, and English translations of some of the more valuable foreign language articles

---

[1] Obviously, the same ground gradient problems that exist in a substation yard should not be present within a building. This will be true provided the floor surface either assures an effective insulation from earth potentials, or else is effectively equivalent to a conductive plate or close mesh grid that is always at substation ground potential, including the building structure and fixtures.

Therefore, even in a wholly indoor substation it may be essential to consider some of the possible hazards from perimeter gradients (at building entrances) and from transferred potentials described in Section 6. Furthermore, in the case of indoor gas-insulated facilities, the effect of circulating enclosure currents may be of concern, as discussed in Section 8.

The concept and use of safety criteria are described in Sections 1 - 6, practical aspects of designing a grounding system are covered in Sections 7 - 11, and procedures and evaluation techniques for the grounding system assessment (in terms of safety criteria) are described in Sections 12 - 18. Supporting material is organized in Appendixes A - J.

Some of the special hazards encountered in gas-insulated substations and techniques for analyzing these problems are discussed in Section 8 and Appendix D.

No attempt is made to cover the grounding problems peculiar to dc substations. A quantitative analysis of the effects of lightning surges is also beyond the scope of this guide. However, the listed references contain further information on both subjects.

A grounding system designed as recommended herein will, nonetheless, provide a high degree of protection against steep wave front surges entering the station and passing to earth through its ground electrodes.[2]

1.2 Relation to Other Standards. This guide only briefly discusses the field tests required to evaluate soil resistivity. The procedures for measuring the resistance of the installed grounding system, the surface gradients, and the continuity of the grid conductors are described in more detail in ANSI/IEEE Std 81-1983 [3].[3]

ANSI/IEEE Std 142-1982 [5] generally takes its recommended grounding practices from this guide. Also known as "The IEEE Green Book," it covers some of the practical aspects of grounding in more detail, such as equipment grounding, cable routing to avoid induced ground currents, cable sheath grounding, static and lightning protection, indoor installations, etc. However, as such, it refers to the previous edition of this guide.

IEEE Std 367-1979 [9] provides a detailed explanation of the asymmetrical current phenomenon and of the fault current division, which to a large degree parallels that given herein. Of course, the reader should be aware that the ground potential rise calculated for the purpose of telecommunication protection and relaying applications is based on a somewhat different set of assumptions concerning the maximum grid current, in comparison with those used for the purposes of this guide.

Finally, a guide that is presently being developed by the IEEE Power Generation Committee's Working Group on Generator Station Grounding Practices will give more detailed information on specific generating plant grounding problems. These include stack and cooling tower grounding, building grounds, etc. See 1.5.

1.3 Key Definitions. Most of the definitions given herein pertain solely to the applications of this guide, though those approved or standardized by other bodies

---

[2] The greater impedance offered to steep front surges will somewhat increase the voltage drop in ground leads to the grid system, and decrease the effectiveness of the more distant parts of the grid. Offsetting this in large degree is the fact that the human body apparently can tolerate far greater current magnitudes in the case of lightning surges than in the case of 50 or 60 Hz currents.

[3] The numbers in brackets correspond to those of the references listed in 1.4.

are used whenever possible. No further reference will be made to any of the key definitons stated below, unless necessary for clarity. All other definitions are placed within the text of individual sections. An alphabetical index of all definitions used is given in Appendix F. For additional definitions refer to ANSI/IEEE Std 100-1984 [4].

In spite of the fact that grounding concepts are generally well understood, some design solutions are more a reflection of a routine rather than of a well-conceived design approach. For this reason, the following point is made prior to the first definition: No ground current or ground fault current will flow into the earth unless at least one ground return circuit exists. This return circuit enables the current produced by a source, however distant, to return through the earth to that source. The definition of a *ground return circuit* is primary, and all other *ground* related definitions are secondary, following from it.

ground return circuit. A circuit in which the earth or an equivalent conducting body is utilized to complete the circuit and allow current circulation from or to its current source.

ground. A conducting connection, whether intentional or accidental, by which an electric circuit or equipment is connected to the earth or to some conducting body of relatively large extent that serves in place of the earth.

grounded. A system, circuit, or apparatus referred to is provided with ground for the purposes of establishing a ground return circuit and for maintaining its potential at approximately the potential of earth.

ground current. A current flowing into or out of the earth or its equivalent serving as a ground.

initial symmetrical ground fault current. The maximum rms value of symmetrical fault current after the instant of a ground fault initiation. As such, it represents the rms value of the symmetrical component in the first half-cycle of a current wave that develops after the instant of fault at time zero. Generally,

$$I_{f(0+)} = 3I_0'' \qquad \text{(Eq 1)}$$

where

$I_{f(0+)}$ = initial symmetrical ground fault current

I_0'' = rms value of zero-sequence symmetrical current that develops immediately after the instant of fault initiation, that is, reflecting the subtransient reactances of rotating machines contributing to the fault

NOTE: Elsewhere in the guide, this initial symmetrical fault current is shown in an abbreviated notation, as I_f, or is referred to only as $3I_0$. The underlying reason for the latter notation is that, for purposes of this guide, the initial symmetrical fault current is assumed to remain constant for the entire duration of the fault.

decrement factor. An adjustment factor used in conjunction with the initial symmetrical ground fault current parameter in safety-oriented grounding calculations. It allows us to obtain an rms equivalent of the asymmetrical current wave for a given fault duration, accounting for the effect of initial dc offset and its attenuation during the fault.

effective asymmetrical fault current. The rms value of asymmetrical current wave, integrated over the entire interval of fault duration. (See Fig 1.)

NOTE: In terms of this guide, it can be expressed as

$$I_F = D_f(t_f)\, I_f \qquad \text{(Eq 2)}$$

where

I_F = effective asymmetrical fault current in A
I_f = (initial) symmetrical ground fault current in A
$D_f(t_f)$ = decrement factor accounting for the effect of a dc offset during the subtransient period of fault current wave on an equivalent time basis of the entire fault duration, t_f, for t_f given in s

1.4 References. The following publications shall be used in conjunction with this standard.

[1] ANSI C2-1984, National Electrical Safety Code (note particularly Rule 93C).[4]

[2] ANSI/ASTM D448-80, Specifications for Standard Sizes of Coarse Aggregate for Highway Construction.[5]

[3] ANSI/IEEE Std 81-1983, IEEE Guide for Measuring Earth Resistivity, Ground Impedance, and Earth Surface Potentials of a Ground System — Part I: Normal Measurements.

[4] ANSI/IEEE Std 100-1984, IEEE Standard Dictionary of Electrical and Electronics Terms.

[5] ANSI/IEEE Std 142-1982, IEEE Recommended Practice for Grounding of Industrial and Commercial Power Systems.

[6] ANSI/IEEE Std 837-1984, IEEE Standard for Qualifying Permanent Connections Used in Substation Grounding.

[7] ANSI/IEEE C37.010-1979, IEEE Application Guide for AC High-Voltage Circuit Breakers Rated on a Symmetrical Current Basis.

[8] ANSI/IEEE C57.12.00-1980, IEEE Standard General Requirements for Liquid-Immersed Distribution, Power and Regulating Transformers.

---

[4] ANSI publications are available from the Sales Department, American National Standards Institute, 1430 Broadway, New York, NY, 10018.

[5] ANSI/ASTM publications are available from the Sales Department, American National Standards Institute, 1430 Broadway, New York, NY, 10018 and the Sales Department, American Society of Testing and Materials, 1916 Race St, Philadelphia, PA 19103.

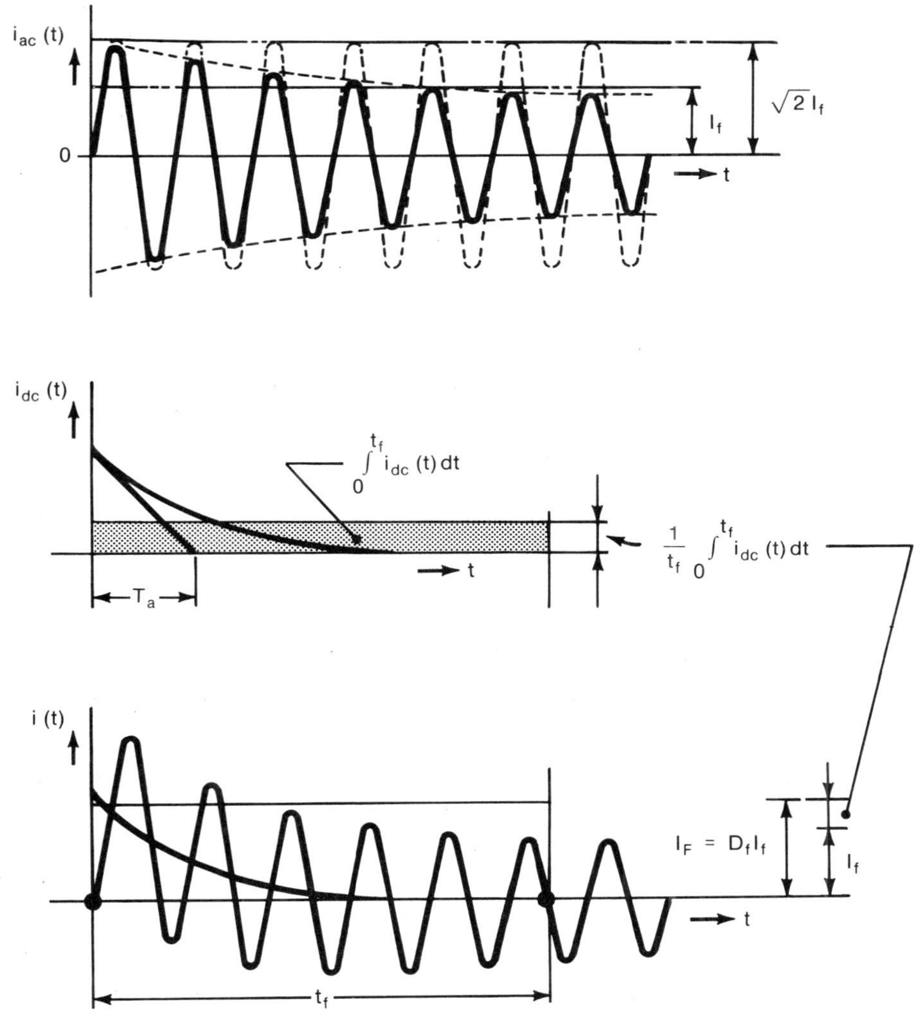

Fig 1
Relationship Between Actual Values of Fault Current and Values of I_F, I_f, and D_f for Fault Duration t_f

[9] IEEE Std 367-1979, IEEE Guide for the Maximum Electric Power Station Ground Potential Rise and Induced Voltage from a Power Fault.[6]

[10] IEEE Std 590-1977, IEEE Cable Plowing Guide.

[6] IEEE publications are available from the Institute of Electrical and Electronics Engineers, Service Center, 445 Hoes Lane, Piscataway, NJ 08854.

2. Safety in Grounding

2.1 Basic Problem. In principle, a safe grounding design has two objectives:

(1) To provide means to carry electric currents into the earth under normal and fault conditions without exceeding any operating and equipment limits or adversely affecting continuity of service

(2) To assure that a person in the vicinity of grounded facilities is not exposed to the danger of critical electric shock

A practical approach to safe grounding thus concerns and strives for controlling the interaction of two grounding systems:

(1) The intentional ground, consisting of ground electrodes buried at some depth below the earth surface

(2) The accidental ground, temporarily established by a person exposed to a potential gradient in the vicinity of a grounded facility

People often assume that any object grounded, however crudely, can be safely touched. This misconception probably contributed to accidents in the past, as a low station ground resistance is not, in itself, a guarantee of safety. There is no simple relation between the resistance of the ground system as a whole and the maximum shock current to which a person might be exposed. Therefore, a station of relatively low ground resistance may be dangerous under some circumstances[7], while another station with very high resistance may be safe or can be made safe by careful design.

For instance, if a substation is supplied from an overhead line with no shield or neutral wire, a low grid resistance is important. A substantial part of the total ground fault current enters the earth causing an often steep rise of the local ground potential; Fig 2(a).

If a shield wire, gas-insulated bus, or underground cable feeder, etc, is used, a part of the fault current returns through this metallic path directly to the source. Since this metallic link provides a low impedance parallel path to the return circuit, the rise of local ground potential is ultimately of lesser magnitude; Fig 2(b).

In either case, the effect of that portion of fault current that enters the earth within the station area should be further analyzed. If the geometry, location of ground electrodes, local soil characteristics, and other factors contribute to an excessive potential gradient at the earth surface, the grounding system may be inadequate despite its capacity to carry the fault current in magnitudes and durations permitted by protective relays.

[7] One exception is the case in which the product of the maximum grid current I_G flowing in the earth via the grounding system resistance R_g results in a voltage low enough to be contacted safely.

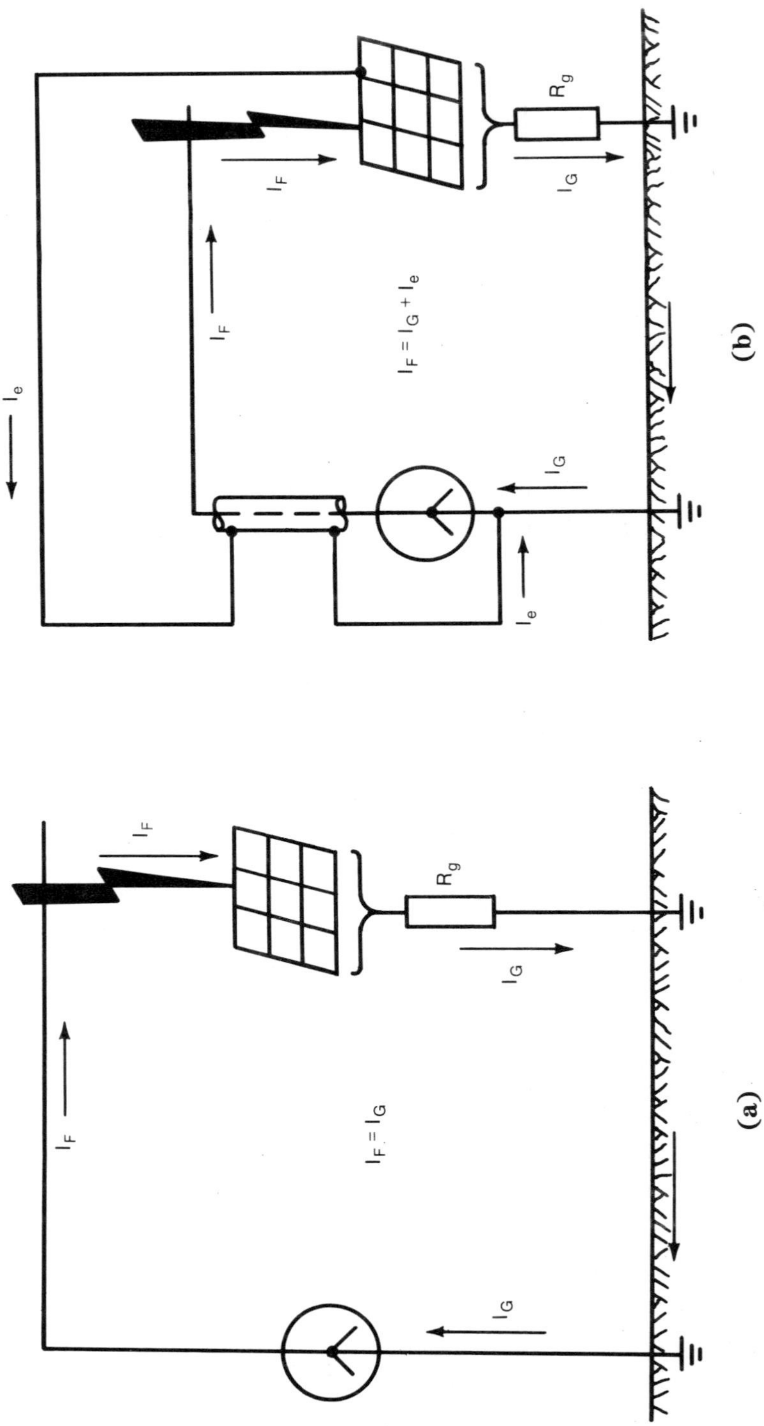

Fig 2
Typical Faulted Substation With and Without Multiple Ground Return Paths

The following Sections 3-6 cover in detail those principal assumptions and criteria that enable us to evaluate all necessary factors in protecting the most precious element of the accidental circuit, human life.

2.2 Conditions of Danger. During typical ground fault conditions, the flow of current to earth will produce potential gradients within and around a substation. Figure 3 shows this effect for a station with a simple rectangular grounding grid in homogeneous soil.

Fig 3
Equipotential Contours of a Typical Grounding Grid With and Without Ground Rods

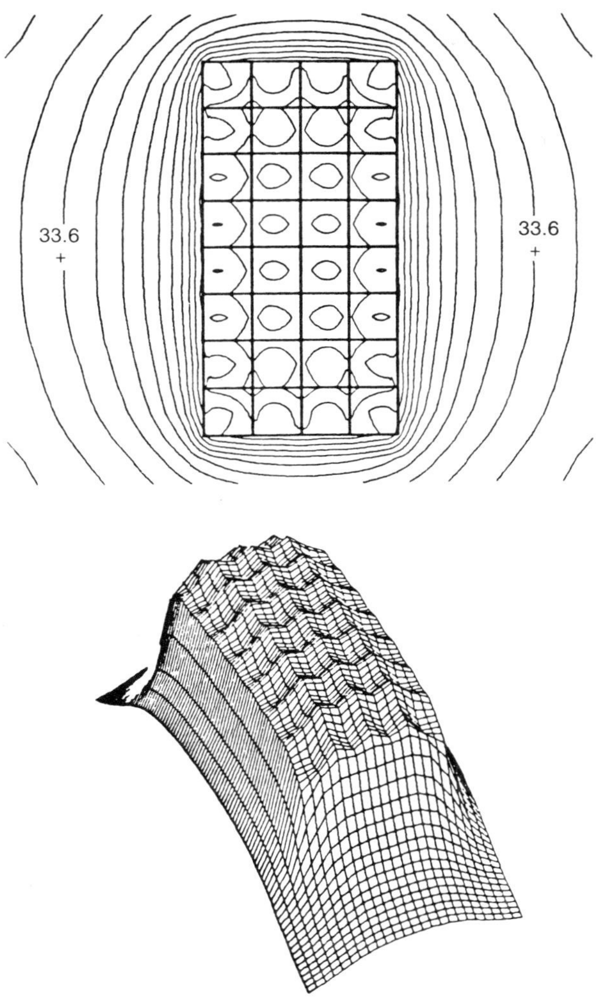

Unless proper precautions are taken in design, the maximum potential gradients along the earth surface may be of sufficient magnitude during ground fault conditions to endanger a person in the area. Moreover, dangerous potential differences may develop between structures or equipment frames that are grounded and the nearby earth.

The circumstances that make electric shock accidents possible are:

(1) Relatively high fault current to ground in relation to the area of ground system and its resistance to remote earth

(2) Soil resistivity and distribution of ground currents such that high potential gradients may occur at points at the earth surface

(3) Presence of an individual at such a point, time, and position that the body is bridging two points of high potential difference

(4) Absence of sufficient contact resistance or other series resistance to limit current through the body to a safe value under the above circumstances

(5) Duration of the fault and body contact, and hence, of the flow of current through a human body for a sufficient time to cause harm at the given current intensity

The relative infrequency of accidents of the type being studied, as compared to accidents of other kinds, is due largely to the low probability of coincidence of all the unfavorable conditions mentioned above.[8]

Nevertheless, some fatalities due to gradients have occurred in the past. Therefore, it is the responsibility of the engineer to lower this possibility.

---

[8] German Grounding Standard DIN 57141 (1977 edition) recognizes this low probability and allows reduction for grounding calculations of a given fault current magnitude by a certain factor. For instance, a 0.7 value is recommended for stations of 110 kV class and above.

3. Range of Tolerable Current

Effects of an electric current passing through the vital parts of a human body depend on the duration, magnitude, and frequency of this current. The most dangerous consequence of such an exposure is a heart condition known as ventricular fibrillation, resulting in immediate arrest of blood circulation.

3.1 Effect of Frequency. Humans are very vulnerable to the effects of electric current at frequencies of 50 or 60 Hz. Currents of approximately 0.1 A can be lethal. Authorities generally agree that the human body can tolerate a slightly higher 25 Hz current and approximately five times higher direct current. At frequencies of 3000 - 10 000 Hz, even higher currents can be tolerated [B28], [B30][9]. In some cases the human body is able to tolerate very high currents due to lightning surges.

Detailed studies of the effects of both direct and oscillatory impulse currents are reported in the literature [B19], [B21].

Information regarding special problems of dc grounding is contained in the 1958 Report of the Conversion Substation Committee [B16]. The hazards of an electric shock produced by the electrostatic effects of overhead transmission lines are reviewed in Part 1 of the 1971 Report of the General Substations Subcommitee [B59]. Additional information on the electrostatic effects of overhead transmission lines can be found in Chapter 8 of the *Transmission Line Reference Book 345 kV and Above* (the EPRI Red Book) [B46].

3.2 Effects of Magnitude and Duration. The most common physiological effects of electric current on the body, stated in order of increasing current magnitude, are perception, muscular contraction, unconsciousness, fibrillation of the heart, respiratory nerve blockage, and burning [B53].

Current of 1 mA is generally recognized as the threshold of perception, that is, the current mangitude at which a person is just able to detect a slight tingling sensation in his hands or fingertips caused by the passing current [B21].

Currents of 1 - 6 mA, often termed let-go currents, though unpleasant to sustain, generally do not impair the ability of a person holding an energized object to control his muscles and release it. Dalziel's classic experiment with 28 women and 134 men provides data indicating an average let-go current of 10.5 mA for women and 16 mA for men, and 6 mA and 9 mA as the respective threshold values [B29].

In a 9 - 25 mA range, currents may be painful and can make it hard or impossible to release energized objects grasped by the hand. For still higher currents

[9] The numbers in brackets with the prefix B refer to those of the bibliography in Chapter 20.

muscular contractions could make breathing difficult. Unlike the cases of respiratory inhibition from the much greater current mentioned next, these effects are not permanent and disappear when the current is interrupted — unless the contraction is very severe and breathing is stopped, not for seconds, but for minutes. Yet even such cases often respond to resuscitation [B25].

It is not until current magnitudes in the range of 60–100 mA are reached that ventricular fibrillation, stoppage of the heart, or inhibition of respiration might occur and cause injury or death. A person trained in cardiopulmonary resuscitation should administer CPR until the victim can be treated at a medical facility [B27].

Hence, this guide emphasizes the importance of the fibrillation threshold. If shock currents can be kept below this value by a carefully designed grounding system, injury or death may be avoided.

As shown by Dalziel and others [B26], [B29], the nonfibrilating current of magnitude I_B at durations ranging from 0.03–3.0 s is related to the energy absorbed by the body as described by the following equation:

$$(I_B)^2\, t_s = S_B \qquad \text{(Eq 3)}$$

where

I_B = rms magnitude of the current through the body
t_s = duration of the current exposure in s
S_B = empirical constant related to the electric shock energy tolerated by a certain percent of a given population

A more detailed discussion of this equation is provided in Section 4.

3.3 Importance of High-Speed Fault Clearing. Considering the significance of fault duration both in terms of Eq 3 and implicitly as an accident-exposure factor, high-speed clearing of ground faults is advantageous for two reasons:

(1) The probability of electric shock is greatly reduced by fast fault clearing time, in contrast to situations in which fault currents could persist for several minutes or possibly hours

(2) Both tests and experience show that the chance of severe injury or death is greatly reduced if the duration of a current flow through the body is very brief; the allowed current value may therefore be based on the clearing time of primary protective devices, or that of the back-up protection

A good case could be made for the former because of the low combined probability that relay malfunctions will coincide with all other adverse factors necessary for an accident, as has already been described in Section 2.

If the probabilistic aspects are neglected, choice of the backup relay clearing times is more conservative since it assures greater safety margins with regard to Eq 3.

An additional incentive to use switching times less than 0.5 s results from the research done by Biegelmeier and Lee [B8]. Their research provides evidence that a human heart becomes increasingly susceptible to ventricular fibrillation when the time of exposure to current is approaching the heartbeat period, but that the danger is much smaller if the time of exposure to current is in the region of 0.06 - 0.3 s.

In reality, high ground gradients from faults are usually infrequent, and shocks from this cause still more so. Futhermore, both events are often of very short duration. Thus, it would not be practical to design against shocks that are merely painful and cause no serious injury, that is, for currents below the fibrillation threshold.

4. Permissible Body Current Limit

The magnitude and duration of the current conducted through a human body at 50 or 60 Hz should be less than those that cause ventricular fibrillation.

4.1 Duration Formula. The duration for which a 50-60 Hz current can be tolerated by most people is related to its magnitude by Eq 3. Based on the results of Dalziel's studies, it is assumed that 99.5% of all persons can safely withstand, without ventricular fibrillation, the passage of a current in magnitude and duration determined by the following formula:

$$I_B = k/\sqrt{t_s} \qquad \text{(Eq 4)}$$

where, in addition to the terms previously defined, for Eq 3,

$$k = \sqrt{S_B}$$

Dalziel found that the shock energy that can be survived by 99.5% of persons weighing approximately 50 kg (110 lbs) results in a value of S_B of 0.0135. Thus, k_{50} = 0.116 and the formula for the allowable body current becomes

$$I_B = 0.116/\sqrt{t_s} \quad \text{for 50 kg body weight} \qquad \text{(Eq 5)}$$

Note that the above equation results in values of 116 mA for 1 s and 367 mA for 0.1 s.

Since Eq 4 is based on tests limited to a 0.03-3.0 s range, it obviously is not valid for very short or long times, and some values of current can be tolerated indefinitely [B20], [B24].

In a perspective of the past 40 years of researching I_B, in 1936 Ferris *et al.* [B49] suggested 100 mA as the fibrillation threshold if shock durations are not specified. The value 100 mA was derived from extensive experience at Columbia University, on animals having body and heart weights comparable to man, for a maximum shock duration of 3 s. Some of the more recent experiments suggest the existence of two distinct thresholds: one where the exposure time is shorter than one heartbeat period and another one for the current exposure longer than one heartbeat. For a 50 kg (110 lbs) adult, Biegelmeier proposes the threshold values at 500 and 50 mA, respectively [B6], [B7]. Other recent works on this subject are by Lee and Kouwenhoven [B27], [B65], [B68].

4.2 Alternative Assumptions. Fibrillation current is actually a function of individual body weight, as illustrated in Fig 4. This shows the relationship between the critical current and body weight for several species of animals (calves, dogs, sheep, pigs), and a 0.5% common threshold region for mammals.

In the 1961 edition of this guide, constants S_B and k in Eqs 3 and 4, given as 0.0272 and 0.165, respectively, had been assumed valid for 99.5% of all men weighing approximately 70.3 kg (155 lbs). Dalziel's more recent studies (in 1968), on which Eq 4 is based, lead to the alternate value of k = 0.157 and S_B = 0.0246 as being applicable to persons weighing 70 kg (155 lbs) [B22], [B24]. Thus,

$$I_B = 0.157/\sqrt{t_s} \quad \text{for 70 kg body weight} \qquad \text{(Eq 6)}$$

The maximum 3 s nonfibrillating current of 91 mA at the 70 kg base is still below the threshold of fibrillation of 107 mA at 50 kg (110 lbs), as shown in Fig 4.

Fig 4
Fibrillating Current Versus Body Weight for Various Animals Based on a Three-Second Shock

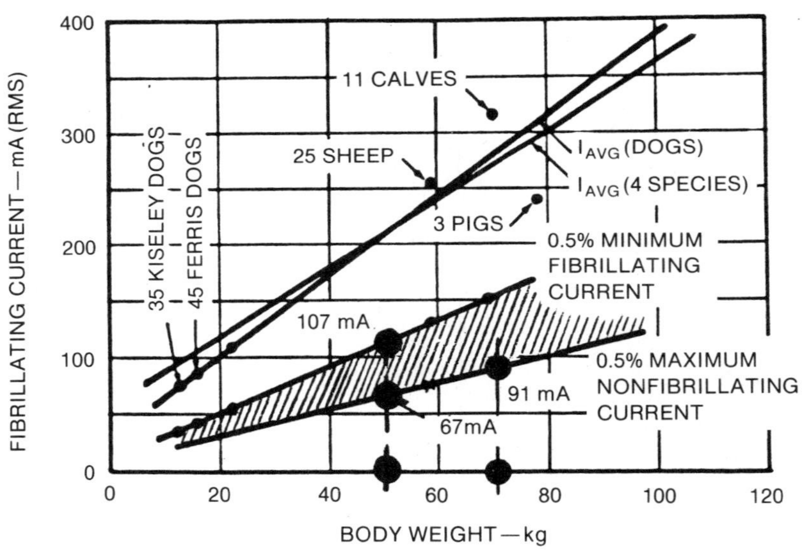

VALUE OF CONSTANT k FOR EFFECTIVE RMS VALUES OF I_B ($k = I_B \sqrt{t_s}$):

$k_{70} = 0.091\sqrt{3} = 0.157$
$k_{50} = 0.067\sqrt{3} = 0.116$
$k_{50} = 0.107\sqrt{3} = 0.185$
FIBRILLATION

Users of this guide may select $k = 0.157$ provided that the average population weight can be expected to be at least 70 kg.[10]

Equation 4 indicates that much higher currents can be allowed where fast operating protective devices can be relied upon to limit fault duration. A judgment decision is needed as to whether to use the clearing time of regular high-speed relays, or that of the back-up protection, as the basis for calculation.

4.3 Note on Reclosing. Reclosure after a ground fault is common in modern operating practice. In such circumstances, a person might be subjected to the first shock, which would not permanently injure him, but would upset and disturb him temporarily. Next, a single fast automatic reclosure could result in a second shock, initiated within less than 0.5 s from the start of the first. It is this second shock, occurring after a relatively short interval of time before the person has recovered, that might cause a serious accident. With manual reclosure, the possibility of exposure to a second shock is reduced since the reclosing time interval may be substantially greater.

The cumulative effect of two or more closely spaced shocks has not been thoroughly evaluated, but a reasonable allowance can be made by using the sum of individual shock durations as the time of a single exposure. This is discussed in more detail in 13.4.

[10] Typically, these conditions can be met in places that are not accessible to the public, such as in switchyards protected by fences or walls, etc. Depending on specific circumstances, an assessment should be made if a 50 kg criterion (Eq 5) ought to be used for areas outside the fence.

5. Accidental Ground Circuit

5.1 Resistance of the Human Body. For dc and ac at normal power frequency, the human body can be represented by a noninductive resistance. The resistance is between extremities, that is, from one hand to both feet, or from one foot to the other one. In either case, the value of this resistance is difficult to establish. The resistance of the internal body tissues, not including skin, is approximately 300 Ω, whereas values of body resistance including skin, ranging from 500 - 3000 Ω, have been suggested in the literature [B20], [B53], [B54], [B64], [B83].

As already mentioned in 3.2, Dalziel conducted extensive tests to determine safe let-go currents, with hands and feet wet, in salt water. Values obtained using 60 Hz for men were as follows: the current was 9.0 mA; corresponding voltages were, hand-to-hand, 21.0 V, and, hand-to-feet, 10.2 V. Hence, the ac resistance for a hand-to-hand contact is equal to 21.0/0.009 or 2330 Ω and the resistance hand-to-feet equals 10.2/0.009 or 1130 Ω, based on this experiment [B29].

For higher voltages (above 1 kV) and currents (above 5 A), the human resistance is decreased by damage or puncture of the skin at the point of contact. However, a wet hand contact resistance may be very low at any voltage. The resistance of shoes is uncertain, though it may be very low for damp leather.

Thus, for the purposes of this guide:

(1) Hand and shoe contact resistances will be assumed as equal to zero

(2) A value of 1000 Ω is selected for the calculations that follow as representing the resistance of a human body from hand-to-both-feet and also from hand-to-hand, or from one foot to the other foot:

$$R_B = 1000\ \Omega \qquad \text{(Eq 7)}$$

5.2 Current Paths Through the Body. It should be remembered that the choice of a 1000 Ω resistance value relates to paths such as those between the hand and one or both feet, where a major part of the current passes through parts of the body containing vital organs, including the heart. It is generally agreed that current flowing from one foot to the other is far less dangerous. Referring to tests done in Germany, Loucks mentioned that much higher foot-to-foot than hand-to-foot currents had to be used to produce the same current in the heart region, stating that the ratio is as high as 25:1 [B69].

Based on these conclusions, resistance values greater than 1000 Ω could possibly be allowed, where a path from one foot to the other foot is concerned. However, the following factors should be considered.

(1) A voltage between the two feet, painful but not fatal, might result in a fall that could cause a current flow through the chest area. The degree of this danger would further depend on the fault duration and the possibility of another, successive fault — perhaps on reclosure.

(2) A person might be working or resting in a prone position when a fault occurs.

It is apparent that the dangers from foot-to-foot contact are far less than from the other type. However, since deaths have occurred from the former, it is a hazard that should not be ignored [B10], [B66].

5.3 Accidental Circuit Equivalents. Using the value of tolerable body current established by Eq 5 or 6 and the appropriate circuit constants, it is possible to determine the tolerable voltage between any two critical points of contact.

Let it be noted that for the accidental circuit equivalent, the following notation applies:

I_A = current through the accidental circuit
R_A = total effective resistance of the accidental circuit
I_B = permissible body current, defined by Eq 5 or 6

Obviously,

$I_A < I_B$ is always required for safety

Since the body resistance is assumed constant, to require $I_A < I_B$ is equivalent to saying that fibrillation may be prevented by keeping the total watts-seconds (Ws) of energy absorbed in the body during a shock below a certain value. This value is 0.0135 Ws for k_{50} = 0.116 A, and 0.0246 Ws for k_{70} = 0.157 A, respectively. Thus, it can be seen that Dalziel's formula actually represents the relationship between shock current magnitude and duration for a constant shock energy.

Resistance of the accidental circuit R_A is a function of the body resistance R_B and the footing resistance R_F (resistance of the ground just beneath the feet). The footing resistance may affect appreciably the value of R_A, a fact that may be most helpful in some difficult situations. For the purposes of circuit analysis, the human foot is usually represented as a conducting metallic disk and the contact resistance of shoes and socks is neglected. As shown by Sunde [B98], the self and mutual resistances for two metallic disks of radius b, separated by a distance d_F on the surface of a homogeneous earth of resistivity ρ, are:

$$R_{\text{foot}} = \rho/(4b) \quad \text{(Eq 8)}$$

$$R_{M\text{foot}} = \rho/(2\,\pi\, d_{\text{foot}}) \quad \text{(Eq 9)}$$

where

R_{foot} = self-resistance of each foot to remote earth in Ω
$R_{M\text{foot}}$ = mutual resistance between the feet in Ω
b = equivalent radius of a foot in m
d_{foot} = separation distance of the feet in m

The resistances of the ground beneath the two feet in series and in parallel are:

$$R_{2Fs} = 2\ (R_{\text{foot}} - R_{M\text{foot}}) \qquad \text{(Eq 10)}$$

$$R_{2Fp} = \tfrac{1}{2}\ (R_{\text{foot}} + R_{M\text{foot}}) \qquad \text{(Eq 11)}$$

where, in addition to the symbols described above,

R_{2Fs} = resistance of two feet in series

R_{2Fp} = resistance of two feet in parallel

Figure 5 defines the circuit equivalent of a foot-to-foot contact. Here the potential U, shunted by the body, is the maximum potential difference between two accessible points on the ground surface, separated by the distance of one pace. The equivalent circuit resistance for the step potential circuit is given by Eq 12:

$$R_A = R_B + 2\ (R_{\text{foot}} - R_{M\text{foot}}) \qquad \text{(Eq 12)}$$

Next, the equivalent circuit for a hand-to-two-feet contact is illustrated in Fig 6.

The equivalent circuit resistance for the touch potential circuit is given by Eq 13:

$$R_A = R_B + \tfrac{1}{2}\ (R_{\text{foot}} + R_{M\text{foot}}) \qquad \text{(Eq 13)}$$

Bibliographic references [B10], [B43], [B59], and [B66] choose a 0.08 m (3 in) radius for the disk representing one foot and neglect the mutual resistance term.

Fig 5
Step Voltage Circuit

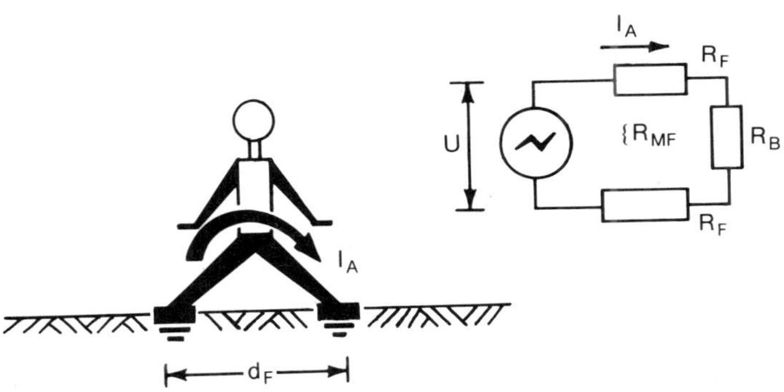

d_F = 1 m
R_A = $R_B + 2R_F - 2R_{MF}$
I_A = U/R_A
R_B = 1000 Ω

where

I_A = the current of accidental circuit
R_A = the total resistance of accidental circuit

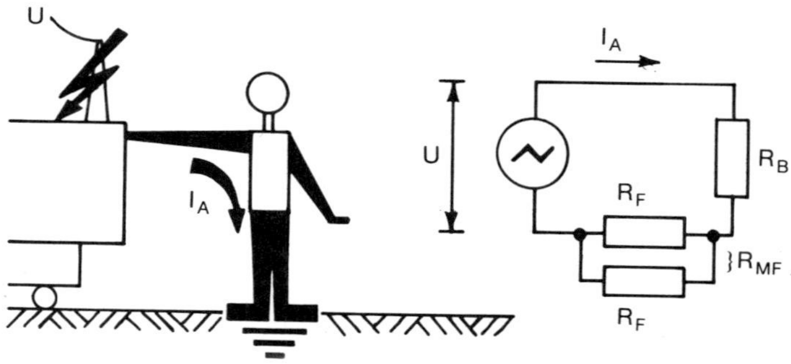

Fig 6
Touch Voltage Circuit

With only slight approximation, equations for the series and parallel resistances of two feet can be obtained in numerical form and expressed in terms of ρ, as shown below:

$$R_{2Fs} = 6(\rho) \qquad \text{(Eq 14)}$$

$$R_{2Fp} = 1½(\rho) \qquad \text{(Eq 15)}$$

Therefore, for all practical purposes, the resistance of a foot is equal to 3ρ. Equation 14 is used when computing the body current resulting from step voltages and Eq 15 applies when calculating the body current produced by a mesh or touch potential with both feet *buried* at zero depth in the soil near the surface.

For example, if $\rho = 2000$ Ω-m, Eqs 14 and 15 yield 12 000 and 3000 Ω for the series and parallel resistances, respectively.

A more exact calculation of the self and mutual resistances using 1 m separation yields $R_{2Fs} = 11863$ Ω and $R_{2Fp} = 3284$ Ω. Use of a value of $d_{foot} = 1$ m is conservative in calculating R_{2Fs}. Though it might produce a slightly higher value of resistance than would a smaller separation between the feet, the resulting step voltage also is much higher with a larger separation than it would be with a smaller one, and that would be the dominant effect on body current.

The large separation is also conservative in computing R_{2Fp} because it produces a lower resistance than a small separation would.

5.4 Effect of a Thin Surface Layer of Crushed Rock. Equations 8 and 9 are derived based on the assumption of uniform soil resistivity. However, a 0.08–0.15 m (3–6 in) layer of crushed rock is often spread on the earth's surface above the ground grid to increase the contact resistance between the soil and the feet of people in the substation. The crushed rock also improves the surface for the movement of equipment and vehicles in the substation. The area covered by this crushed rock layer is generally of sufficient size to validate the assumption of the feet being in contact with a material of uniform resistivity in the lateral direction. However, the relatively shallow depth of the crushed rock as compared to the equivalent radius of the foot precludes the assumption of uniform resistivity in the vertical direction when computing the self and mutual resistances of the feet.

If the underlying soil has a lower resistivity than the crushed rock, only some grid current will go upward into the thin upper layer of crushed rock, and the surface voltage will be very nearly the same as that without the rock layer. The current through the body will be lowered considerably with the addition of the crushed rock surface because of the greater contact resistance between the earth and the feet. However, this resistance may be considerably less than that of a crushed rock layer of great thickness (that is, thick enough to assume uniform resistivity in all directions). How much less depends on the relative values of earth and crushed rock resistivities and in the thickness of the rock layer. A typical case described in the literature shows that the effective resistance of a 0.25 m layer of limestone having a 5000 Ω-m (wet) resistivity is roughly equivalent to 75% of its nominal value if the resistivity of the earth's soil is 250 Ω-m [B58].

The following equations for R_{foot} and $R_{M\text{foot}}$ are derived from [B58]:

$$R_{\text{foot}} = \frac{\rho_1}{4b} F(X_1) \tag{Eq 16}$$

$$R_{M\text{foot}} = \frac{\rho_1}{2\pi d_{\text{foot}}} F(X_2) \tag{Eq 17}$$

In Eqs 16 and 17, b and d_{foot} are defined in 5.3 and $F(X)$ is a function based on the feet spacing and the relative values of the earth and crushed rock resistivities:

$$F(X) = 1 + 2 \sum_{n=1}^{\infty} Q \tag{Eq 18}$$

$$Q = \frac{K^n}{\sqrt{1+(2nX)^2}} \tag{Eq 19}$$

$$K = \frac{\rho - \rho_s}{\rho + \rho_s} \tag{Eq 20}$$

where

ρ_s = crushed rock resistivity in Ω-m
ρ = earth resistivity in Ω-m
$X = X_1 = h_s/b$ for R_{foot}
$X = X_2 = h_s/d_{\text{foot}}$ for $R_{M\text{foot}}$
h_s = thickness of the crushed rock surface layer in m

These equations could also be derived by applying the method of images to Sunde's equations in [B98]. However, since the quantity $F(X)$ is rather tedious to evaluate without a computer or programmable calculator, these values have been

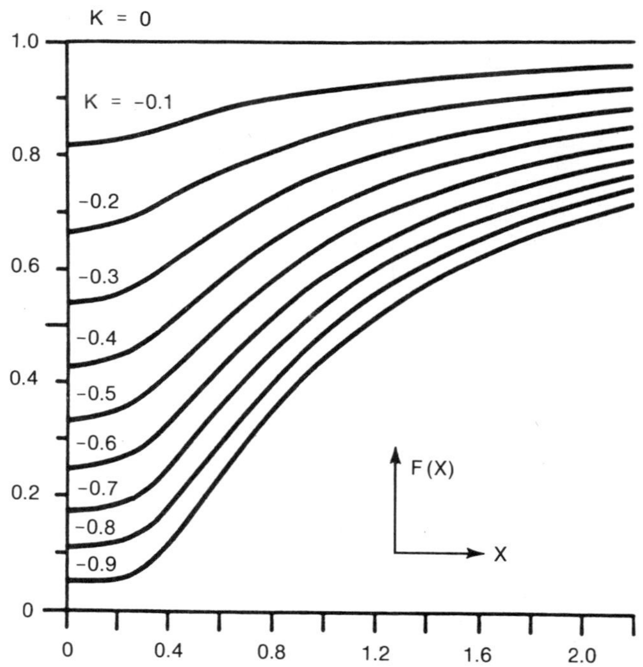

Fig 7
Function $F(X)$ Versus X and Reflection Factor K

precalculated and graphed for a wide range of values (X) and factor K, as shown in Fig 7.

Example 1. Let a layer of surface material be 0.1 m (4 in) thick, and have the nominal resistivity of 2000 Ω-m; the underlying soil resistivity is 222 Ω-m, b = 0.08 m, and d_{foot} = 1 m. From these data it follows that $K = -0.80$, $X_1 = 1.25$, and $X_2 = 0.1$. Using Fig 7, one can find $F(X_1) = 0.57$ and $F(X_2) = 0.11$. Substitute into Eqs 16 and 17: $R_{\text{foot}} = 3562\ \Omega$ and $R_{M\text{foot}} = 35\ \Omega$. Finally, using Eqs 10 and 11 one obtains $R_{2Fs} = 7054\ \Omega$ and $R_{2Fp} = 1798\ \Omega$.

In order to simplify the above procedure for routine use, the mutual resistance term can be neglected and b assumed always equal to 0.08 m. On this basis, the equations for the series and parallel resistances of two feet can alternatively be expressed in a form that is analogous to that of Eqs 14 and 15, used for uniform soil:

$$R_{2Fs} = 6.0\ C_s\ (h_s, K)\ \rho_s \qquad \text{(Eq 21)}$$

and

$$R_{2Fp} = 1.5\ C_s\ (h_s, K)\ \rho_s \qquad \text{(Eq 22)}$$

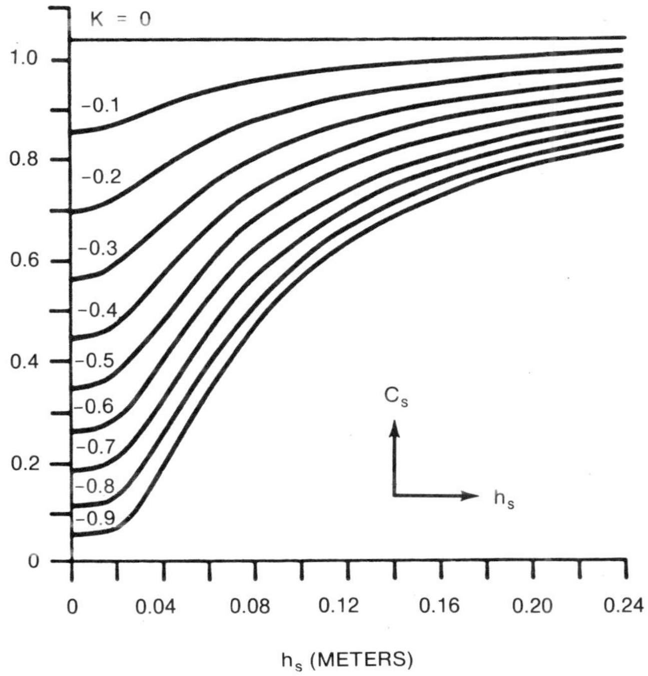

Fig 8
Reduction Factor C_s As a Function of Reflection Factor K and Crushed Rock Layer Thickness h_s

where

C_s = reduction factor for derating the nominal value of surface layer resistivity determined as follows:

C_s = 1 for crushed stone resistivity equal to soil resistivity

Otherwise,[11]

$$C_s = \frac{1}{0.96}\left[1 + 2\sum_{n=1}^{\infty} \frac{K^n}{\sqrt{1 + (2n\, h_s/0.08)^2}}\right]$$

For the latter case of $C_s < 1$, in which C_s is a function of (h_s, K) and which distinguishes Eqs 21 and 22 from Eqs 14 and 15, the values of C_s are plotted in Fig 8.

---

[11] Simple alternative approaches, based on the equivalent hemisphere concept, such as

$$C_s \approx 1 - a\left[\frac{1 - \frac{\rho}{\rho_s}}{2h_s + a}\right]$$

; a = 0.106 m, which avoids the infinite summation series, are also possible; refer to pp 14-15 of [B100] and to Jackson's discussion of Sverak's equations on p 19 of the same reference.

Example 2. For the same resistivity data used in Example 1, and also assuming again h_s = 0.1 m, and K = -0.8, factor C_s can be found from Fig 7. Here, C_s is determined to be approximately 0.6. R_{2Fs} is 7200 Ω and R_{2Fp} is 1800 Ω.

The converse of the derating principle is also true. If the underlying soil has a higher resistivity than the crushed rock, a substantial portion of the grid current will go upward into the thin layer of crushed rock. However, unlike the case described above, the surface potentials will be altered substantially, due to this concentration of current near the surface. Although Eqs 16 and 17 are valid for computing the self and mutual resistances of the feet even if the soil has a higher resistivity than the crushed rock, these equations do not account for the alteration of the surface potentials. Thus, the effective resistivity of the crushed rock should not be upgraded without taking into account this change in surface potential. This problem can best be solved by using multilayer soil analysis (see Section 11).

6. Criteria of Permissible Potential Difference

6.1 Typical Shock Situations. Figures 9 and 10 show four basic situations involving a person and grounded facilities during a fault. For a foot-to-foot contact the accidental circuit equivalent is that of Fig 5, and its driving voltage U is equal to E_s (step voltage). For the three remaining examples of a hand-to-both-feet contact, Fig 6 applies, and U is equal to E_t (touch voltage), E_m (mesh voltage), or E_{trrd} (transferred voltage), respectively.

During a fault, the earth becomes saturated by the currents emanating from the grid and other permanent ground electrodes buried below the earth surface. The resulting potential gradients have a primary effect on the value of U.

For clarity, let the following be defined:

ground potential rise (GPR). The maximum voltage that a station grounding grid may attain relative to a distant grounding point assumed to be at the potential of remote earth.

NOTE: Under normal conditions, the grounded electrical equipment operates at near zero ground potential, that is, the potential of a grounded neutral conductor is nearly identical to the potential of remote earth. During a ground fault, the portion of fault current that is conducted by a station grounding grid into the earth causes the rise of the grid potential with respect to remote earth. This voltage rise, GPR, is proportional to the magnitude of the grid current, and to the grid resistance.

step voltage. The difference in surface potential experienced by a person bridging a distance of 1 m with his feet without contacting any other grounded object.

touch voltage. The potential difference between the ground potential rise (GPR) and the surface potential at the point where a person is standing, while at the same time having his hands in contact with a grounded structure.

NOTE: In a conventional substation, the worst touch voltage is usually found to be the potential difference between a hand and the feet at a point of maximum reach distance. However, in the case of a metal-to-metal contact from hand-to-hand, or from hand-to-feet, which is of concern in the gas-insulated substations, both situations should be investigated for the possible worst reach condition, including both hands; Fig 11.

mesh voltage. The maximum touch voltage to be found within a mesh of a ground grid.

transferred voltage. A special case of the touch voltage where a voltage is transferred into or out of the substation.

Typically, the case of transferred voltage occurs when a person standing within the station area touches a conductor grounded at a remote point or a person

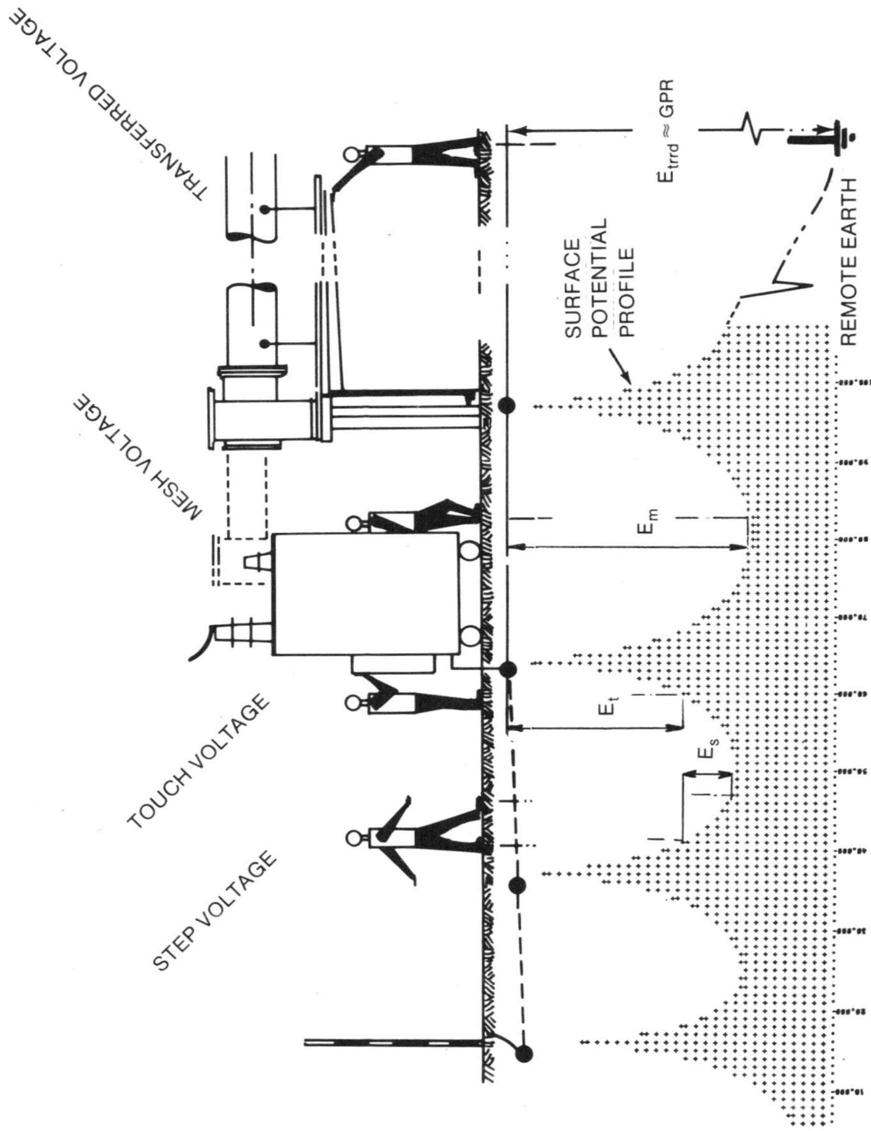

Fig 9
Basic Shock Situations

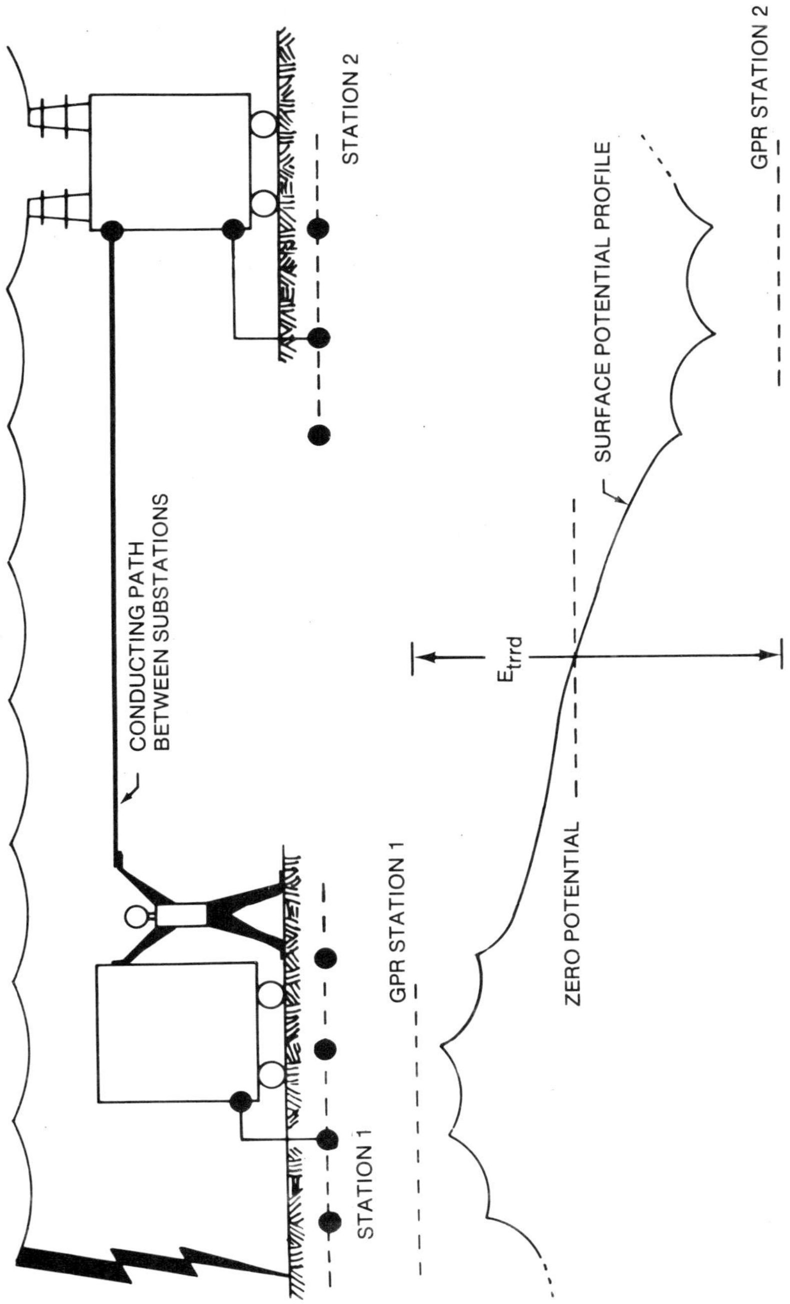

Fig 10
Typical Situation of External Transferred Potential

standing at a remote point touches a conductor connected to the station grounding grid. During fault conditions, the resulting potential to ground may equal or exceed the full GPR of a grounding grid discharging the fault current, rather than the fraction of this total voltage encountered in the *ordinary* touch contact situations (see Fig 10). In fact, as discussed in Section 15, the transferred voltage may exceed the sum of the GPR's of both substations, due to induced voltages on communication circuits, static or neutral wires, pipes, etc. It is impractical, and often impossible, to design a ground grid based on the touch voltage caused by external transferred voltages. Hazards from these external transferred voltages are best avoided by using isolating or neutralizing devices and by treating and clearly labeling these circuits, pipes, etc., as being equivalent to *live* lines.

6.2 Step and Touch Voltage Criteria. The safety of a person depends on preventing the critical amount of shock energy from being absorbed before the fault is cleared and the system de-energized. The maximum driving voltage of any accidental circuit should not exceed the limits defined below. For step voltage the limit is

$$E_{\text{step}} = (R_B + R_{2Fs})\, I_B \qquad \text{(Eq 23)}$$

Combining Eqs 23, 21, 7, and either 5 or 6,

$$E_{\text{step}50} = (1000 + 6C_s(h_s,K)\,\rho_s)\,0.116/\sqrt{t_s} \qquad \text{(Eq 24)}$$

or

$$E_{\text{step}70} = (1000 + 6C_s(h_s,K)\,\rho_s)\,0.157/\sqrt{t_s} \qquad \text{(Eq 24a)}$$

The actual step voltage, E_s, should be less than the maximum allowable step voltage, E_{step}, to ensure safety. Similarly, the touch voltage limit is

$$E_{\text{touch}} = (R_B + R_{2Fp})\, I_B \qquad \text{(Eq 25)}$$

Combining Eqs 25, 22, 7, and either 5 or 6,

$$E_{\text{touch}50} = (1000 + 1.5C_s(h_s,K)\,\rho_s)\,0.116/\sqrt{t_s} \qquad \text{(Eq 26)}$$

or

$$E_{\text{touch}70} = (1000 + 1.5C_s(h_s,K)\,\rho_s)\,0.157/\sqrt{t_s} \qquad \text{(Eq 26a)}$$

where

C_s = 1 for no protective surface layer or is determined from Fig 8 if a protective surface layer of high resistivity and small thickness is used
ρ_s = the resistivity of the surface material in Ω-m
t_s = duration of shock current in s

The actual touch voltage, mesh voltage, or transferred voltage should be less than the maximum allowable touch voltage, E_{touch}, to ensure safety.

6.3 Typical Shock Situations for Gas-Insulated Substations. In the grounding analysis of gas-insulated substations (GIS), the touch voltage considerations present several unique problems. Unlike conventional facilities, the GIS equipment features a metal sheath enclosing gas-insulated switchgear and inner high-voltage buses. Each bus is completely contained within its enclosure and the enclosures are grounded. Since a voltage is induced in the outer sheath whenever a current flows in the coaxial busbar, certain parts of the enclosure might be at different potentials with respect to the station ground. To evaluate the maximum voltage occurring on the bus enclosure during a fault, it is necessary to determine the inductance of the outer sheath to ground, the inductance of the inner conductor, and the mutual inductances for a given phase configuration of individual buses.

A person touching the outer sheath of a GIS might be exposed to voltages resulting from two basic fault conditions:

(1) An internal fault within the gas-insulated bus system, such as a flashover between the bus conductor and the inner wall of the enclosure

(2) A fault external to the GIS in which a fault currrent flows through the GIS bus and induces currents in the enclosures

Since the person may stand on a grounded metal grating and the accidental circuit may involve a hand-to-hand and hand-to-feet current path, the analysis of GIS grounding necessitates consideration of an additional problem — that of the permissible touch voltage for a metal-to-metal contact; Fig 11.

Most GIS manufacturers consider the enclosure properly designed and adequately grounded if the potential difference between individual enclosures and, with respect to the voltage level of other grounded structures, does not exceed 65-130 V during a fault. As shown below, the substitution of $\rho = 0$ in the foot

Fig 11
Typical Metal-to-Metal Touch Situations in GIS

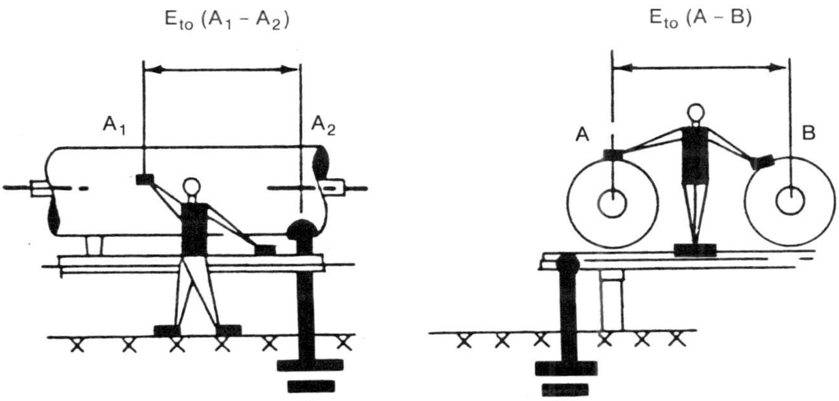

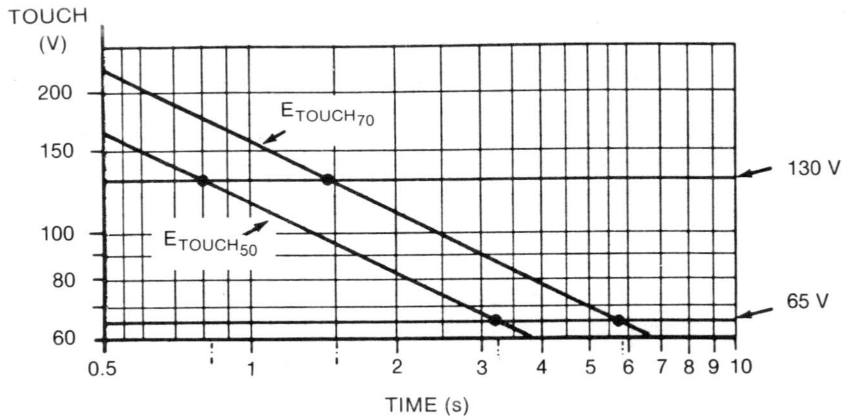

Fig 12
Touch Voltage Limits for Metal-to-Metal Contact and a Typical Range of Enclosure Voltages to Ground

resistance terms of Eqs 26 and 26a reveals that this voltage range corresponds to fault times ranging from 0.8–3.2 s if a 50 kg criterion is used, and ranging from 1.46–5.8 s for the assumption of a 70 kg body. This relationship is, however, better perceived in the graphical form of Fig 12, which also helps to grasp the related problem of sufficient safety margins.

The reduced equations for a metal-to-metal contact are:

$$E'_{touch_{50}} = 116/\sqrt{t_s} \qquad \text{(Eq 27)}$$

and

$$E'_{touch_{70}} = 157/\sqrt{t_s} \qquad \text{(Eq 28)}$$

The fault conditions and the corresponding circuit equivalents for determining or verifying the critical safety design parameters of GIS grounding will be discussed in more detail in Section 8 and Appendix D.

6.4 Effect of Sustained Ground Currents. After the safe step and touch voltage limits are established, the grounding system can then be designed based on the available short-circuit current and overall clearing time. The designer should also consider sustained low-level (below setting of protective relays) fault magnitudes that may be above the let-go current threshold. Some sustained faults above the let-go current, but below the fibrillation threshold, may cause asphyxiation from prolonged contraction of the chest muscles. However, it would not be practical to design against lesser shocks that are painful, but cause no permanent injury.

7. Principal Design Considerations

7.1 General Concept. A grounding system should be installed in a manner that will limit the effect of ground potential gradients to such voltage and current levels that will not endanger the safety of people or equipment under normal and fault conditions, as well as assure continuity of service. Related definitions are:

ground electrode. A conductor imbedded in the earth and used for collecting ground current from or dissipating ground current into the earth.

grounding grid. A system of horizontal ground electrodes that consists of a number of interconnected, bare conductors buried in the earth, providing a common ground for electrical devices or metallic structures, usually in one specific location.

NOTE: Grids buried horizontally near the earth's surface are also effective in controlling the surface potential gradients. A typical grid usually is supplemented by a number of ground rods and may be further connected to auxiliary ground electrodes, to lower its resistance with respect to remote earth.

ground mat. A solid metallic plate or a system of closely spaced bare conductors that are connected to and often placed in shallow depths above a ground grid or elsewhere at the earth surface, in order to obtain an extra protective measure minimizing the danger of the exposure to high step or touch voltages in a critical operating area or places that are frequently used by people. Grounded metal gratings, placed on or above the soil surface or wire mesh placed directly under the crushed rock, are common forms of a ground mat.

grounding system. Comprises all interconnected grounding facilities in a specific area.

In the discussion that follows, it is assumed that the system of ground electrodes has the form of a grid of horizontally buried conductors, supplemented by a number of vertical ground rods connected to the grid. Based on two surveys, the first reported in an AIEE application guide in 1954 [B1], and the second published in 1980 [B33], this concept represents the prevailing practice of most utilities both in the USA and in other countries.

Some of the reasons for using the combined system of vertical rods and horizontal conductors are as follows:

(1) In substations a single electrode is, by itself, inadequate in providing a safe grounding system. In turn, when several electrodes, such as ground rods, are connected to each other and to all equipment neutrals, frames, and structures that are to be grounded, the result is, essentially, a grid arrangement of ground elec-

trodes, regardless of the original objective. If the connecting links happen to be buried in a soil of good conductivity, this network alone may represent an excellent grounding system. Partly for this reason, some utilities depend on the use of a grid alone. However, ground rods are of a particular value, as explained next.

(2) If the magnitude of current dissipated into the earth is high, it seldom is possible to install a grid with resistance so low as to assure that the rise of a ground potential will not generate surface gradients unsafe for human contact. Then, the hazard can be eliminated only by control of local potentials through the entire area. A system that combines a horizontal grid and a number of vertical ground rods penetrating lower soils has the following advantages:

(a) While horizontal (grid) conductors are most effective in reducing the danger of high step and touch voltages on the earth's surface, provided that the grid is installed in a shallow depth usually 0.3-0.5 m (12-18 in) below grade, sufficiently long ground rods will stabilize the performance of such a combined system. For many installations this is important — because of freezing or drying out, the resistivity of upper soil layers could vary with seasons, while the resistivity of lower soil layers remains nearly constant.

(b) Rods penetrating the lower resistivity soil are far more effective in dissipating fault currents whenever a two- or multilayer soil is encountered and the upper soil layer has higher resistivity than the lower layers. For many gas-insulated substations (GIS) and other space-limited installations, this conditon becomes in fact the most desirable one to occur, or to be achieved by the appropriate design means (extra long ground rods, grounding wells, etc).

(c) If the rods are installed predominately along the grid perimeter in high-to-low or uniform soil conditions, the rods will considerably moderate the steep increase of the surface gradient near the peripheral meshes. Specific details of this arrangement will be further discussed in Section 14, since these are pertinent to the use of simplified methods in determining the voltage gradient at the earth's surface.

7.2 Primary and Auxiliary Ground Electrodes. In general, most grounding systems utilize two groups of ground electrodes: those specifically designed for grounding purposes and the electrodes that comprise various underground metal structures installed for purposes other than grounding, as defined further:

primary ground electrode. A ground electrode specifically designed or adapted for discharging the ground fault current into the ground, often in a specific discharge pattern, as required (or implicitly called for) by the grounding system design.

auxiliary ground electrode. A ground electrode with certain design or operating constraints. Its primary function may be other than conducting the ground fault current into the earth.

Grounding grids, counterpoise conductors, ground rods, and wells are typical primary electrodes. Underground metal structures and reinforcing bars encased in concrete, if connected to the grounding grid, are typical auxiliary electrodes that may have a limited current carrying capability.

7.3 Basic Aspects of Grid Design. Conceptual analysis of a grid system usually starts with inspection of the station layout plan, showing all major equipment and structures. In order to establish the basic ideas and concepts, the following points may serve as guidelines for starting a typical grounding grid design:

(1) A continuous conductor loop should surround the perimeter to enclose as much area as practical. This measure helps to avoid high current concentration and hence high gradients both in the grid area and near the projecting cable ends. Enclosing more area also reduces the resistance of the grounding grid.

(2) Within the loop, conductors should be laid in parallel lines and, where practical, along the structures or rows of equipment, to provide for short ground connections.

(3) A typical grid system for a substation may include 4/0 bare copper conductors buried 1.3 - 0.5 m (12 - 18 in) below grade, spaced 3 - 7 m (10 - 20 ft) apart, in a grid pattern. At cross-connections, the conductors would be securely bonded together. Ground rods may be at the grid corners and at each second junction point along the perimeter. Ground rods may also be installed at major equipment. In multilayer or very resistive soils, it might be useful to use longer rods. (Lengths exceeding 100 ft have been used by some utilities.)

(4) This grid system would be extended over the entire substation switchyard and often beyond the fence line. Multiple ground leads or larger sized conductors would be used where high concentrations of current may occur, such as at a neutral-to-ground connection of generators, capacitor banks, or transformers.

(5) The ratio of the sides of the mesh usually is from 1:1 to 1:3, unless a precise (computer-aided) analysis warrants more extreme values. Frequent cross-connections have a relatively small effect on lowering the resistance of a grid. Their primary role is to assure adequate control of the surface potentials. The cross-connections are also useful in securing multiple paths for the fault current, minimizing the voltage drop in the grid itself, and providing a certain measure of redundancy in the case of a conductor failure.

7.4 Design in Difficult Conditions. In areas where the soil resistivity is rather high or the substation space is at a premium, it may not be possible to obtain a low impedance grounding system by spreading the grid electrodes over a large area, as is done in more favorable conditions. Such a situation is typical of many GIS installations, occupying only a fraction of the land area normally used for conventional equipment. This often makes the control of surface gradients difficult. Some of the solutions include:

(1) Connection(s) of remote ground grid(s) and adjacent grounding facilities; a combined system utilizing separate installations in buildings, underground vaults, etc. A predominant use of remote ground electrodes requires careful consideration of transferred potentials, surge arrester locations, and other critical points. A significant voltage drop may develop betwen the local and remote grounding facilities.

(2) Use of deep-driven ground rods and drilled ground wells, in combination with a chemical treatment of soils, or use of bentonite clays for backfilling.

(3) Use of counterpoise wire mats. In exposed areas, it is feasible to combine both an insulating material and fabricated mats made of wire mesh, expanded metal, or gratings; first to equalize the gradient field near the surface and then to reduce conductance from the surface to the underlying metal structures. A typical counterpoise mesh might consist of copper-clad steel wires of AWG No 6 size, arranged in a 0.6 · 0.6 m (24 · 24 in) grid pattern, installed 0.05 - 0.15 m (2 - 6 in) below the earth's surface and overlaying the main grounding grid, which is installed in greater depth, usually between 0.3 - 0.5 m (12 - 18 in).

(4) Where feasible, controlled use of other available means to lower the overall resistance of a ground system, such as connecting static wires and neutrals to the ground (see 13.3). Typical is the use of metallic objects on the site that qualify for and can serve as auxiliary ground electrodes, or as ground ties to other systems. Consequences of such applications, of course, have to be carefully evaluated.

(5) Wherever practical, a nearby deposit of low resistivity material of sufficient volume can be used to install an extra (satellite) grid. This satellite grid, when sufficiently connected to the main grid, will lower the overall resistance and, thus, the ground potential rise of the grounding grid. The nearby low resistivity material may be a clay deposit or it may be a part of some large structure, such as the concrete mass of hydroelectric dam [B108].

7.5 Connections to Grid. Conductors of adequate ampacity and mechanical strength should be used for the connections between:

(1) All ground electrodes, such as grounding grids, rodbeds, ground wells, and, where applicable, metal, water, or gas pipes, water well casings, etc

(2) All above-ground conductive metal parts that might accidentally become energized, such as metal structures, machine frames, metal housings of conventional or gas-insulated switchgear, transformer tanks, guards, etc

(3) All fault current sources such as surge arresters, capacitor banks or coupling capacitors, transformers, and, where appropriate, machine neutrals, secondary lighting, and power circuits

Copper cables or straps are usually employed for these ground connections. However, transformer tanks are sometimes used as part of a ground path for surge arresters thereon. Similarly, most steel or aluminum structures may be used for the ground path if it can be established that their conductance, including that of any joints, is and can be maintained as equivalent to that of the conductor that would normally be installed. Where this practice is followed, any paint films that might otherwise introduce a highly resistive joint should be removed and a suitable joint compound applied or other effective means, such as jumpers across the joints, taken to prevent subsequent deterioration of the joint. In the case of GIS installations, extra attention should be paid to the possibility of unwanted circulation of induced currents. Section 8 covers the subject in more detail.

Equal division of currents between multiple ground leads at cross-connections or similar junction points should not be assumed.

All accessible ground leads should be inspected on a periodic basis. Exothermic-weld, brazed, or pressure type connectors can be used for underground connections (see 9.5). Soldered connections should be avoided because of the possibility of failure under high fault currents.

Open circuits, even in exposed locations, can escape detection, and it obviously is impractical to inspect buried portions of the grounding network once it is installed. Those facilities that are most likely to supply or carry a high current, such as transformer and circuit breaker tanks, switch frames, and arrester pads, should always be connected to the grid with more than one ground lead. The leads should preferably be run in opposite directions to eliminate common mode failure.[12]

[12] One possible exception is grounding of the secondaries of potential and current transformers. The grounding of such devices usually must be restricted to a single point to avoid any parallel path that could cause undesirable circulation of currents affecting the performance of relays and metering devices.

8. Special Considerations for Gas-Insulated Substations (GIS)

8.1 GIS Characteristics. Gas-insulated substations (GIS) are subjected to the same magnitude of ground fault current and require the same low impedance grounding as conventional substations.

gas-insulated substation. A compact, multicomponent assembly, enclosed in a grounded metallic housing in which the primary insulating medium is a compressed gas, and that normally consists of buses, switchgear, and associated equipment (subassemblies).

Typically, the GIS installation necessitates 10–25% of the land area required for conventional equipment. Because of this smaller area it may be difficult to obtain adequate grounding solely by conventional methods. Particular attention should be given to the bonding of the metallic enclosures of the GIS assembly, as these enclosures carry induced currents of significant magnitude, which must be confined to specific paths. In this respect, grounding recommendations by the manufacturer of a given GIS usually need to be strictly followed.

8.2 Enclosures and Circulating Currents. The shielding effectiveness of the bus enclosure is determined by its impedance, which governs the circulation of induced currents.

enclosure currents. Currents that result from the voltages induced in the metallic enclosure by the current(s) flowing in the enclosed conductor(s).

continuous enclosure. A bus enclosure in which the consecutive sections of the housing along the same phase conductor are bonded together to provide an electrically continuous current path throughout the entire enclosure length. Cross-bondings, connecting the other phase enclosures, are made only at the extremities of the installation and at a few selected intermediate points.

noncontinuous enclosure. A bus enclosure with the consecutive sections of the housing of the same phase conductor electrically isolated (or insulated from each other), so that no current can flow beyond each enclosure section.

With separate enclosures for each phase, the magnitude and direction of the enclosure current is influenced by the size of the enclosure and the phase spacing between the buses, as well as by the method of interconnecting the enclosures.

In a *continuous* enclosure design, a voltage is induced in an enclosure by the current in the conductor that it surrounds, producing a longitudinal current flow in the enclosure. When a continuity of all phase enclosures is maintained through short connections at both ends, the enclosure current is only slightly less than that flowing in the inner bus in the opposite direction. This current returns through the housing (enclosures) of adjacent phases when the load is equalized between phases. The magnetizing current lags the enclosure current by approximately 90 degrees; the flux is mainly contained within the enclosure.

In a *noncontinuous* enclosure design, there are no external return paths for enclosure currents. Thus the voltage induced in a noncontinuous enclosure by the current of an inner bus(es) that it surrounds cannot produce any longitudinal current flow. Also, voltages might be induced in each enclosure by the currents in the conductors not enclosed by it. Nonuniform voltages result, causing local current flows in each isolated enclosure section, with the currents flowing in nonuniform patterns. Because of these properties, the noncontinuous design is generally considered less advantageous that that of the continuous type. As such, it is not currently used by the industry.

8.3 Grounding of Enclosures. Normally, the continuous type enclosures provide a return path for induced currents so that the conductor and enclosure form a concentric pair with effective external shielding of the field internal to the enclosure. However, under asymmetrical faults, the dc component is not shielded and causes an external voltage drop due to enclosure resistance.

Frequent bonding and grounding of GIS enclosures is the best solution to minimize hazardous touch and step voltages within the GIS area. Additional measures[13] include the use of conductive platforms (ground mats) that are connected to GIS structures and grounded. Servicing platforms usually are an integral part of the GIS design supplied by the manufacturer.

main ground bus. A conductor or system of conductors provided for connecting all designated metallic components of the GIS to a station grounding system.

To limit the undesirable effects caused by circulating currents, the following requirements should be met:

(1) All metallic enclosures should normally operate at ground voltage level.

(2) When grounded at the designated points, the bus enclosure design should ensure that no significant voltage differences exist between individual enclosure sections and that neither the supporting structures nor any part of the grounding systems is adversely influenced by the flow of induced currents.

(3) To avoid the circulation of enclosure currents beyond regular return path within the GIS assembly, power cable sheath grounds should be tied to the grounding system via connections that are separated from the GIS enclosures. To facilitate this isolation, the design of cable terminations (potheads) should be such that an isolating air gap or proper insulation elements are provided.

---

[13] However, despite all measures described, the presence of circulating currents can cause different parts of the GIS metal housing to have a slightly different potential to ground. Though the resulting voltage differences are small and generally of no concern with regard to a shock hazard, accidental metallic bridging of adjacent enclosures can cause annoying sparks.

(4) Enclosure return currents also cannot be permitted to flow through any externally mounted current transformers.

8.4 Cooperation Between GIS Manufacturer and User. Usually, it is the GIS manufacturer who defines clearly what constitutes the main ground bus of the GIS and specifies what is required of the user for connecting the GIS assembly to the station ground. Ample documentation is necessary to assure that none of the proposed connections from the main ground bus to the grounding grid will interfere with the required enclosure current path or any other operational feature of the GIS design. That may be especially pertinent if the main ground bus consists of a system of interconnections between the GIS components and structures, and no separate busbar (continuous common ground bus loop) is furnished.

Usually the GIS manufacturer also provides, or is responsible for:

(1) Providing the subassembly-to-subassembly bonding to assure safe voltage gradients between all intentionally grounded parts of the GIS assembly and between those parts and the main ground bus of the GIS

(2) Furnishing readily accessible connectors of sufficient mechanical strength to withstand electromagnetic forces and normal abuse, and that are capable of carrying the anticipated maximum fault current in that portion of the circuit without overheating

(3) Providing ground pads or connectors, or both, allowing, at least, for two paths to ground from the main ground bus, or from each metallic enclosure and auxiliary piece of GIS equipment designated for a connection to the station ground if the main ground bus of the GIS assembly does not actually exist

(4) Recommending proper procedures for connections between dissimilar metals, typically between a copper cable or a similar ground conductor and aluminum enclosures

The user usually provides information on the sources of fault current and the expected magnitudes and durations that ought to be considered. Moreover, he should be of assistance to the GIS manufacturer in reviewing all proposed grounding provisions in order to assure proper interfacing of:

(1) Connections for the neutral current of grounded equipment or apparatus and for dissipating surges caused by lightning and switching within the GIS

(2) Devices for dissipating lightning and switching surge currents external to the GIS assembly

(3) Requirements of protective relaying, and satisfying the provisions necessary for telephone and communication facilities

(4) Ground connections to all GIS supporting frames and structures, metallic sheaths, and installation of shielding for cable terminations where applicable

(5) Connections to all those pads or connectors furnished by the GIS manufacturer

(6) Safe voltage for step and touch, under both normal and abnormal operating conditions external to the GIS assembly

(7) Compliance with the grounding specifications, related to correct grounding practices, as mutually agreed to by the GIS manufacturer and the user

8.5 Other Special Aspects of GIS Grounding

8.5.1 Precautions should be undertaken to prevent excessive currents from being induced into adjacent frames, structures, or reinforcing steel, and to avoid establishment of current loops via other station equipment, such as transformers or separate switchgear. If there is the possibility of undesirable current loops via ground connections, or if any sustained current path might partially close or pass through grounded structures, the substation grounding scheme and the physical layout should be carefully reviewed with the GIS manufacturer.

8.5.2 Equal care is needed in the proximity of discontinuities in enclosure grounding paths at the transformer connections to GIS and at the interface points to conventional switchgear to prevent circulating currents in the circuit breaker and transformer tank steel.

8.5.3 Where applicable, all isolating elements should be able to withstand the full potential difference that may occur between the locally grounded system and that external to the GIS. For instance, the isolation of high-pressure oil pipe cables from the GIS grounding system often involves difficulties. Although the individual HV or EHV potheads may provide adequate separation from the external grounds — by the virtue of a design that usually includes the use of base plate insulators made of high voltage rated porcelain or fiberglass, problems sometimes arise if the same level of insulation is also expected at other interface points. One typical problem area is the auxiliary piping between the oil chamber of individual GIS potheads and the oil diffusion chamber at the end of a pipe cable that frequently branches to a variety of oil pressure monitoring instruments and alarm devices [B55]. There the isolation of metal parts is often achieved by the means of ceramic or plastic inserts. Adequate creepage distance should be ensured.

In these and similar circumstances,[14] a close cooperation with the GIS manufacturer in the early stages of the design is very important.

8.6 Notes on Grounding of GIS Foundations. Since the earth path of ground currents is strongly affected by the relative position of conductive objects that are in the ground, more attention should be paid to those portions of the GIS grounding system that include discontinuities, or where the design requires an abrupt change in the pattern of ground electrodes. The following circumstances are of concern.

8.6.1 In the limited space of GIS substations, a substantial part of the substation area is often occupied by concrete foundations, which may cause irregularities in a current discharge path. In this respect, a simple monolithic concrete steel reinforced slab is advantageous both as an auxiliary grounding device and for seismic reasons.

---

[14] Transient ground potential rise (TGPR) resulting from insulation breakdown or disconnect switch operation in GIS is a potential problem area. The direct effect of TGPR on humans may not be fatal, but its secondary effect on unsuspecting personnel should be of concern to the design engineer and the manufacturer [B40], [B50].

8.6.2 If a continuous floor slab is used, a good adjunct measure is to tie its reinforcing steel mesh to the common ground bus (main ground bus) so that both the GIS enclosures and the structural steel in and above the foundation will be at approximately the same potential level. The assumption is that this measure should produce a better ground and the reinforcing bars — being considerably closer together than the wires of a typical ground grid — should produce more even potentials within the floor and at the surface.[15]

8.6.3 The reinforcing bars and other metal in GIS foundations can act as auxiliary ground electrodes and may be so used provided that under no circumstances the discharge of current would result in a damage of concrete because of local overheating or a gradual erosion of the concrete-steel bonds. For further details, refer to 12.6.

8.7 Touch Voltage Criteria for GIS. Although the GIS manufacturer generally designs the equipment to meet the already mentioned requirements for safe operation and usually performs most, if not all, calculations that are necessary for determining the sheath voltages and currents during faults, there still are circumstances when the user has to ascertain that the entire installation is safe. Having this possibility in mind, some of the critical aspects of interconnecting the GIS with a grounding system are briefly discussed next.

A certain paradox, inherent to the GIS design, may occur when one tries to determine the best concept of GIS grounding. In contrast to the general wisdom that a large ground connection necessarily equals a good grounding practice, the circulating currents generated in the GIS enclosures during a fault should also be taken into account. To be considered are: (1) where these currents will circulate and (2) where and to what degree the design engineer or GIS manufacturer, or both, prefer these currents to circulate.

Typically in a continuous enclosure design, the path of enclosure currents includes some structural members of the GIS frame and the enclosures themselves. With each phase enclosure tied to the enclosures of adjacent phases at both ends, several loops are formed. Since a cross section of the mentioned structural members is usually much smaller than that of the enclosure and comparable to that of the grounding straps that connect the GIS assembly to a ground grid (and for that matter, also to the reinforcing bars of the concrete foundation), several questions need to be asked:

(1) If the currents divide and flow via all available metallic paths, what ratio is to be expected between the currents circulating within the GIS assembly and those circulating via a ground connection?

(2) How much current circulating via a ground connection loop is too much?

---

[15] It might be argued that the concrete slab, being a fairly good conductor itself, could produce a more uniform voltage at the floor level if no current would flow into the reinforcing bars from the ground system. If the bars are connected, the electrical field in the earth between the bars of the slab and the underlying grid would be zero. (As both mats are at the same potential, hardly any current would flow out of the bars into the concrete and toward the ground grid.) Therefore, the concrete floor with reinforcing bars will produce a substantially uniform potential field across the floor surface.

(3) Should not the GIS be so designed as to be safe if eventually no circulating current would (at least for an external fault) circulate via ground connections?

(4) And finally, just how much of a grounding is needed for the best balance between operational and safety-related requirements?

Presently, there are no clear-cut answers and solutions to the above questions. Some manufacturers prefer to supply a special ground bus (main ground bus) as a part of the GIS package, with clearly designated ground connection points. Others do not use any main ground bus at all, but simply designate certain points on the enclosure as grounding pads and let the utility complete the grounding.

In either case, it becomes necessary to limit the body current to some value in a milliampere range, while the fault currents that are of concern range from hundreds to thousands of amperes. Thus, one can safely assume that the full potential difference existing prior to a contact would not change while forcing the current through an alternate path including the body. Then the case of a person touching the GIS sheath metal can be reduced to the problem of finding the voltage drop between two points of contact along one or between two enclosures and a common ground. For the hand-to-feet contact made by a person standing on a nonmetallic surface (concrete slab or soil layer above the grounding grid, etc), only a minor modification of the application criterion of Eqs 26 and 26a is required in order to take into account the maximum inductive voltage drop occurring within the GIS assembly.

The touch voltage criterion for GIS is:

$$\sqrt{(E_m, E_t)^2 + (E'_{\text{to max}})^2} < E_{\text{touch}} \qquad \text{(Eq 29)}$$

where

E_m, E_t = calculated values of mesh or touch voltage, as determined for the point underneath a person's feet

$E'_{\text{to max}}$ = (predominantly inductive) maximum value of metal-to-metal voltage difference on and between GIS enclosures, or between these enclosures and the supporting structures, including any horizontal or vertical members for which the GIS assembly is designed

A simplified method for estimating the magnitude of enclosure currents during the external or internal faults shown in Fig 13 is provided in Appendix D.

In practical situations, as shown in Fig 13, a multiplicity of return paths and a considerable cross-coupling occurs. For this reason the simplified models apply only to a certain degree. This, however, makes the calculation of longitudinally induced currents difficult and for some remote external faults often outright impractical, as too many parameters remain undefined. Furthermore, because of a great variety in possible physical arrangements of the GIS assembly, it also is difficult to give specific guidelines. As a rule, the GIS manufacturers do detail calculations of this type for determining the basic design parameters, such as spacing and location of bonds, etc. Therefore, given the limited scope of this document, it is suggested to use the simplified procedure shown in Appendix D only as

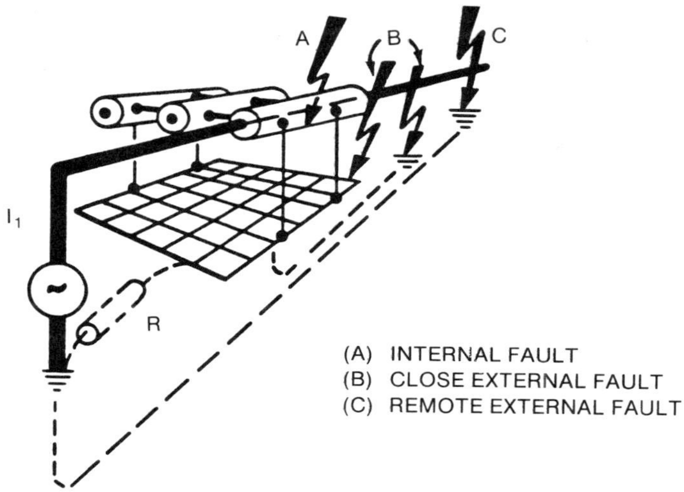

Fig 13
Typical Faults in GIS

a means to obtain a rough estimate, if the lack of manufacturer's data makes it necessary.

Recommendations

(1) In applying the touch voltage criterion Eq 29, the following facts should be considered. While the case of an internal fault with ground return requires the addition of the resistive and inductive voltage drop to the resistive drop representing the difference of potentials between the station ground and the point beneath a person's feet, this generally is not necessary for faults external to the GIS. For an external line-to-ground fault, the voltages induced on the sheath should be checked for a hand-to-hand metal-to-metal contact, but the calculation of step and touch voltages at the earth's surface is the same as that for conventional installations (that is, the inductive term $E'_{to\ max}$ in Eq 29 is zero).

(2) In evaluating the magnitude of induced voltages caused by faults external to the GIS, only the case of a close fault [case (B) in Fig 13] need be analyzed.[16]

[16] For remote external faults, the conditions within GIS will be less severe. For instance, if the method of Appendix D is used and the GIS is tied to an overhead transmission system, the assumption of equivalent depth $S_o = 2h$ (where h represents the depth of grid burial) would no longer be valid for the fault current return circuit. The circuit equation would have to be based on an appropriate equivalent penetration depth, such as that given by Semlyen [B38], [B94].

9. Selection of Conductors and Joints

In assessing which conductor material and what conductor size or what maximum allowable temperature limit need to be applied in individual design situations, the final choice should always reflect the following considerations.

9.1 Basic Requirements. Each element of a grounding system, including grid conductors, joints, connecting leads, and all primary grounding electrodes, should be so designed that for the expected design life of the installation, the element will:

(1) Have sufficient conductivity, so that it will not contribute substantially to local voltage differences

(2) Resist fusing and mechanical deterioration under the most adverse combination of a fault current magnitude and duration

(3) Be mechanically reliable and rugged to a high degree, especially on locations exposed to corrosion or physical abuse

The first requirement for selecting a conductor with sufficient conductivity is usually fulfilled when the other two requirements for current-carrying ability and mechanical strength are satisfied.

9.2 Choice of Material and Related Corrosion Problems. Copper is by far the most common material used for grounding in the USA. Copper conductors, in addition to their high conductivity, have the advantage of being resistant to underground corrosion since copper is cathodic with respect to other metals that are likely to be buried in the vicinity. Copper-clad steel is usually used for ground rods and occasionally for grounding grids. Use of copper, or to a somewhat lesser degree, of copper-clad steel, therefore assures that the integrity of an underground network will be maintained for years, so long as the conductors are of adequate size and not damaged.

Nevertheless, since a grid of copper or copper-clad steel forms a galvanic cell with buried steel structures, pipes, and any of the lead-based alloys that might be present in cable sheaths, it is also likely to hasten the corrosion of the latter. Tinning of copper has been tried by some utilities; that reduces the *cell* potential with respect to steel and zinc by about 50% and practically eliminates this potential with respect to lead (tin being only slightly sacrificial to lead). The disadvantages of using tinned copper conductor is that it accelerates and concentrates the natural corrosion of the metal in a small area. Other often used methods are:

(1) Insulation of the sacrificial metal surfaces with plastic tape, asphalt compound, or both

(2) Routing of buried metal elements so that any copper-based conductor will cross gas pipes or similar objects made of other metals as nearly as possible at right angles, and then applying insulating coatings to one metal or the other where they are in close proximity

(3) A full cathodic protection of sacrificial metals in the area or, where feasible, use of nonmetallic pipes and conduit

Aluminum has been used for ground grids less frequently. Though at first glance the use of aluminum would seem to be a natural choice for GIS equipment since the enclosures are made of aluminum or aluminum alloys, there are several disadvantages to consider:

(1) Aluminum itself may corrode in certain soils, the layer of corroded material is nonconductive for all practical grounding purposes

(2) Gradual corrision caused by alternating currents may also become a problem under some conditions

Thus, aluminum should be used only after a full investigation of all circumstances, despite the fact that, like steel, it would alleviate the problem of contributing to the corrosion of other buried objects and eliminate most difficulties in maintaining a reliable electric connection between dissimilar metals. If so used, the high-purity electric conductor grades are recommended as being more suitable than most alloys. An all-aluminum cable of the same conductance as an equivalent copper conductor will also have an approximately equal short-time ampacity. This follows from the fact that the respective temperature coefficient, specific heat, and density of copper and aluminum are such as to cause their respective melting points to be reached in approximately the same length of time [B107].

Steel has been used for ground-grid conductors in many European countries and is gradually gaining acceptance in the USA, mainly due to the benefit of eliminating most of the adverse effects of copper already mentioned. Of course, such a design requires that attention is paid to the protection of the grid itself. Application of a galvanized or corrosion-resistant steel, in combination with cathodic protection, is typical [B70].

In GIS, the use of cathodic protection may also be required for other reasons. Common is the protection of facilities that are external to the GIS, such as pressurized oil-pipe cables or lead-shielded cables, etc. Because of the complexity of GIS assemblies, it is essential to consider all aspects of corrosion prevention before designing the grounding system. Specific guidelines are difficult to establish since substation conditions may be different due to location and application in the electrical power system.

The subject of underground corrosion and cathodic protection is complex. Many studies have been made and much has been published on this subject. A detailed discussion of these phenomena is beyond the present scope of this guide.

9.3 Minimum Size Formula. A quantitative determination of the short-time temperature rise in a ground conductor can be obtained from Eq 30, taken from the derivation by Sverak [B101]. This equation evaluates the ampacity of any conductor[17] for which the material constants are known, or can be determined by calculation. Material constants of the commonly used grounding materials are listed in Table 1.

$$I = A \sqrt{\left(\frac{\mathrm{TCAP} \cdot 10^{-4}}{t_c\, \alpha_r\, \rho_r}\right) \ln \left(\frac{K_0 + T_m}{K_0 + T_a}\right)} \qquad \text{(Eq 30)}$$

where

I = rms current in kA
A = conductor cross section in mm^2
T_m = maximum allowable temperature in °C
T_a = ambient temperature in °C
T_r = reference temperature for material constants in °C
α_0 = thermal coefficient of resistivity at 0 °C
α_r = thermal coefficient of resistivity at reference temperature T_r
ρ_r = the resistivity of the ground conductor at reference temperature T_a in $\mu\Omega$-cm
K_0 = $1/\alpha_0$, or $(1/d_r) - T_r$
t_c = time of current flow in s
TCAP = thermal capacity factor from Table 1, in J/cm^3/°C (for definition refer to 9.4)

Note that α_r and ρ_r are both to be found for the same reference temperature of r-degrees Celsius. Table 1 provides data for α_r and ρ_r at 20 °C.

If the conductor size is given in circular mils, Eq 30 becomes

$$I = 5.0671 \cdot 10^{-6}\, A \sqrt{\left(\frac{\mathrm{TCAP}}{t_c\, \alpha_r\, \rho_r}\right) \ln \left(\frac{K_0 + T_m}{K_0 + T_a}\right)} \qquad \text{(Eq 31)}$$

Equations 30 and 31, in conjunction with Eq 32 (which defines TCAP), reflect two basic assumptions: (1) all heat will be retained in the conductor, and (2) the

[17] This general equation replaces Onderdonk's formula for copper, used in earlier editions of this guide. As reported in [B101], for the assumption of 1.589 $\mu\Omega/cm^3$ resistivity at 0 °C, TCAP assumed to be 3.4964 J/cm^3/°C, and the temperature coefficient of copper equal to 0.004274 at 0 °C, the substitution of these values into Eq 30 indicate that Onderdonk's formula gives results comparable to the more general formula of Eq 30. Alternately, in Onderdonk's equation shown below, the constant in the denominator would be equal to 32.85 instead of 33, to match Eq 30:

$$I = A \sqrt{\left(\frac{1}{33S}\right) \log_{10} \left(1 + \frac{T_m - T_a}{234 + T_a}\right)}$$

where

I = rms current in A
A = copper cross section in cmils
S = time in s during which current I is applied
T_m = maximum allowable temperature in °C
T_a = ambient temperature in °C

Table 1
Material Constants

Description	Material Conductivity (%)	α_r Factor @ 20 °C	K_0 $(1/\alpha_0)$ @ 0 °C	Fusing Temperature (°C)	ρ_r @ 20 °C (μΩ-cm)	TCAP Factor Effective Value (J/cm^3/°C)
Standard Annealed Soft Copper Wire	100.0	0.00393	234	1083	1.7241	3.422
Commercial Hard Drawn Copper Wire	97.0	0.00381	242	1084	1.7774	3.422
Copper-Clad Steel Core Wire	40.0	0.00378	245	1084/ 1300	4.397	3.846
Copper-Clad Steel Core Wire	30.0	0.00378	245	1084/ 1300	5.862	3.846
Commercial EC Aluminum Wire	61.0	0.00403	228	657	2.862	2.556
Aluminum Alloy Wire 5005	53.5	0.00353	263	660	3.2226	2.598
Aluminum Alloy Wire 6201	52.5	0.00347	268	660	3.2840	2.598
Aluminum-Clad Steel Core Wire	20.3	0.00360	258	660/ 1300	8.4805	2.670
Zinc-Coated Steel Core Wire	8.5	0.00320	293	419/ 1300	20.1	3.931
Stainless Steel No 304	2.4	0.00130	749	1400	72.0	4.032

product of specific heat (SH) and specific weight (SW), TCAP, is approximately constant since SH increases and SW decreases at about the same rate. For most metals, these premises are applicable over a reasonably wide temperature range, as long as the fault duration is within a few seconds. The error is always on the conservative side.[18]

9.4 Alternate Formulations. When working with materials that are not listed in Table 1, most engineering handbooks provide enough information, including the specific heat and weight data, for determining TCAP.

Specific heat (cal/gram/°C) and specific weight (gram/cm^3) are related to the thermal capacity per unit volume in (Ws/cm^3) as follows:

$$(\text{cal/gram/°C}) \cdot (\text{gram/cm}^3) = 4.184\ (\text{Ws/cm}^3/°\text{C})$$

$$1\ \text{Ws} = 1\ \text{J}$$

[18] Readers interested in more information on this subject may find a comprehensive treatment of the fusing problem by a different method, a comparision of several methods of calculation, as well as a large number of further references in [B75].

Hence, TCAP is defined by

$$TCAP = 4.184 \cdot SH \cdot SW \tag{Eq 32}$$

where

SH = specific heat in cal/gram/°C
SW = specific weight in gram/cm^3

Once TCAP is determined, Eqs 30 or 31 can be used to determine the ampacity of the conductor.

Equations 30 and 31 can be arranged to give required conductor size as a function of conductor current.

$$A_{mm^2} = I \sqrt{\frac{\dfrac{t_c \, \alpha_r \, \rho_r \cdot 10^4}{TCAP}}{\ln\left[1 + \left(\dfrac{T_m - T_a}{K_0 + T_a}\right)\right]}} \tag{Eq 33}$$

$$A_{cmils} = 1973.52 \; I \sqrt{\frac{\dfrac{t_c \, \alpha_r \, \rho_r \cdot 10^4}{TCAP}}{\ln\left[1 + \left(\dfrac{T_m - T_a}{K_0 + T_a}\right)\right]}} \tag{Eq 34}$$

Example. The 1976 edition of this guide provided tabulated data on the minimum size for copper and copper-clad steel (CCS), given in cmils/A. A similar tabulation can be made, using Eq 34 and Table 1, to get data for 30% and 40% copper-clad steel, and for 100% and 97% copper conductors. For instance, to calculate the 1 s size of a 30% copper-clad steel cable, one gets:

$t_c = 1.0$, $\alpha_{20} = 0.00378$, $\rho_{20} = 5.862$, TCAP = 3.846, $T_m = 1084$, $T_a = 40$, $K_0 = 245$

Thus, for I = 1 A (0.001 kA)

$$A_{cmils} = 1.97352 \sqrt{\frac{57.16}{1.5397}}$$

$$A_{cmils} = 12.02$$

Figure 14 and Table 2 provide a quick reference for most common materials assuming the following design parameters:

(1) Ambient temperature of 40 °C

(2) Conductor fusing temperature limit, as given in Table 1

(3) Maximum allowable temperature for brazed joints, 450 °C, as discussed in 9.5

(4) Maximum allowable temperature for critical cables and bolted joints, 250 °C as discussed in 9.5

Fig 14
Nomogram for Conductor Sizing

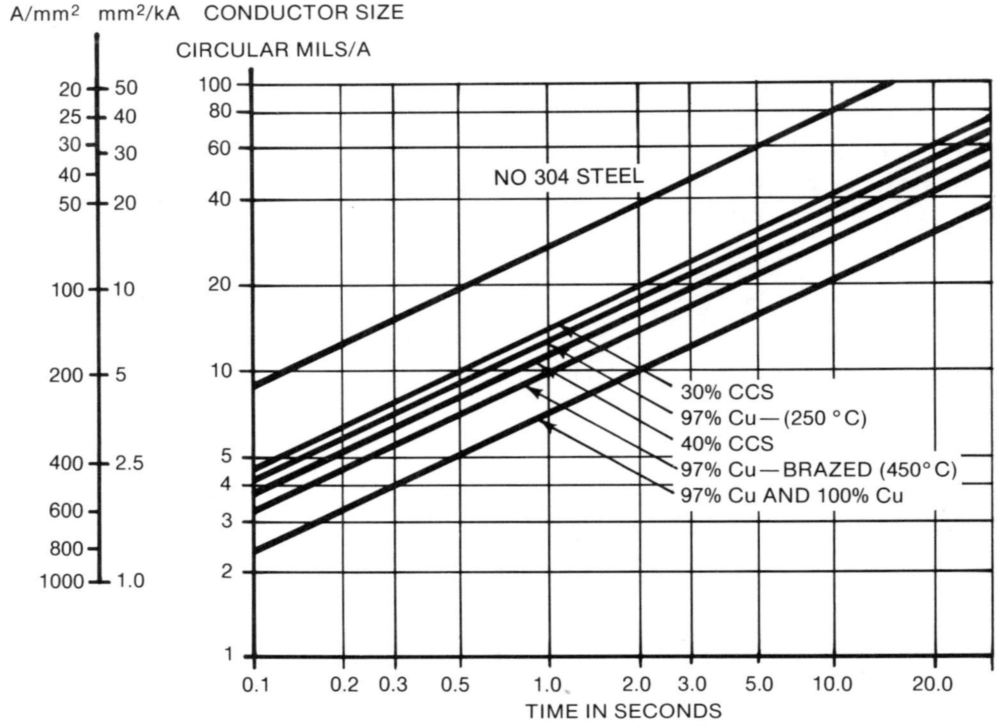

Table 2
Minimum per Unit Conductor Sizes (cmils/A)

Fault Time (s)	100% Cu Only	97% Cu Only	40% CCS Only	30% CCS Only	97% Cu/Temperature Limits (450 °C)	(250 °C)
30.0	38.4	38.7	57.0	65.8	51.1	64.5
4.0	14.0	14.2	20.8	24.0	18.7	23.5
1.0	7.0	7.1	10.4	12.0	9.3	11.8
0.5	4.9	5.0	7.4	8.5	6.6	8.3

9.5 Selection of Joints. All joints that connect various parts of the grounding network into an electrically continuous system of apparatus, conductors, and ground electrodes should be evaluated in terms of conductivity, thermal capacity, mechanical strength, and reliability.

An obvious consideration is to ensure that the connection will withstand expected mechanical stresses without any significant deterioration due to corrosion metal fatigue, and electromagnetic forces for many years.

Electromagnetic forces produced by a high fault current can be severe; copper cables were observed to stretch in staged fault tests when temperatures approached the fusing limit of the tested conductor. Also, where overhead ground wires are installed in tension, some reduction in strength (due to annealing) should be anticipated.

The most common methods of making ground connections utilize exothermic welds, brazed joints, and pressure type connectors. Provided these connectors are properly designed and installed, some of the guidelines of their applications are:

(1) If, for mechanical reasons, annealing of a conductor is a consideration, it may be prudent not to exceed 250 °C regardless of the type of connection used.

(2) The temperature limit of 450 °C is a reasonable value for brazed connections, considering that in practice many copper-based eutectic brazing alloys will start to melt at temperatures less than 600 °C.

(3) Exothermic welded joints will intimately join the cable with a connector or material that has about the same fusing temperature, so that the entire connection can be viewed and rated as being an integral part of one homogeneous conductor.

(4) Pressure type connectors exist in a variety of types and makes. The bolted, wedge, and compression types are most common. In general, pressure type connectors operate at lower temperatures than the conductor. Due to a heat-sink effect caused by the presence of a relatively large connector, the conductor may fuse before the joint fails. Presently, ANSI/IEEE Std 837-1984 [6] provides detailed information on the application and testing of these connectors. If there is uncertainty or a lack of test data, it is reasonably conservative to design for temperatures within a 250-350 °C range. In the previous editions of this guide, a 250 °C limit was suggested for bolted connections.[19]

9.6 Additional Sizing Factors. As a rule, the designer should take precautions to ensure that the temperature of any conductor will not exceed the maximum allowable temperature of the lowest rated component, or some other limitation, such as:

(1) Low temperature due to special circumstances. Typically, conductors near flammable materials could be subjected to more stringent limitations.

(2) Environmental factors. A possible exposure to corrosive environment should be carefully examined. If a gradual degradation of the grounding system could occur during the planned design life, extra allowances should be made in this respect.

The down leads to the grid conductor may be subjected to the total fault current into the grid, while the grid conductor subdivides this current so that each conductor segment in the grid is only subjected to some fraction of the total grid

[19] Research into the origins of this rating has indicated that the main reason for this value was to avoid annealing of hard-drawn copper conductors in above-grade applications [B96].

current. Thus, the down leads may be required to be larger than the grid conductor in order to have sufficient ampacity for this total grid current.

The National Electrical Safety Code ANSI C2-1984 [1] specifies AWG No 6 copper or AWG No 4 aluminum as the minimum size for surge arrester ground leads.

Conductors that are used as ground leads conducting the lightning current seldom require further consideration. The size of a conductor, which is selected according to its fault current duty, usually is also adequate for carrying short-time surges caused by lightning. There is no recorded evidence of a copper conductor larger than AWG No 10 ever being fused because of the passage of lightning current [B5].

9.7 Final Choice of Conductor Size. In practice, the requirements on mechanical reliability will set a minimum conductor size. The earlier AIEE and IEEE guides recommended minimum sizes of 1/0 and 2/0 copper for brazed and bolted joints, respectively. A recent international survey has shown that about 66% of the questioned utilities use No 4/0 copper conductor for grids and approximately 16% prefer to use conductors as large as 500 kcmils. On the other hand, close to 25% of utilities report the use of copper conductors as small as 1/0 without any mechanical problems [B33].

While it might seem proper for the designer to establish minimum sizes in light of local conditions, the need for conservatism deserves consideration. Some of the specific reasons are:

(1) Relay malfunctions and human errors can result in fault duration in excess of primary clearing times. The back-up clearing time is usually adequate for sizing the conductor. For small substations, this may approach 3 s or longer. However, since large substations usually have complex or redundant protection schemes, the fault will generally be cleared in 1 s or less.

(2) The ultimate value of current used to determine the conductor size should take into account the possibility of future growth. It is less costly to include an adequate margin in conductor size during the initial design than to try to reinforce a number of ground leads at a later date.

10. Soil Characteristics

10.1 Soil as a Grounding Medium. The behavior of a ground electrode buried in soil can be analyzed by means of the circuit in Fig 15. As shown, most soils behave both as a conductor of resistance *r*, and as a dielectric. Except for high-frequency and steep-front waves penetrating a very resistive soil material, the charging current is negligible in comparison to the *leakage* current, and the earth can be represented by a pure resistance.

10.2 Effect of Voltage Gradient. The soil resistivity is not affected by a voltage gradient unless the latter exceeds a certain critical value. The value somewhat varies with the soil material, but it usually has the magnitude of several kilovolts per centimeter. Once exceeded, arcs would develop at the electrode surface and progress into the earth so as to increase the effective size of the electrode, until the gradients are reduced to values that the soil material can withstand. This condition is illustrated by the presence of gaps in Fig 15. Since the substation grounding system normally is designed to comply with far more stringent criteria of step and touch voltage limits, the gradient can always be assumed to be below the critical range.

10.3 Effect of Current Magnitude. Soil resistivity in the vicinity of ground electrodes may be affected by current flowing from the electrodes into the surrounding soil. The thermal characteristics and the moisture content of the soil will determine if a current of a given magnitude and duration will cause significant drying and thus increase the effective soil resistivity. A conservative value of current density, as given by Armstrong [B3], is not to exceed 200 A/m^2 for 1 s.

10.4 Effect of Moisture Temperature and Chemical Content. Electrical conduction in soils is essentially electrolytic. For this reason the resistivity of most soils rises abruptly whenever the moisture content accounts for less than 15% of the soil weight. The amount of water further depends upon the grain size, compactness, and variability of the grain sizes. However, as shown in Fig 16, curve 2, the resistivity is little affected once the moisture content exceeds 22% [5].[20]

The effect of temperature on soil resistivity is nearly negligible for temperatures above the freezing point. At 0 °C, the water in the soil starts to freeze and the resistivity increases rapidly. Curve 3 shows this typical variation for a clay soil containing 15.2% of moisture by weight [5].

[20] See, specifically, p 122, Tables 6 and 7 of [5].

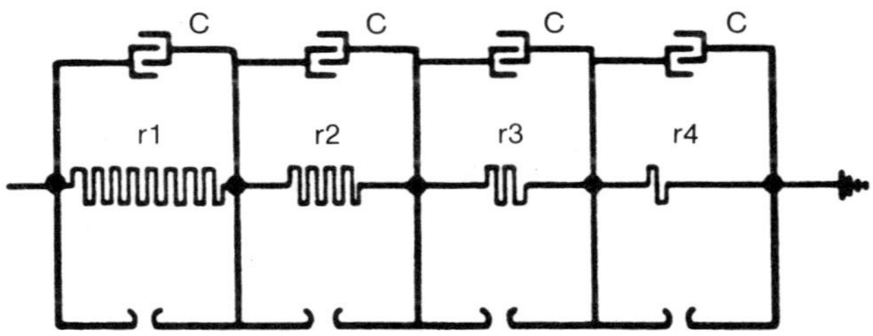

Fig 15
Soil Model

The composition and the amount of soluble salts, acids, or alkali present in the soil may considerably affect its resistivity. Curve 1 of Fig 16 illustrates a typical effect of salt (sodium chloride) on the resistivity of a soil containing 30% of moisture by weight [B106].

Figure 16 should not be used for calculation purposes. To determine the actual soil resistivity, tests such as those described in ANSI/IEEE Std 81-1983 [3] should be performed at the site.

10.5 Use of Crushed-Stone Layer. Gravel or crushed rock coverings, usually about 0.08-0.15 m (3-6 in) in depth, are very useful in retarding the evaporation of moisture and thus in limiting the drying of topsoil layers during prolonged dry weather periods. Also, as discussed in 5.4, covering the surface with a material of high resistivity is very valuable in reducing shock currents. The value of this layer in reducing shock currents is not always fully realized. Tests by Bodier at a station in France showed that the river gravel used as yard surfacing when moistened had a resistivity of 5000 Ω-m. A layer 4-6 in thick decreased the *danger factor* (ratio of body to short-circuit current) by a ratio of 10:1, as compared to the natural moist ground. Tests by Langer in Germany compared body currents when touching a hydrant while standing on *wet coarse gravel* of 6000 Ω-m resistivity with body currents while standing on *dry* sod. The current in the case of *dry* sod was of the order of 20 times the value for *wet coarse gravel.* Tests reported by Elek provide further confirmation of these benefits [B10], [B43], [B66].

In basing calculations on the use of a layer of clean crushed rock or gravel, consideration should be given to the possibility that insulation may become impaired in part through filling of voids by compression of the lowest ballast layers into the soil beneath by material from subsequent excavations, if not carefully removed, and in some areas, by settlement of airborne dust.

The range of resistivity values for a crushed-stone layer depends on many factors, some of which are kinds of stone, size, condition of stone (that is, clean or with fines), amount and type of moisture content, atmospheric contamination, etc. Table 3 indicates that the resistivity of the water with which the rock is wet has considerable influence on the measured resistivity of the crushed-stone layer. Thus, crushed stone subjected to sea spray may have substantially lower resis-

Fig 16
Effects of Moisture, Temperature, and Salt upon Soil Resistivity

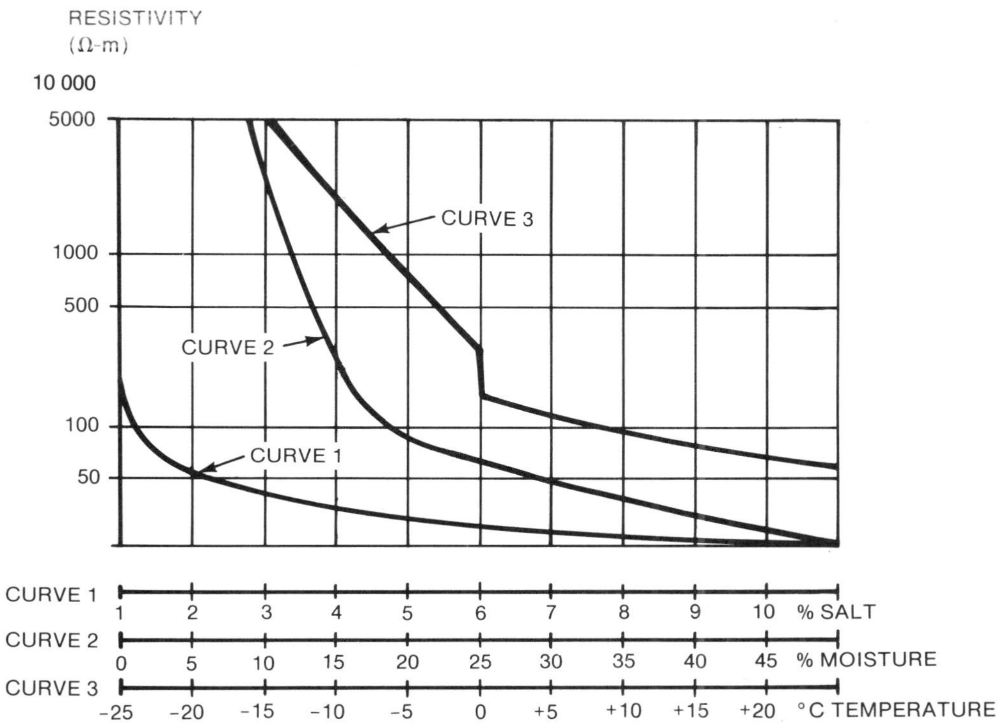

Table 3
Typical Crushed-Stone Resistivities

Description of Rock Sample	Resistivity of Sample (Ω-m) Dry	Wetted with Ground Water	Wetted with Salt Water
Crusher run granite (with fines) [B61]	$141.8 \cdot 10^6$	1318.7	705.0
#57 clean granite[a] [B61]	$192.5 \cdot 10^6$	8106.8	2166.5
Clean limestone[b] [B61]	$7.3 \cdot 10^6$ - $68.5 \cdot 10^6$	2094.8 - 2912.4	1274.8 - 1470.8
Gravel (type and size unknown) [B2]	$1.22 \cdot 10^6$	8534.4	24.4
Crushed rock (type and size unknown) [B2]	$18.3 \cdot 10^6$	4267.2	121.9

[a] Standard size designation from ANSI/ASTM D448-80 [2], approximately ¾-1.0 in.
[b] Nonstandard size = actual gradings as follows: 100% passing 1.0 in screen, 85-95% passing ¾ in screen, 15-25% passing ½ in screen, 5-10% passing ⅜ in screen, 0-2% passing #4 mesh.

11. Soil Structure and Selection of Soil Model

11.1 Investigation of Soil Structure. Field investigation of a station site is most essential for determining both the general soil composition and obtaining some basic ideas as to its homogeneity. Usually, excavations and other civil engineering work are already in progress at or near the site where the ground system will be located. The boring test samples and other geological investigations often provide useful information on the presence of various layers and the nature of soil material, leading at least to some ideas as to its resistivity and the range of values at the site.

11.2 Classification of Soils and Ranges of Resistivity. A number of tables exist in the literature, showing the ranges of resistivity for various soils and rocks. The tabulation from Rüdenberg [B88] has the advantage of extreme simplicity. More detailed data are available in engineering handbooks and publications [B98], [B110]. See Table 4.

11.3 Resistivity Measurements. Estimates based on soil classification yield only a rough approximation of the resistivity. Actual resistivity tests therefore are imperative. These should be made at a number of places within the site. Station sites where the soil may possess uniform resistivity throughout the entire area and to a considerable depth are seldom found. Typically, there are several layers, each having a different resistivity. Most often lateral changes also occur, but in comparison to the vertical ones, these changes usually are more gradual. Soil resistivity tests should be made to determine any important variation of resistivity with depth. As a rule, the number of such readings taken should be greater where the variations are large, especially if some readings are so high as to suggest a possible safety problem.

If the resistivity varies appreciably with depth, it is often desirable to use an increased range of probe spacings. The idea is that a fairly accurate estimate for still greater spacings can be determined by extrapolation. This is possible because, as the probe spacing is increased, the test source current penetrates more and more distant areas, in both vertical and horizontal directions, regardless of how much the current path is distorted due to the varying soil conditions [B71].

A number of measuring techniques are described in detail in ANSI/IEEE Std 81-1983 [3]. The Wenner's four-pin method is the most commonly used technique. In brief, four probes are driven into the earth along a straight line, at equal distances A apart, driven to a depth B. The voltage between the two inner (poten-

Table 4
Range of Earth Resistivity

Type of Earth	Average Resistivity
Wet organic soil	10 Ω-m
Moist soil	10^2 Ω-m
Dry soil	10^3 Ω-m
Bedrock	10^4 Ω-m

tial) electrodes is then measured and divided by the current between the two outer (current) electrodes to give a value of mutual resistance R. Then,

$$\rho = \frac{4\,\pi\,AR}{1 + \dfrac{2A}{\sqrt{A^2 + 4\,B^2}} - \dfrac{A}{\sqrt{A^2 + B^2}}} \qquad \text{(Eq 35)}$$

where

ρ = resistivity of the soil in Ω-m
R = resistance in ohms resulting from dividing the voltage between the potential probes by the current flowing between the current electrodes
A = distance between adjacent electrodes in m
B = depth of the electrodes in m

If B is small compared to A, as is the case of probes penetrating the ground a short distance only, the above equation can be reduced to

$$\rho = 2\pi AR \qquad \text{(Eq 36)}$$

The current tends to flow near the surface for small probe spacings, whereas more of the current penetrates deeper soils for large spacings. Thus, it is customary to assume that the resistivity measured for a given probe spacing A represents the apparent resistivity of the soil to a depth of A. Thus, Eq 35 can be used to determine the apparent resistivity (ρ_a) at a depth A.

Palmer modified the above method to give greater sensitivity for large probe spacings, as described in ANSI/IEEE Std 81-1983 [3], [B84].

Resistivity measurement records should include temperature data and information on the moisture condition of the soil at the time of measurement. All data available on buried conductors already known or suspected to be in the area studied should also be recorded.

Buried bare conductors in contact with the soil can invalidate readings made by the method described if they are close enough to alter the test current flow pattern. For this reason, the soil resistivity measurements are of little value in an area where grid conductors have already been installed, except, perhaps, for shallow-depth measurements in or near the center of a very large mesh rectangle. In such cases, a few approximate readings might be taken in a short distance outside the grid, with the probes so placed as to minimize the effect of the grid on the current flow pattern. Though not conclusive as to conditions inside the grid, such readings may be used for approximation, especially if there is reason to believe that the soil in the entire area is reasonably homogeneous.

11.4 Uniform Soil Assumption. The derivation of the above equations for soil measurements is based on the assumption that the soil resistivity is uniform. This requires that the soil resistivity is constant both laterally and with depth to infinity. Obviously, this is never the case. However, this assumption can be made without significant error if the soil is essentially uniform (both horizontally and vertically) to a distance (measured from the edge of the grid) or approximately 3 - 5 times the diagonal dimension of the grid. The uniform soil assumption can be used with less accuracy when the resistivity varies slightly with depth by using the apparent resistivity (ρ_a), as described in 11.3.

11.5 Nonuniform Soil Assumptions. More exact theoretical approaches to situations where resistivity varies markedly with depth are suggested by Sunde, and in some of the books on geophysical prospecting to which he refers. For example, it is often possible, from field readings taken with a wide range of probe spacings, to deduce a stratification of the earth into two or more layers of appropriate thickness that will account for the actual test variations [3], [B98].

11.5.1 Two-Layer Soil Model. Remarks below are limited to the assumption of the simplest soil stratification, that is, it is anticipated that a two-layer model is reasonably valid for the actual soil conditions and the range of resistivity variations found on the site. Luckily, in practice it is often possible to satisfy these requirements without risking a serious calculation error.

In principle, as shown in Sunde [B98]:

(1) A grounding system in a two-layer soil environment behaves differently in comparison with the same system in uniform soil. Generally, for a grounding system in uniform soil or in two-layer soil with ρ_1 less than ρ_2 (upper layer soil resistivity less than lower layer resistivity), the current density is higher in the conductors at the outer edges of the grounding grid. In two-layer soil with ρ_1 greater than ρ_2 (the soil in the upper layer is more resistive than the lower layer soil), the current density is more uniform over all the conductors of the grounding system. This is caused by the tendency of the leakage current to go downward into the layer of lower resistivity, rather than up and outward to the more resistive upper layer. More recent studies by Thapar and Gross [B105] and Dawalibi *et al.* [B37] provide a wealth of information on this subject.

(2) The equations that govern the performance of a grounding system buried in multilayer soil can be obtained by solving Laplace's equations for a point current source, or by the method of images, which gives identical results. The use of either method in determining the earth potential caused by a point current source results in an infinite series of terms representing the contributions of each consequent image of the point current source, as shown in Fig 16 and described below. Exact formulation of the equations that include these effects is given in numerous references, for instance, [B35], [B57], and [B98].

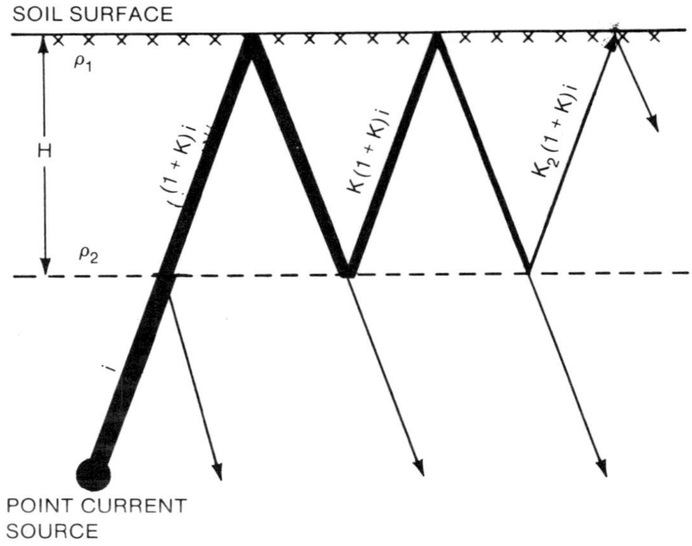

Fig 17
Reflections of Current in Two-Layer Soil With Current Source in Lower Soil

The abrupt changes in resistivity at the boundaries of each soil layer can be described by means of a reflection factor. This reflection factor K is defined as

$$K = \frac{\rho_2 - \rho_1}{\rho_1 + \rho_2} \tag{Eq 37}$$

where ρ_1 and ρ_2 are the resistivity values of the upper and lower layers of soil, respectively. In Fig 17, an observer in the lower layer of soil would see a current source of magnitude i, an image reflected from the subsoil interface of magnitude $-Ki$, and an infinite series of images reflected from the topsoil surface of magnitudes $K^n(1-K)^2$, where n goes from 0 to infinity. An observer in the top layer would see a source and its reflection at the top soil surface, both with apparent magnitude $(1+K)i$, and an infinite series of pairs of reflections having magnitudes $K^n\,(1+K)i$, where n goes from 1 to infinity. These reflections would be at successively greater heights and depths. A similar figure could be drawn to represent the case of a current source in the topsoil layer.

While the most accurate representation of a grounding system should certainly be based on the actual variations of soil resistivity present at the substation site, it will rarely be economically justifiable or technically feasible to model all these variations. However, in most cases, the representation of a ground electrode based on an equivalent two-layer earth model is sufficient for designing a safe grounding system.

ANSI/IEEE Std 81-1983 [3] provides convenient methods for determining the equivalent resistivities of the upper and lower layers of soil and the height of the upper layer for such a model.

11.5.2 Comparison of Uniform and Two-Layer Soil Model. This two-layer model approach has been found to be much more accurate than the uniform soil model. Some of the reasons are:

(1) Variations in soil resistivity have considerable influence on the performance of most grounding systems, affecting both the value of ground resistance and ground potential rise, and the step and touch surface voltages. In general, for negative values of K (upper layer more resistive than lower layer), the resistance is less than that of the same grounding system in uniform soil with resistivity ρ_1. In contrast, for positive values of K, the resistance is generally higher than that in uniform soil with resistivity ρ_1. A similar relationship exists for the step and touch voltages produced on the surface of a two-layer earth versus that on the surface of uniform soil. For negative values of K, the step and touch voltages are generally lower than the voltages for the same grounding system in uniform soil of resistivity ρ_1. Also, for positive values of K, the step and touch voltages are generally higher than in uniform soil.[21]

(2) Other parameters, such as the surface layer height H, also affect the differences in the performance of ground electrodes in a two-layer environment and in uniform soil conditions. The general rule is that when the upper layer height H becomes significantly larger than the electrode's own dimensions, the performance of the electrode approaches the performance of the same electrode in uniform soil of resistivity ρ_1.

(3) Also, it must be recognized that the above characteristics are based on the premise of a constant fault current source. The actual currents in the grounding system will change from case to case as a function of ρ_1 and ρ_2, reflecting the local changes relative to all other ground fault current paths predetermined by the fault location. This current division is discussed in Section 13. Therefore, in certain cases some of the assumptions given above may not always hold true.

Since the use of two-layer or multilayer models necessitates the application of digital computers or similar means having large memory space available, it is impractical to insist on the use of multilayer models for all grounding studies. For design applications involving relatively simple grounding arrangements of electrodes buried in a reasonably uniform soil, the approximate methods provided elsewhere in the guide will be suitable for obtaining a realistic design with adequate safety margins. However, for designs involving a large grounded area, odd-shaped grids, etc, or where the resistivity of soil is clearly very nonuniform, the engineer responsible for the design should decide which method to use and whether or not a multilayer model is warranted, based on all the information available [B112].

---

[21] As discussed in 10.5, it is a common practice to have a thin layer of crushed rock overlaying the grounded area of a substation. It could appear that such a high resistivity layer, having the layer height H, much less than the depth of the grounding system, might worsen both the step and touch voltage. However, this is not the case. The crushed rock surfacing is used to increase the contact resistance between a person's foot and the earth's surface. Thus, for a given maximum allowable body current, considerably higher step and touch voltages can be allowed if a crushed rock surfacing is present.

12. Evaluation of Ground Resistance

12.1 Usual Requirements. An ideal grounding should provide a near zero resistance to remote earth. In practice, the ground potential rise at the station site increases proportionally to the fault current; the higher the current, the lower value of a total system resistance thus has to be obtained. For most transmission and other large substations, the ground resistance should be about 1 Ω or less. In smaller distribution substations the usually acceptable range is from 1-5 Ω, depending on the local conditions.

12.2 Simplified Calculations. Estimation of the total resistance to remote ground is one of the first steps in determining the size and basic layout of a grounding system. At first glance this may appear difficult: the grounding system is not yet designed and so its resistance, depending on the design, is unknown. Fortunately, the station resistance depends primarily on the area to be occupied by the ground system, which is usually known in the early design stage.

Thus, as a first approximation, a minimum value of the substation grounding resistance in uniform soil can be estimated by means of the formula of a circular metal plate at zero depth once the soil resistivity has been determined:

$$R_g = \frac{\rho}{4}\sqrt{\frac{\pi}{A}} \quad \text{(Eq 38)}$$

where

R = station ground resistance in Ω
ρ = average earth resistivity in Ω-m
A = the area occupied by the ground grid in m^2

Next, an upper limit of the substation resistivity can be obtained by adding a second term to the above formula, as proposed by Laurent [B67] and Niemann [B80]:

$$R_g = \frac{\rho}{4}\sqrt{\frac{\pi}{A}} + \frac{\rho}{L} \quad \text{(Eq 39)}$$

where L is the total[22] buried length of conductors (in m).

[22] In the case of a grid-rod combination in uniform soil, a combined length of horizontal conductors and ground rods will yield a slightly conservative estimate of L, since ground rods usually are more effective on a per unit length basis.

The second term recognizes the fact that the resistance of any actual grounding system that consists of a number of conductors is higher than that of a solid metallic plate, and that the difference will decrease with the increasing length of buried conductors, approaching 0 for infinite L, when the condition of a solid plate is reached.

Equations 38 and 39 can be used with reasonable accuracy for grid depths less than 0.25 m. For grid depths between 0.25 and 2.5 m, correction for the grid depth is required. Using Sverak's approximation,

$$R_g = \rho \left[\frac{1}{L} + \frac{1}{\sqrt{20A}} \left(1 + \frac{1}{1 + h\sqrt{20/A}} \right) \right] \qquad \text{(Eq 40)}$$

where h is the depth of the grid. For grids without ground rods, this formula has been tested to yield results that are practically identical to those obtained with Eq 42 of Schwarz, described in 12.3 (see also [B100]).

The following tabulation from Kinyon's report [B63] offers some idea of how the calculated and actual measured resistance for five different substations compare. Equation 39 was used to compute the grid resistance. See Table 5.

Recommendations

(1) Equation 38 should be used only when a value of substation resistance is desired for estimating the maximum fault current.

(2) Equation 39 or 40 should be helpful in estimating the substation ground potential rise for a preliminary design evaluation, to determine the approximate length of buried conductors needed for control of the step and touch voltages.

Table 5
Typical Grid Resistances

Parameter	SUB 1	SUB 2	SUB 3	SUB 4	SUB 5
Soil Texture	Sand & Gravel	Sandy Loam	Sand & Clay	Sand & Gravel	Soil & Clay
Resistivity (Ω-m)*	2000	800	200	1300	28
Grid area (ft^2)	15 159	60 939	18 849	15 759	61 479
Buried length (ft)	3120	9500	1775	3820	3000
R_g (calculated Ω)	25.7	4.97	2.55	16.15	0.19
R_g (measured Ω)	39.0	4.10	3.65	18.2	0.21

* An average value of all measured resistivity values is frequently substituted for the *uniform soil* resistivity in Eq 39. If this average resistivity is used, Eq 39 usually produces a resistance that is higher than the value that would result from a direct resistance measurement. The calculated and measured resistance values shown above do not reflect this trend, because Kinyon based his calculations on the "... lowest average value of resistivity measured on the site." Readers are referred to Kinyon's report for further discussion on his choice of resistivity values used in Table 5 [B63].

(3) For better estimates of the ground resistance of grids with ground rods, equations such as the Schwarz formula, described next, should preferably be used.

12.3 Schwarz's Formula. Total resistance of a system consisting of a combination of horizontal (grid) and vertical (rods) electrodes is lower than the resistance of either component alone, but still higher than that of their parallel combination. The total resistance is [B89], [B98]:

$$R_g = \frac{R_1 R_2 - R_{12}^2}{R_1 + R_2 - 2R_{12}} \quad \text{(Eq 41)}$$

where

R_1 = resistance of grid conductors
R_2 = resistance of all ground rods (rodbed)
R_{12} = mutual resistance between the group of grid conductors and group of ground rods

Schwarz developed a set of convenient formulas, defining R_1, R_2, and R_{12} in terms of the basic design parameters, assuming uniform soil conditions [B92].

However, in practice, it is often desirable to drive ground rods deep into the ground to reach more conductive soils. In the equations below, the expressions for R_2 and R_{12} have been modified for such a possibility.

$$R_1 = (\rho_1/\pi l_1)\,(\ln(2l_1/h') + K_1\,(l_1/\sqrt{A}) - K_2) \quad \text{(Eq 42)}$$

$$R_2 = (\rho_a/2n\pi l_2)\,[\ln(8l_2/d_2) - 1 + 2K_1\,(l_2/\sqrt{A})\,(\sqrt{n} - 1)^2] \quad \text{(Eq 43)}$$

$$R_{12} = (\rho_a/\pi l_1)\,[\ln(2l_1/l_2) + K_1\,(l_1/\sqrt{A}) - K_2 + 1] \quad \text{(Eq 44)}$$

where

ρ_1 = soil resistivity encountered by grid conductors buried at depth h in Ω-m
ρ_a = apparent soil resistivity as seen by a ground rod in Ω-m,
H = thickness of the upper layer soil in m
ρ_2 = soil resistivity from depth H downward in Ω-m
l_1 = total length of grid conductors in m
l_2 = average length of a ground rod in m
h = depth of grid burial in m
h' = $\sqrt{d_1 h}$ for conductors buried at depth h, or 0.5 d_1 for conductors at $h = 0$ (on earth's surface)
A = area covered by a grid of dimensions $a \cdot b$ in m^2
n = number of ground rods placed in area A
K_1, K_2 = constants related to the geometry of the system [Fig 18(a) and (b)]
d_1 = diameter of grid conductor in m
d_2 = diameter of ground rods in m
a = short-side grid length in m
b = long-side length in m

The preceding Eqs 42, 43, and 44 are valid for a two-layer soil environment, with upper layer thickness H, in which ground rods penetrate the more conductive lower layer. In such a case, that is for $\rho_1 \geq \rho_2$, where the grid is buried in the upper layer ρ_1, but the ground rods are partly in ρ_1 and partly in ρ_2, R_2, and R_{12} are calculated with the use of an apparent soil resistivity seen by the ground rods, ρ_a, defined as follows:

$$\rho_a = l_2 (\rho_1 \rho_2)/(\rho_2 H + \rho_1 (l_2 - H)) \qquad \text{(Eq 45)}$$

if the top of each ground rod is flush with the earth's surface. The practical field testing and derivation of this expression is described in [B9].

For the more usual case of the rod's top being in the same depth as the grid,

$$\rho_a = l_2 (\rho_1 \rho_2) / (\rho_2 (H-h) + \rho_1 (l_2 + h-H)) \qquad \text{(Eq 46)}$$

For uniform soils, $\rho_2 = \rho_1$.

If the difference between ρ_1 and ρ_2 is not too great (preferably ρ_2 not lower than 0.2 ρ_1), and the first layer thickness H is at least 0.1 b, the resulting equations are reasonably accurate for most practical calculations and relatively easy to use [B78], [B79].[23] Moreover, the ability to work with separate expressions for a grid and a set of rods becomes advantageous in simplified calculations.

A slight problem with the application of these equations was that the factors (coefficients) K_1 and K_2 had been originally presented by Schwarz only in a graphical form.

However, given the near-linear character of these curves, it is possible to use a linearized form of $y = px + q$ to obtain K_1 and K_2 within the range of values shown in Fig 18(a) and (b); or to linearly interpolate between several points taken from the original curve. Finally, in computer applications, it might be worthwhile to use the more elaborate expressions derived by Kercel [B62].

He also provides valuable information on the use of Schwarz's formula for determining the minimum buried length of conductors to attain a specified resistance for one, two, or three interconnected ground grids in uniform soil.

12.4 Note on Resistance of Primary Electrodes. In general, the resistance of any primary electrode depends on the soil resistivity and the size and type of arrangement of all individual conductors comprising the ground electrode. In more complex arrangements involving criss-crossed wires and a large number of rods in the same area, the mutual resistance between individual elements plays an important role. More literature exists on this aspect of accurate evaluation of grounding systems than any other. For studies utilizing computers, numerous references are available.[24]

12.5 Chemical Treatment of Soils and Use of Bentonite. It is often impossible to achieve the desired reduction in ground resistance by adding more grid con-

[23] For more difficult two-layer soil conditions, Nahman provides more extensive corrections.

ductors or ground rods. An alternate solution is to effectively increase the diameter of the electrode by modifying the soil surrounding the electrode. As the current density and ground resistance are inversely proportional to the diameter of an electrode, the inner shell of soil closest to the metal normally comprises the bulk of the electrode resistance to remote earth. This phenomenon is often utilized to an advantage, as follows:

(1) Use of sodium chloride, magnesium, and copper sulfates, or calcium chloride, to increase the conductivity of the soil immediately surrounding an electrode.

(2) Use of bentonite, a natural clay containing the mineral montmorillionite, which was formed by volcanic action years ago. It is noncorrosive, stable, and has a resistivity of 2.5 Ω-m, at 300% moisture. The low resistivity results mainly from an electrolytic process between water, Na_20 (soda), K_20 (potash), CaO (lime), MgO (magnesia), and other mineral salts that ionize forming a strong electrolyte with ph ranging from 8-10. Unlike a salt bed, this electrolyte will not gradually leach out, as it is part of the clay itself. Provided with a sufficient amount of water, it swells up to 13 times its dry volume and will adhere to nearly any surface it touches. If exposed to direct sunlight, it tends to seal itself off, preventing the drying process from penetrating deeper.

Due to its hygroscopic nature, it acts as a drying agent drawing any available moisture from the surrounding environment. Bentonite needs water to obtain and maintain its beneficial characteristics. Its initial moisture content is obtained at installation when the slurry is prepared. Once installed, bentonite relies on the presence of ground moisture to maintain its characteristics. Most soils have sufficient ground moisture so that drying out is not a concern. The hygroscopic nature of bentonite will take advantage of the available water to maintain its *as installed* condition. However, for this same reason it will not function well in a very dry environment. In dry soils it will not be able to maintain its moisture content, and it will shrink away from the electrode, increasing the electrode resistance. When sufficient soil moisture is available, this limitation disappears and bentonite is an excellent backfill material that allows a substantial reduction of the resistance of ground rods in highly resistive soils [B60].

12.6 Concrete-Encased Electrodes. Concrete, being hygroscopic, attracts moisture. Buried in soil, a concrete block behaves as a semiconducting medium with a resistivity of 30-90 Ω-m. This is of particular interest in medium and highly resistive soils, since a wire or metallic rod encased in concrete has lower resistance than a similar electrode buried directly in the earth. This encasement reduces the resistivity of the most critical portion of material surrounding the metal element in much the same manner as does a chemical treatment of soils. However, this

[24] Some of the recent references, such as that of Garrett and Holley [B52], become truly useful when the algorithm for computing the grid resistance is incorporated into a program that already has provisions for a mathematical treatment of large arrays, which are further used in solving the problem of a surface gradient field.

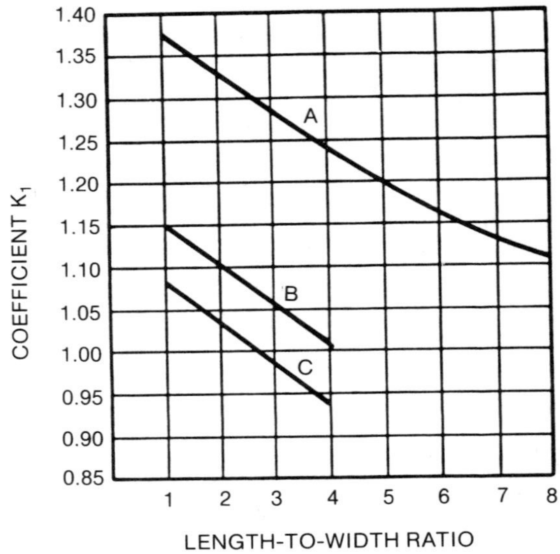

CURVE A — FOR DEPTH h = 0
$y_A = -0.04x + 1.41$
CURVE B — FOR DEPTH h = 1/10 $\sqrt{AREA}$
$y_B = -0.05x + 1.20$
CURVE C — FOR DEPTH h = 1/6 $\sqrt{AREA}$
$y_C = -0.05x + 1.13$

Fig 18
Coefficients K_1 and K_2 of Schwarz's Formula
(a) Coefficient K_1

phenomenon may often be found as being both a design advantage and disadvantage. Some of the reasons are:

(1) On the one hand, it is impractical to build foundations for structures where the inner steel (reinforcing bars) is not electrically connected to the metal of the structure. Even if extreme care were taken with the anchor bolt placement in order to prevent any direct metal-to-metal contact, the semiconductive nature of concrete would provide an equally effective connection.

(2) On the other hand, the presence of a small dc current can cause corrosion of rebar material. Although ac current as such does not produce corrosion, approximately 0.01% of the ac current becomes rectified at the interface of the steel bar and concrete [B86].

(3) As a result, splitting of concrete may occur either due to the above phenomenon (since corroded steel occupies approximately 2.2 times its original volume, producing pressures approaching 5000 lbf/in^2), or due to the passage of a very high current, which would vaporize the moisture in the concrete.

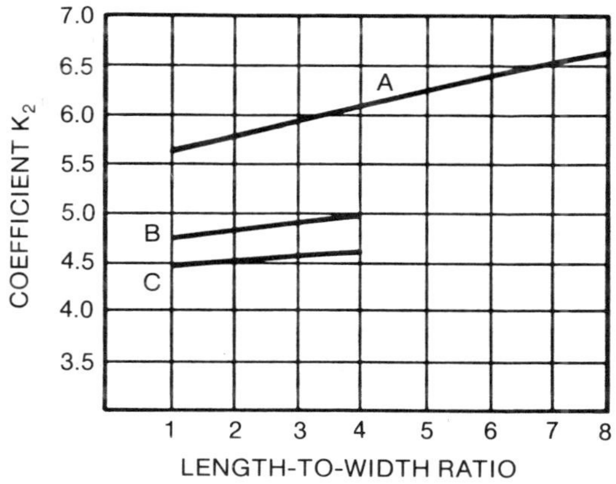

CURVE A — FOR DEPTH h = 0
$y_A = 0.15x + 5.50$
CURVE B — FOR DEPTH h = 1/10 $\sqrt{AREA}$
$y_B = 0.10x + 4.68$
CURVE C — FOR DEPTH h = 1/6 $\sqrt{AREA}$
$y_C = -0.05x + 4.40$

Fig 18 (Continued)
(b) Coefficient K_2

Fortunately, there is a certain threshold potential for dc corrosion, approximately 60 V dc, below which no corrision will occur. A number of field tests concerning the maximum current loading is reported in the literature [B41], [B73], [B12]. The short-time current loading capacity, I_{CE}, of concrete-encased electrodes can be estimated by means of Ollendorf's formula[25] for an indefinitely sustainable current I_∞, adjusted by a 1.4 multiplying factor, or directly from Fig 19.

$$I_{CE} = 1.4\,(I_\infty) = \frac{1.4}{R_z}\sqrt{2\,\lambda_g\,\rho\,(T_\nu\ T_a)} \qquad \text{(Eq 47)}$$

where

λ_g = thermal conductivity of the earth in W/m °C
R_z = ground resistance of the concrete-encased electrode in Ω
ρ = soil resistivity in Ω-m
T_a = ambient temperature in °C
T_ν = maximum allowable temperature to prevent sudden evaporation of moisture in °C
I_∞ = indefinitely sustainable current in A

[25] Ollendorf neglects the cooling effect of evaporated moisture in calculating I_∞ [B82].

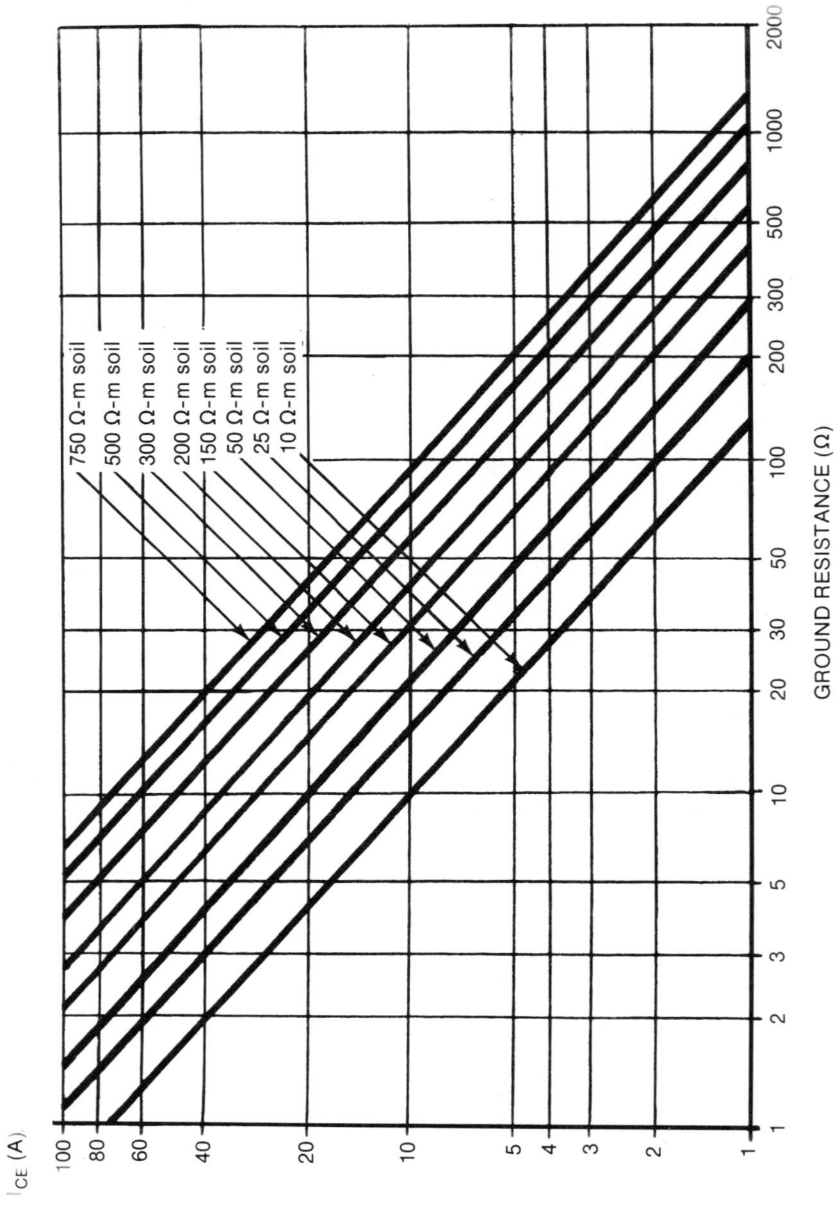

Fig 19
Short-Time Current Loading Capability of Concrete-Encased Ground Electrodes

The applicability of this formula has been verified in [B12], which reports on the results of extensive field testing of concrete poles. In general, if damages are to be prevented, the actual current should be less than the value of I_{CE} found by Eq 47. A 20-25% safety margin is reasonable for most practical applications.

Thus, with proper precautions, the concrete-encased foundations may be used as auxiliary ground electrodes.

Bibliographic reference [B48] uses the following equation for obtaining the effective resistance of a vertical rod encased in concrete, $R_{\text{CE-rod}}$:

$$R_{\text{CE-rod}} = \frac{1}{2\pi l} \left(\rho_c \left[\ln (D/d)\right] + \rho \left[\ln (8l/D)-1\right]\right) \qquad \text{(Eq 48)}$$

where

ρ_c = resistivity of concrete in Ω-m
ρ = resistivity of soil in Ω-m
l = length of a ground rod in m
d = diameter of a ground rod in m
D = diameter of a concrete shell in m

Since the above equation can be related to the commonly used formula for a bare ground rod of length l and diameter d, as shown below,

$$R_{\text{rod}} = \frac{\rho}{2\pi l} \left[\ln (8l/d) -1\right] \qquad \text{(Eq 49)}$$

then Eq 48 can be resolved into

$$R_{\text{CE-rod}} = \frac{1}{2\pi l} \{\rho \left[\ln (8l/D) -1\right] + \rho_c \left[\ln (8l/d) -1\right] - \rho_c \left[\ln (8l/D) -1\right]\} \qquad \text{(Eq 50)}$$

representing a combination of two resistances in series:

(1) Resistance (calculated by Eq 49) of a concrete *rod* of diameter D, directly buried in soil ρ

(2) Resistance of the inner segment of diameter D, containing a metal core of diameter d

Obviously, the latter term is obtained as a difference of the hypothetical resistance values for a rod in concrete, if d and D are entered into the single-medium formula Eq 49, and ρ is replaced by ρ_c.

Such an approach is generally valid for any other electrode having a different shape. Noting, for convenience,

$$R_{\text{SM}} = F(\rho, S_o, G) \qquad \text{(Eq 51)}$$

$$R_{\text{DM}} = F(\rho_c, S_o, G) + F(\rho, S_i, G) \epsilon F(\rho_c, S_i, G) \qquad \text{(Eq 52)}$$

where, in addition to the symbols already mentioned,

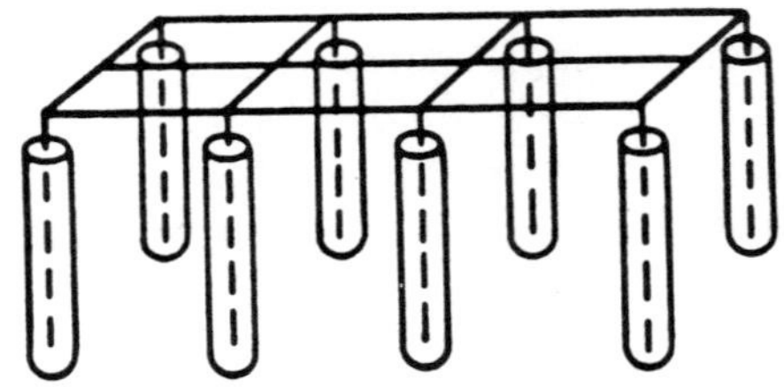

Fig 20
Grid With Encased Vertical Electrodes

R_{SM} = electrode resistance in single medium in Ω
R_{DM} = electrode resistance in dual medium in Ω
S_o = surface area of a given electrode in m^2
S_i = area of interface in m^2
G = a certain geometrical factor characterizing the particular shape of a given electrode

This form is adaptable to a variety of electrodes, buried in soil, and assumed to be surrounded by a concentric shell of a material that has different resistivity than the soil. One possible model of this type, for which Schwarz's formula for a rodbed can easily be modified, is shown in Fig 20.

Recommendations

(1) Connect anchor bolt and angle stubs to the reinforcing steel for a reliable metal-to-metal contact.

(2) Reduce the current duty and dc leakage to allowable levels by making sure that enough primary ground electrodes (grounding grid and ground rods) will conduct most of the fault current.

(3) Bentonite clays may be used in the areas of a high soil resistivity to reduce the resistance of primary grounding. Augering a 150-250 mm (6-10 in) hole and backfilling it with a bentonite slurry around a ground rod is a useful method to prevent the predominance of auxiliary electrodes in dissipating the fault current.

13. Determination of Maximum Grid Current

13.1 Procedure and Related Definitions. The following steps are involved in determining the correct design value of maximum grid current I_G for use in substation grounding calculations:

Step (a). Assess the type and location of those ground faults that are likely to produce the greatest flow of current between the grounding grid and surrounding earth, and hence the greatest rise in grid potential with respect to remote earth (GPR) and largest local surface potential gradients in the substation area.

Step (b). Determine, by computation, the fault current division factor S_f for each of the faults selected in Step (a), and establish the corresponding values of symmetrical grid current I_g.

Step (c). For each fault, based on its duration time t_f, determine the value of decrement factor D_f to allow for the effects of asymmetry of the fault current wave.

Step (d). Select the largest product $D_f I_g$, and hence the worst fault condition, and establish the value of a projection factor C_p to obtain the margin for the future system growth.

Definitions related to the terms used above are as follows.

symmetrical grid current. That portion of the symmetrical ground fault current that flows between the grounding grid and surrounding earth. It may be expressed as

$$I_g = S_f I_f \qquad \text{(Eq 53)}$$

where

I_g = symmetrical grid current in A
I_f = rms value of symmetrical ground fault current in A
S_f = current division factor relating the magnitude of fault current to that of its portion flowing between the grounding grid and surrounding earth

maximum grid current. A design value of the maximum grid current, defined as follows:

$$I_G = C_p D_f I_g \qquad \text{(Eq 54)}$$

where

I_G = maximum grid current in A
D_f = decrement factor for the entire duration of fault t_f, found for t_f given in s
C_p = corrective projection factor accounting for the relative increase of fault currents during the station lifespan; for a zero future system growth $C_p = 1$
I_g = rms symmetrical grid current in A

fault current division factor. A factor representing the inverse of a ratio of the symmetrical fault current to that portion of this current that flows between the grounding grid and surrounding earth.

NOTE: For the purposes of calculating the design value of maximum grid current and symmetrical grid current per definitions of symmetrical grid current and maximum grid current, it may be assumed that the ratio is constant during the entire duration of a given fault. Consequently, this factor may be expressed as

$$S_f = \frac{I_g}{3\,I_0} \qquad \text{(Eq 55)}$$

where

I_g = symmetrical grid current
I_0 = zero-sequence fault current

subtransient reactance. Reactance of a generator at the initiation of a fault. This reactance is used in calculations of the initial symmetrical fault current. The current continuously decreases, but it is assumed to be steady at this value as a first step, lasting approximately 0.05 s after a suddenly applied fault.

transient reactance. Reactance of a generator between the subtransient and synchronous states. This reactance is used for the calculation of the symmetrical fault current during the period between the subtransient and steady states. The current decreases continuously during this period, but is assumed to be steady at this value for approximately 0.25 s.

synchronous reactance. Steady-state reactance of a generator during fault conditions used to calculate the steady-state fault current. The current so calculated excludes the effect of the automatic voltage regulator or governor.

dc offset. Difference between the symmetrical current wave and the actual current wave during a power system transient condition. Mathematically, the actual fault current can be broken into two parts, a symmetrical alternating component and a unidirectional (dc) component, either or both with decreasing magnitudes (usually both). The unidirectional component can be of either polarity, but will not change polarity, and will disappear at some predetermined rate.

dc offset factor. Ratio of the peak fault current to the peak symmetrical value.

***X/R* ratio.** Ratio of the system inductive reactance to resistance. It is proportional to the L/R ratio of time constant, and is, therefore, indicative of the rate of decay of any dc offset. A large X/R ratio corresponds to a large time constant and a slow rate of decay.

13.2 Types of Ground Faults. Many different types of faults may occur in the system. Unfortunately, it may be difficult to determine which fault type and location will result in the greatest flow of current between the ground grid and surrounding earth (current I_G in Figs 21, 22, 23, and 24) since no simple rule applies.

In determining the applicable fault types, consideration should be given to the probability of occurrence of the fault. Multiple simultaneous faults, even though they may result in higher ground current, should not be considered if their probability of occurrence is negligible. It is thus recommended, for practical reasons, that investigation be confined to single-line-to-ground and line-to-line-to-ground faults.

In the case of a line-to-line-to-ground fault, the zero-sequence fault current is

$$I_0 = \frac{E\,(R_2 + \mathrm{j}X_2)}{(R_1 + \mathrm{j}X_1)\,(R_0 + R_2 + 3R_f + \mathrm{j}\,(X_0 + X_2)) + (R_2 + \mathrm{j}X_2)\,(R_0 + 3R_f + \mathrm{j}X_0)} \qquad \text{(Eq 56)}$$

Fig 21
Fault Within Local Station; Local Neutral Grounded

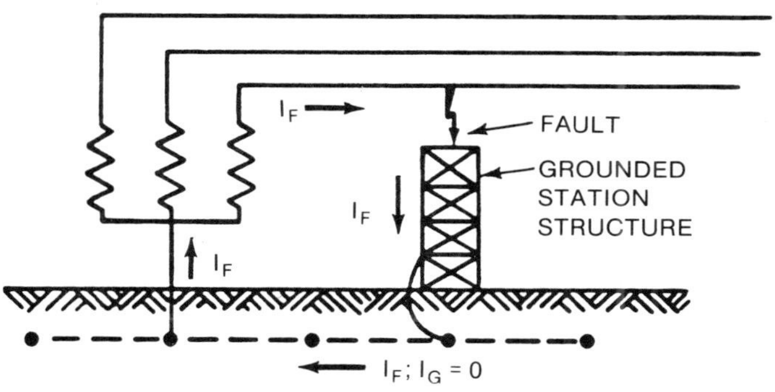

Fig 22
Fault Within Local Station; Neutral Grounded at Remote Location

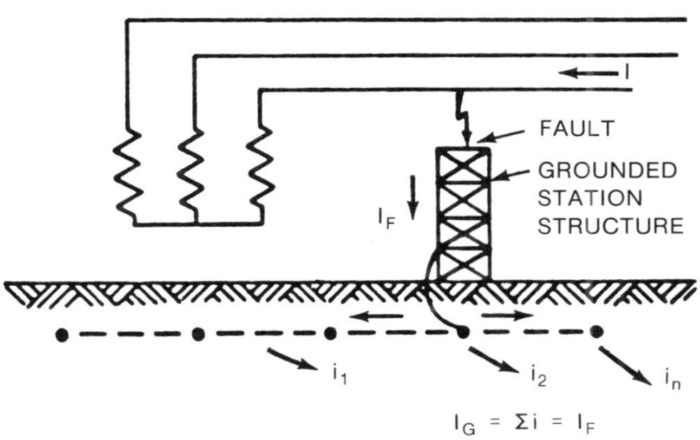

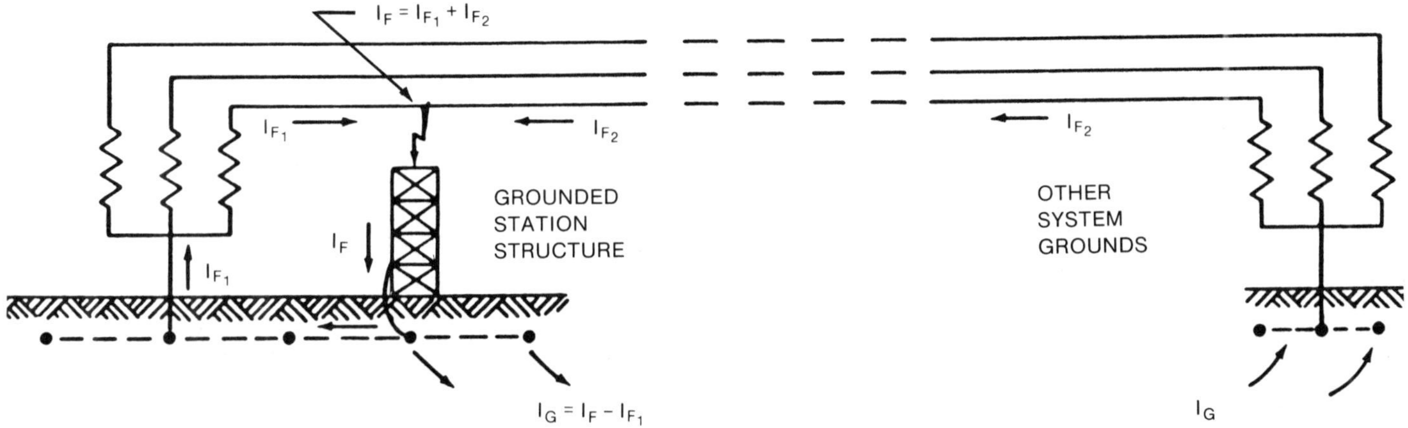

Fig 23
Fault in Station; System Grounded at Local Station and Also at Other Points

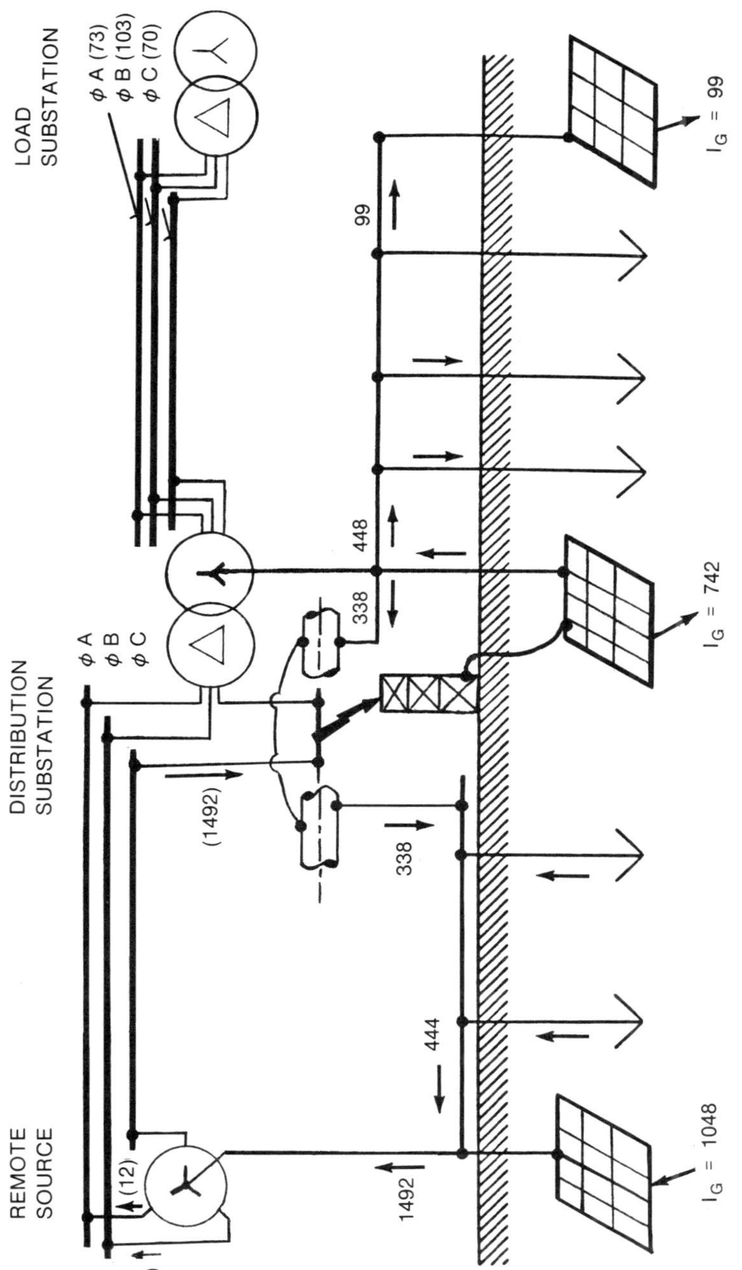

Fig 24
Typical Current Division for a Fault on High Side of Distribution Substation

In the case of a single-line-to-ground fault, the zero-sequence fault current is

$$I_0 = \frac{E}{3R_f + (R_1 + R_2 + R_0) + j\,(X_1 + X_2 + X_0)} \qquad \text{(Eq 57)}$$

In many cases, however, the effect of the resistance terms in the above equation is negligible. For practical purposes, the following simplified equations are sufficiently accurate and more convenient:

Line-to-line-to-ground fault:

$$I_0 = \frac{EX_2}{X_1\,(X_0 + X_2) + X_2\,X_0} \qquad \text{(Eq 58)}$$

Line-to-ground fault:

$$I_0 = \frac{E}{X_1 + X_2 + X_0} \qquad \text{(Eq 59)}$$

In the above equations:

I_0 = symmetrical rms value of zero sequence fault current in A
E = phase-to-neutral potential in V
R_f = estimated minimum resistance of the fault, itself, in Ω (normally it is assumed $R_f = 0$)
R_1 = positive-sequence equivalent system resistance, Ω/phase, computed at fault location
R_2 = negative-sequence equivalent system resistance, Ω/phase, computed at fault location
R_0 = zero-sequence equivalent system resistance, Ω/phase, computed at fault location
X_1 = positive-sequence equivalent system reactance (subtransient), Ω/phase, computed at fault location
X_2 = negative-sequence[26] equivalent system reactance, Ω/phase, computed at fault location
X_0 = zero-sequence equivalent system reactance, Ω/phase, computed at fault location

The values R_1, R_2, R_0, X_1, X_2, and X_0 are computed looking into the system from the point of fault.

13.3 Effect of Station Ground Resistance. In the great majority of cases it is sufficient to derive the maximum grid current I_G, as described in 13.1 and 13.2, by neglecting the system resistance, the station ground resistance, and the resistance at the fault. The error thus introduced is usually small, and is always on the side of safety.

[26] In most calculations it is usually permissible to assume a ratio of X_2/X_1 equal to unity, and hence $X_1 = X_2$, especially if an appreciable percentage of the positive sequence reactance to the point of fault is that of static apparatus and transmission lines.

However, there may be unusual cases where the predicted station ground resistance is so large, in relation to system reactance, that it is worthwhile to take the resistance into account by including it into the more exact Eq 56 or 57. This poses a problem since the station ground system is not yet designed and its resistance is not known. However, the resistance can be estimated by the use of the approximate formulas of 12.2 or 12.3. This estimated resistance generally gives sufficient accuracy for determining the current I_g, and hence I_G.

13.4 Effect of Fault Resistance. If the fault is an insulation breakdown within the local station, the only safe assumption is that the resistance of the fault be assumed zero (Figs 21, 22, 23, and 24).

In the case of a fault outside of the local station area, on a line connected to the station bus (Fig 24), it is permissible, if a conservative minimum value of fault resistance R_f can be assigned, to use this in the ground fault current calculations. This is done by multiplying R_f by 3 and adding it to the other resistance terms as indicated in the denominator of Eqs 56-59. If, however, the actual fault resistance does not maintain a value at least as great as the value of R_f used in the calculations, then the fault resistance should be neglected. Any error from neglecting R_f will, of course, be on the side of safety.

13.5 Effect of Overhead Ground Wires and Neutral Conductors. Where transmission line overhead ground wires or neutral conductors are connected to the station ground, they divert a substantial portion of the ground currents away from the station ground grid. Where this situation exists, the overhead ground wires or neutral conductors can be taken into consideration in the design of the ground grid.

It should be realized that connecting the station ground to overhead ground wires or neutral conductors, or both, and through them to transmission line towers, will usually have the overall effect of increasing the hazard at tower bases, while lessening it at the substation. This is due to the fact that each of the nearby towers will share in each voltage rise of the substation ground mat whatever the cause, instead of being affected only by a local insulation failure or flashover at one of the towers. Conversely, when such a tower fault does occur, the effect of the connected station ground system should decrease the magnitude of gradients near the tower bases.

13.6 Effect of Direct Buried Pipes and Cables. Buried cables with their sheaths or armor in effective contact with the ground, and buried metallic pipes will have a somewhat similar effect when they are bonded to the station ground system, but extend beyond its perimeter.

Buried cables with their sheaths or armor in effective contact with the earth, and buried metallic pipes bonded to the station ground system and extending beyond its perimeter will have an effect similar to that of overhead ground wires and neutrals. By conducting part of the ground current away from the station before it enters the ground, the potential rise of the grid during the fault, and the local gradients in the station will be somewhat lessened. As discussed later in Section 15, however, external hazards may sometimes be introduced [B11], [B88].

Because of the complexities and uncertainties in the pattern of current flow, the effect is often difficult to calculate. Some guidelines to the computation of the input impedance of such current paths leaving the substation are supplied by Rüdenberg [B59] and Laurent [B42]. A more recent study of this problem is presented in EPRI publication EL-904 [B77], which provides methods for computing the impedance of both above-ground and buried pipes. From these values an approximate calculation can determine the division of ground current between these paths, the station ground system, and any overhead ground wires that are present and connected.

13.7 Worst Fault Type and Location — Step (a). The worst fault type for a given grounding system is the one resulting in the highest value of the maximum grid current I_G. Since this current is proportional to the zero-sequence current and the current division factor, and since the current division is almost independent of the fault type, *the worst fault type can be defined as the one resulting in the highest zero-sequence current flow into the earth*, $3I_0$. In a given location, a single-line-to-ground fault will be the worst fault type if $Z_0 Z_1 > Z_2^2$ at the point of fault, and a line-to-line-to-ground fault will be the worst type if $Z_1 Z_0 < Z_2^2$. In the usual case where Z_2 is assumed equal to Z_1, the above comparisons reduce to $Z_0 > Z_1$ and $Z_0 < Z_1$, respectively.

The question of the fault location producing the maximum grid current I_G involves several considerations. The worst fault location may be either on the high voltage side or on the low voltage side, and in either case may be either inside the station or outside on a line, at a certain distance from the station. (A fault is classified as inside the station if it is related to a metallic structure that is electrically connected to the substation grounding grid via negligible impedance.) There are no universal rules for the determination of the worst fault location. The following discussion relates to some, but by no means all, possibilities.

For distribution substations with the transformer *grounded only on the distribution side*, the worst fault location for I_G usually occurs on the high side terminals of the transformer. However, if the source of ground fault current on the high side is weak, or if a parallel operation of several transformers results in a strong ground fault current source on the low side, the worst fault location may be found somewhere on the distribution circuit.

For faults on the low side terminals of such a secondary grounded transformer, the transformer's contribution to the fault circulates in the station grid conductor with negligible leakage current into the earth and, thus, has no effect in the substation ground potential rise (GPR), as shown in Fig 21.

For faults outside the substation on a distribution feeder (far enough to be at remote earth with respect to the ground grid), a large portion of the fault current will return to its source (that is, the transformer neutral) via the station grid, thus contributing to the substation GPR.

In transmission substations with three-winding transformers or autotransformers, the problem is more complex. The worst fault location for I_G may occur on either the high or low side of the transformer; both locations should be checked. In either case, it can be assumed that the worst fault location is at the terminals of the transformer inside the substation, if the system contribution to the fault current is larger than that of the transformers in the substation. Conversely, the worst fault location may be outside the substation on a transmission line, if the transformer contribution dominates.

Exceptions to the above generalities exist. Therefore, for a specific system, several candidates for the worst fault location should be considered. For each candidate, the applicable value of zero-sequence current I_0 should be established in this step.

13.8 Computation of Current Division — Step (b). For the assumption of a sustained flow of the initial zero-sequence current, that is, for the entire fault duration $I_0 = I_0'' = \text{const}$, the symmetrical grid current can be expressed as

$$I_g = S_f\,(3I_0) \tag{Eq 60}$$

To determine I_g, the current division factor S_f must be computed.

The process of computing consists of deriving an equivalent representation of the overhead ground wires, neutrals, etc, connected to the grid and then solving the equivalent to determine what fraction of the total fault current flows between the grid and earth, and what fraction flows through the ground wires or neutrals. S_f is dependent on many parameters, some of which are:

(1) Location of the fault, as described in 13.7

(2) Magnitude of station ground grid resistance, to be discussed in 13.6

(3) Buried pipes and cables in the vicinity of or directly connected to, or both, the station ground system, as discussed in 13.6

(4) Overhead ground wires, neutrals, or other ground return paths, as discussed in 13.5

Because of S_f, the symmetrical grid current I_g, and therefore also I_G, are closely related to the location of the fault. If the additional ground paths of item (3) and (4) above are neglected, the current division ratio can be computed using traditional symmetrical components. However, the current I_g computed using such a method may be overly pessimistic, even if the future system expansion is taken into consideration.

Since overhead transmission lines are present at most substations, the remaining discussion refers only to overhead ground wires and neutral conductors, although the principles involved also apply to buried pipes, cables, or any other conducting path connected to the grid.

High-voltage transmission lines are commonly provided with overhead static wires, either throughout their length or for short distances from each substation. They may be grounded at each tower along the line or they may be insulated from the towers and used for communication purposes. There are many bibliographic sources that provide assistance in determining the effective impedance of a static wire as seen from the fault point [B13], [B14], [B15], [B18], [B39], [B67], [B85], [B109]. Many of these methods may, however, be difficult to apply by the design engineer. Since it is beyond the scope of this guide to discuss in detail the applicability of each method to all possible system configurations, only a brief description of some of the more recent methods will be given.

Endrenyi presents an approach in which, for a series of identical spans, the tower impedances and overhead ground wires or neutrals are reduced to an equivalent lumped impedance Z_∞ to derive several formulas for various types of fault location [B44], [B45]. Except for estimating purposes, Endrenyi recommends including the mutuals between multiple ground conductors and introduces the coupling factor, μ, to account for the mutual impedance between the neutral conductors and the phase conductors. This technique is developed further by Verma and Mukhedkar [B109].

In the cascaded matrix method of Sebo [B93], an impedance matrix is derived for each span of the line, and the individual span matrices are cascaded into a resulting matrix representing the entire line. This technique allows one to take into account all self and mutual impedances (except between the tower footing grounds), and the location and type of fault. A correction for the end effects of the line is suggested, using a modified screening factor.

With some limitations in applicability and accuracy, the span-by-span calculation technique can be considerably simplified. A typical approach, in which all mutual couplings between the neutral conductor and phase conductors and between neutral conductors are ignored, has been recently described by Garrett [B51]. In this technique, each neutral conductor is modeled by the impedance of each span and the equivalent ground impedance of each tower to form a network resembling a ladder. This ladder network is then reduced, using simple network reduction techniques, to an input impedance as seen from the fault point. The input impedance of each circuit is combined with the grid resistance and three times this resulting value is included in the zero-sequence equivalent fault impedance. The current division factor S_f is computed by applying Kirchoff's current law to obtain the current division between the grid resistance and the input impedance of each circuit. Although this, or similar approximate approaches, is limited in applicability and accuracy, in many cases it may provide a reasonable estimate of the influence of overhead ground wires and neutrals on both the resistance of the grounding system and the current division ratio.

Dawalibi [B32] provides algorithms for deriving simple equations to solve for the currents in the grid and in each tower. These equations are obtained from one or both ends of each line and do not require the large computer storage requirements of the techniques that model each span individually. Dawalibi also addresses the effects of the soil structure (that is, multilayer earth resistivities) on the self and mutual impedances of the conductors and on the current division ratio.

Meliopoulos *et al.* [B72] introduce an equivalent conductor to represent the effects of earth using Carson's formula. Every span in each line is modeled and the resulting network is solved for current flows. From this solution, the current division ratio is computed. The number of lines and substations modeled are limited only by the computer used to solve the network.

Obviously, the techniques that model the static wires, phase conductors, towers, etc, in detail will give the best evaluation of the current division ratio S_f. However, the approximate methods discussed above have been compared with the detailed methods and found to give comparable answers for many simple examples. Thus, the choice of the method used to determine S_f will depend on the complexity of the system connected to the substation and the desired degree of accuracy. A simple example follows, showing the results of three of the methods described above. In the following example, Endrenyi's method is used and compared with the results of other methods.

Example 1. Figure 25 shows a three-terminal network with one transmission line between the substation and each of the three energized terminals. The three terminals and transmission lines are identical. Each transmission line has 100 sections, each 0.5 km in length. The various impedances for each line section, tower footing resistance, terminal resistance, and grid resistance are:

Z_{ai} = 0.1 + j0.425 Ω/section
Z_{gi} = 3.5 + j0.65 Ω/section
Z_{mi} = 0.025 + j0.190 Ω/section
R_i = 10.0 + j0.0 Ω
R_{T1} = R_{T2} = R_{T3} = 3.0 + j0.0 Ω
R_g = 2.5 + j0.0 Ω

where

Z_{ai} = self impedance of faulted phase conductor/span
Z_{gi} = self impedance of overhead static wire or neutral/span
Z_{mi} = mutual impedance between phase and neutral conductor/span
R_i = impedance to remote earth of tower i
R_{T1}, R_{T2}, R_{T3} = terminal ground resistances
R_g = station ground resistance to remote earth

The soil is assumed to be uniform with a resistivity of 1000 Ω-m.

Fig 25
Example System for Computation of Current Division Factor S_f

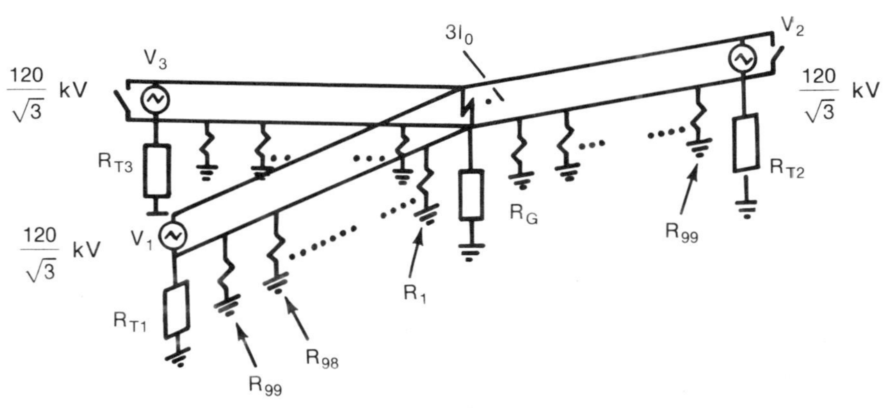

As shown in the figure, a single line-to-ground fault occurs at the substation from the phase conductor bus to the substation neutral.

Using Endrenyi's method described in [B44], the equivalent impedance of the overhead static wire for each line (as seen from the fault point) is

$$\begin{aligned} Z_{li} &= (0.5)\, Z_{gi} + \sqrt{Z_{gi}\, R_i} \\ &= (0.5)\, 3.56 + \sqrt{(3.56)(10)} \\ &= 7.75\ \Omega \end{aligned}$$

For three static wires, the combined equivalent of the static wire network is

$$7.75 / 3 = 2.58\ \Omega$$

The equivalent fault impedance is approximately

$$\begin{aligned} Z_f &= \frac{100}{3}\,(0.1 + \mathrm{j}0.425) + \frac{(2.58)(2.5)}{2.58 + 2.5} \\ &= 4.60 + \mathrm{j}\,14.17\ \Omega \end{aligned}$$

Thus, the total fault current ($3I_0$) is

$$\begin{aligned} 3I_0 &= \frac{120\,000 / \sqrt{3} + \mathrm{j}\,0.0}{4.60 + \mathrm{j}\,14.17} \\ &= 4650.4\ \mathrm{A} \end{aligned}$$

Since the current division factor, S_f, is

$$S_f = \frac{2.58}{2.58 + 2.5} = 0.508$$

the current I_g is found as

$$\begin{aligned} I_g &= S\,(3I_0) = (0.508)(4650.4) \\ &= 2362.4\ \mathrm{A} \end{aligned}$$

Using Dawalibi's algorithm [B32], the total fault currrent $3I_0$ is 4637 A. Approximately 50% (I_g = 2318 A) of the fault current flows through grid to remote earth, so the current division factor equals 0.50. Using Garrett's simplified approach and ignoring mutual coupling, the total fault current is 4627 A. Approximately 51.7% (I_g = 2392 A) of the fault current flows through the grid; S_f equals 0.517.

As shown above, the approximate and detailed methods are in close agreement for this example. However, for more complex systems, with both local and remote ground sources and with dissimilar lines and sources, the results may not be in close agreement.

13.9 Effect of Asymmetry — Step (c). The maximum grid current, I_G, as described in 13.1 and 13.2, is the maximum asymmetrical ac current that will flow between the grounding grid and surrounding earth. This asymmetrical current, defined by Eq 54, includes the symmetrical ac current, I_g, as well as a correction for a dc component. The dc component decays exponentially and is known as the dc offset current. Since the design of a grounding grid must consider the asymmetrical current, a *decrement factor*, D_f, will be derived to take into account the effect of dc current offset.

In general, the asymmetrical fault current includes the subtransient, transient and steady-state ac components, and the dc offset current component. Both the subtransient and transient ac components and the dc offset decay exponentially, each having a different attenuation rate. Thus, as a periodic function of time t, the asymmetrical fault current may be expressed as

$$i_f(t) = \sqrt{2}\,E\left[Y_{ac}(t)\cos(\omega t) + \frac{1}{Z''}\,e^{-t/T_a}\right] \quad \text{(Eq 61)}$$

where

$i_f(t)$ = asymmetrical fault current at any instant t, t in s
E = prefault rms voltage, line-to-neutral
$Y_{ac}(t)$ = equivalent ac system admittance, decreasing with the elapsed time from the initiation of fault; if identical rotating machines are solely anticipated, $Y_{ac}(t)$ is approximately equal to:

$$\left(\frac{1}{Z''} - \frac{1}{Z'}\right) e^{-t/T''} + \left(\frac{1}{Z'} - \frac{1}{Z}\right) e^{-t/T'} + \frac{1}{Z}$$

for:

T'' = machine subtransient time constant in s
T' = machine transient time constant in s
T_a = dc offset time constant in s
Z'' = system subtransient impedance at fault location, with machines represented by their subtransient reactances
Z' = system transient impedance at fault location, with machines represented by their transient reactances
Z = system steady-state impedance at fault location, with machines represented by their synchronous reactances
ω = system frequency in rads/s

Typical values for T'' range from 0.02 - 0.04 s, while typical values for T' range from 0.20 - 0.05 s. T_a is determined by the system X/R ratio at the fault location.

In reality, short circuits occur at random with respect to the voltage wave. However, the shock contact may exist at the moment the fault is initiated. Hence, to allow for the most severe condition, it is necessary to assume that the maximum possible dc offset will be present at the moment of an accidental shock contact.

For fault durations of approximately two cycles or less, the ac component of the asymmetrical wave is very nearly determined by the subtransient impedance. For fault durations of approximately 6 - 60 cycles, the ac component is very nearly determined by the transient impedance.

However, the previous analysis notwithstanding, in typical applications of this guide it is assumed that the ac component *does not decay* with time, but remains at its initial subtransient value. On this basis, in the expression for the fault current decreasing with time, Eq 61, $Y_{ac}(t)$ remains at its initial value $1/Z''$ regardless of the fault duration, and the expression can be rewritten as

$$i_f(t) = \frac{\sqrt{2}\,E}{Z''}\left[\cos(\omega t) + e^{-t/T_a}\right] \qquad \text{(Eq 62)}$$

Finally, since the experimental data in the fibrillation threshold are based on the energy content of a symmetrical sine wave of constant amplitude, it is necessary to establish an equivalent rms value of the asymmetrical current wave for the maximum time of possible shock exposure. This value, in accordance with the definition of the effective asymmetrical fault current I_F, can be determined by integration of Eq 62 over the entire duration interval of fault t_f in s

$$I_F = \sqrt{\frac{1}{t_f}\int_0^{t_f}\left[i_f(t)\right]^2 dt} \qquad \text{(Eq 63)}$$

where

I_F = effective rms value of approximate asymmetrical current for the entire duration of a fault
t_f = time duration of fault in s
t = time (variable) after the initiation of fault in s

Evaluating the integral of Eq 63 in terms of Eq 62, and recognizing that the initial symmetrical fault current $I_f = E/Z''$, it follows that

$$I_F = I_f\sqrt{\frac{2}{t_f}\int_0^{t_f}\left[\cos(\omega t) + e^{-t/T_a}\right]^2 dt} \qquad \text{(Eq 64)}$$

Therefore, the decrement factor D_f is determined by the ratio I_F/I_f, yielding

$$D_f = \sqrt{1 + \frac{T_a}{t_f}\left(1 - e^{-t_f/T_a}\right)} \qquad \text{(Eq 65)}$$

where

t_f = fault duration in s
T_a = equivalent system subtransient time constant in s ($T_a = X''/\omega R''$; for 60 Hz $T_a = X''/120\pi R''$)

The X''/R'' ratio to be used here is the system X/R ratio at the fault location for a given fault type. This X/R ratio is usually approximated using the X and R components of the system subtransient impedance.

Table 6
Typical Values of D_f

Fault Duration t_f (s)	Cycles (60 Hz ac)	Decrement Factor D_f
0.008	½	1.65
0.1	6	1.25
0.25	15	1.10
0.5 or more	30 or more	1.0

Equation 65 can be used to compute the decrement factor for specific X/R ratios and fault durations. Typical values of the decrement factor with an assumed X/R ratio of 20 are shown in Table 6.

For relatively long fault durations, the effect of the dc offset current can be assumed to be more than compensated by the decay of the subtransient component of ac current. A decrement factor of 1.0 is, therefore, conservative for fault durations of 30 cycles or more.

For closely spaced successive shocks (possibly from reclosures), past editions of this guide suggested a decrement factor computed using the shortest single fault duration, even if the time "t_s" used elsewhere in the calculations is based on the sum of the individual shock durations. However, the preceding discussion of the asymmetrical fault current decrement factor suggests that the use of the shortest fault duration in conjunction with the longest shock duration, or sum of the shock durations, may result in an over-designed grounding system. This is especially true for faults of intermediate duration (that is, 6–30 cycles), where the decrement factor is relatively large and the ac component of current is assumed to remain at its subtransient value. Crawford and Griffith [B17] suggest that the shock duration and fault duration be assumed identical, which will result in sufficient grid design for cases involving no automatic reclosures or successive (high-speed) shocks. However, since little or no testing has been done on the effects of repetitive shocks separated by only a few cycles, the design engineer should judge whether or not he should use the longest shock duration for time "t_s" elsewhere in the calculations and the shortest fault duration for the time "t_f" in computing the decrement factor with Eq 65.

NOTE: It is important that the values of the decrement factor given in Table 6 not be confused with the multiplying factors given by ANSI/IEEE C37.010-1979 [7]. The decrement factor is D_f, and is used to determine the effective current during a given time interval after inception of a fault, whereas the multiplying factors given by ANSI/IEEE C37.010-1979 [7] are used to determine the rms current at the end of this interval. Because of the decay of ac and dc transient components with time, the decrement factors determined by Eq 65 are slightly higher than the factors given by ANSI/IEEE C37.010-1979 [7] for short fault and shock durations.

13.10 Effect of Future Changes—Step (d). It is a common experience for maximum fault currents at a given location to increase as system capacity is added or new connections are made to the grid. While an increase in system capacity will increase the maximum expected fault current I_F, new connections may increase or decrease the maximum grid current I_G. One case in which the grid current may decrease with new connections is when new transmission lines are added with ground or neutral wires, or both. In general, if no margin for increase in I_G is included in the original ground system design, the design may become unsafe. Also, subsequent additions will usually be much less convenient and more expensive to install. Allowance for an increase in I_G can be made by decreasing the value of system impedance used in the calculations; or simply by multiplying the value of calculated fault current by an appropriate factor, C_p; $C_p > 1$. It has been a widely accepted practice to assume the total fault current, I_F, between the grid and surrounding earth (that is, ignoring any current division) in an attempt to allow for system growth. While this assumption would be overly pessimistic for present year conditions, it may not exceed the current I_G computed considering current division and system growth. If the system growth is taken into account and current division is ignored, the resulting grid will be overdesigned. An estimate of the future system conditions can be obtained by including all system additions forecasted.

Caution should be exercised when future changes involve such design changes as disconnection of overhead ground wires coming into the substations. Such changes may have an effect on ground fault currents, resulting in an inadequate grounding system. However, future changes such as additions of incoming overhead ground wires may decrease the current division ratio, resulting in the existing ground system being, in effect, overdesigned.

14. Design of Grounding System

14.1 Design Criteria. As stated in 2.1, there are two main design goals to be achieved by any substation ground system under normal as well as fault conditions. These are (1) to provide means to dissipate electric currents into the earth without exceeding any operating and equipment limits, and (2) to assure that a person in the vicinity of grounded facilities is not exposed to the danger of critical electric shock.

The design procedures described in the following sections are aimed at achieving safety from dangerous step and touch voltages within a substation. It is pointed out in 6.1 that it is possible for transferred potentials to exceed the GPR of the substation during fault conditions. Section 15 discusses some of the methods used to protect personnel and equipment from these transferred potentials. Thus, the design procedure described here is based on assuring safety from dangerous step and touch voltages within, and immediately outside, the substation fenced area. Since the mesh voltage is the worst possible touch voltage inside the substation (excluding transferred potentials), the mesh voltage will be used as the basis of this design procedure.

Step voltages are inherently less dangerous than mesh voltages. If, however, safety within the grounded area is achieved with the assistance of a high resistivity surface layer (crushed rock), which does not extend outside the fence, then step voltages may be dangerous. In any event, the computed step voltages should be compared with the permissible step voltage after a grid has been designed that satisfies the touch voltage criterion.

For equally spaced ground grids, the mesh voltage will increase along meshes from the center to the corner of the grid. The rate of this increase will depend on the size of the grid, number and location of ground rods, spacing of parallel conductors, diameter and depth of the conductors, and the resistivity profile of the soil. In a computer study of three typical grounding grids in uniform soil resistivity, the data shown in Table 7 were obtained. These grids were all symmetrically shaped square grids with no ground rods and equal parallel conductor spacing. The corner E_m was computed at the center of the corner mesh. The actual worst case E_m occurs slightly off-center (toward the corner of the grid), but is only slightly higher than the E_m at the center of the mesh.

As indicated in Table 7, the corner mesh voltage is generally much higher than that in the center mesh. This will be true unless the grid is unsymmetrical (that is, has projections, or is L-shaped, etc), has ground rods located on or near the perimeter, or has extremely nonuniform conductor spacings. Thus, in the simpli-

Table 7
Typical Ratio of Corner-to-Center Mesh Voltage

Grid Number	Grid Size (Meshes · Meshes)	E_m Corner/Center
1	10 · 10	2.71
2	20 · 20	5.55
3	30 · 30	8.85

fied equations for the mesh voltage E_m given in 14.5, only the mesh voltage at the center of the corner mesh is used as the basis of the design procedure.[27] Analysis based on computer programs, described on 14.9, may use this approximate corner mesh voltage, the actual corner mesh voltage, or the actual worst-case touch voltage found anywhere within the grounded area as the basis of the design procedure. In either case, the initial criterion for a safe design is to limit the computed mesh or touch voltage to below the tolerable touch voltage from Eq 26 or 26a.

14.2 Critical Parameters. The following site-dependent parameters have been found to have substantial impact on the grid design: maximum grid current (I_G), fault duration (t_f), shock duration (t_s), soil resistivity (ρ), high resistivity surface material (ρ_s), and grid geometry. Several parameters define the geometry of the grid, but the area of the grounding system, the conductor spacing, and the depth of the ground grid have the most impact on the mesh voltage, while parameters such as the conductor diameter and the thickness of the surfacing material have less impact [B1], [B33], [B36], [B97]. A brief discussion or review of the critical parameters is given below.

14.2.1 Maximum Grid Current (I_G). The evaluation of the maximum design value of ground fault current that flows through the substation grounding grid into the earth, I_G, has been described in Section 13. In determining the maximum current I_G, by means of Eq 54, consideration should be given to the resistance of the ground grid, division of the ground fault current between the alternate return paths and the grid, the decrement factor, and the future expansion of the power system.

14.2.2 Fault Duration (t_f) and Shock Duration (t_s). The fault duration and shock duration are normally assumed equal, unless the fault duration is the sum of successive shocks, such as from reclosures (Section 13). The selection of t_f should reflect fast clearing time for transmission substations and slow clearing times for distribution and industrial substations. The choices t_f and t_s should result in the most pessimistic combination of fault current decrement factor and allowable body current. Typical values for t_f and t_s range from 0.25 - 1.0 s. Sections 3.2 - 4.3 and 13.4 give more detailed information on the selection of t_f and t_s.

14.2.3 Soil Resistivity (ρ). The grid resistance and the voltage gradients within a substation are directly dependent on the soil resistivity. Since in reality

[27] Unless otherwise specified, the remainder of the guide will use the term mesh voltage (E_m) to mean the touch voltage at the center of the corner mesh.

soil resistivity will vary horizontally as well as vertically, sufficient data must be gathered for a substation yard. Section 11.3 describes the widely used Wenner technique to measure the soil resistivity [B110], [B71].

Since the simplified equations for E_m and E_s given in 14.5 assume uniform resistivity soil, the equations can employ only a single value for the resistivity. There is no simple method to determine a value from the field test data that can yield an accurate ground grid analysis using these simplified equations. However, the following points may provide the user with general guidelines.

(1) The soil can be considered uniform if the difference between two extreme resistivity values of the field test data is 30% or less. In this case a simple average of all the resistivity values can be used in Eqs 70 and 73.

(2) When an equivalent two-layer soil model is determined (see 11.5.2) and the ground system is in the upper layer, the value of ρ_1 (upper-layer soil resistivity) can be used in the simplified equations. As stated in 11.5.2, for negative values of reflection factor K, the grids designed using uniform soil analysis will have higher step and touch voltages than the grids designed with the equivalent two-layer model, if ρ_1 is used as the soil resistivity in Eqs 70 and 73.

14.2.4 Resistivity of Surface Layer (ρ_s). A thin surface layer of crushed rock helps in limiting the body current by adding resistance to the equivalent body resistance. Values from 1000-5000 Ω-m have been used for ρ_s. Refer to 5.4 and 10.5 for more details on the application of this parameter.

14.2.5 Grid Geometry. In general, the limitation on the physical parameters of a ground grid are based on economics and the physical limitations of the installation of the grid. The economic limitation is obvious: it is impractical to install a copper plate grounding system. Section 16 describes some of the limitations encountered in the installation of a grid. For example, the digging of the trenches into which the conductor material is laid limits the conductor spacing to approximately 2 m or more. Typical conductor spacings range from 3-15 m, while typical grid depths range from 0.5-1.5 m. For the typical conductors ranging from AWG 2/0 to 500 kcmils, the conductor diameter has negligible effect on the mesh voltage. The area of the grounding system is the single most important geometrical factor in determining the resistances of the grid. The larger the area grounded, the lower the grid resistance and, thus, the lower the GPR and mesh voltage.

14.3 Index of Design Parameters. Table 8 contains a summary of the design parameters used in the design procedure.

14.4 Design Procedure. The block diagram of Fig 25 illustrates the sequences of steps to design the ground grid. The parameters shown in the block diagram are identified in the index presented in Table 8 of 14.3.

(1) The property map and general location plan of the substation should provide good estimates of the area to be grounded. A soil resistivity test, described in Section 11, will determine the soil resistivity profile and the soil model needed (that is, uniform or two-layer model).

(2) The conductor size is determined by equations given in 9.3 and 9.4. The fault current $3I_0$ should be the maximum expected future fault current that will

Table 8
Index of Design Parameters

Symbol	Description	Reference Sections
$3I_0$	Symmetrical fault current in substation for conductor sizing in A	9.3, 9.4, 13.2, 13.4
I_G	Maximum grid current that flows between ground grid and surrounding earth (incuding dc offset) in A	13.4
ρ	Soil resistivity in Ω-m	11
ρ_s	Surface layer resistivity in Ω-m	5.4, 10.5
h_s	Surface layer thickness	5.4
C_p	Current projection factor for future system growth	13.1, 13.10
C_s	Surface layer resistivity derating factor	5.4
t_c	Duration of fault current for sizing ground conductor in s	9.3, 9.4, 9.6
t_f	Duration of fault current for determining decrement factor in s	13.9
t_s	Duration of shock for determining allowable body current in s	3.2-4.3
h	Depth of ground grid conductors in m	12.2, 12.3
d	Diameter of grid conductor in m	9.3, 9.4, 9.6
A	Total area enclosed by ground grid in m^2	12.2, 12.3
D	Spacing between parallel conductors in m	14.4, 14.5
D_f	Decrement factor for determining I_G	13.1, 13.9, 14.2
n	Number of parallel conductors in one direction	14.4, 14.5
K_m	Spacing factor for mesh voltage, simplified method	14.4, 14.5
K_s	Spacing factor for step voltage, simplified method	14.4, 14.5
K_i	Corrected factor for grid geometry, simplified method	14.4, 14.5
K_{ii}	Corrective weighting factor that adjusts the effects of inner conductors on the corner mesh, simplified method	14.5, A
K_h	Corrective weighting factor that emphasizes the effects of grid depth, simplified method	14.5, A
L	Total length of grounding system conductor, including grid and ground rods in m	14.5
R_g	Resistance of ground system in Ω	12.1-12.4
E_m	Mesh voltage at the center of the corner mesh for simplified method in V	14.4-14.5
E_s	Step voltage between a point above the outer corner of the grid and a point 1 m diagonally outside the grid for simplified method in V	14.4-14.5
$E_{touch_{50}}$	Tolerable touch voltage for human with 50 kg body weight in V	6.2, 6.3
$E_{touch_{70}}$	Tolerable touch voltage for human with 70 kg body weight in V	6.2, 6.3
$E_{step_{50}}$	Tolerable step voltage for human with 50 kg body weight in V	6.2
$E_{step_{70}}$	Tolerable step voltage for human with 70 kg body weight in V	6.2

be conducted by any conductor in the grounding system, and the time t_f should reflect the maximum possible clearing time (including back-up).

(3) The tolerable touch and step voltages are determined by equations given in 6.2 and 6.3. The choice of time t_s is based on the judgement of the design engineer, with guidance from 3.2-4.3.

(4) The preliminary design should include a conductor loop surrounding the entire grounded area, plus adequate cross conductors to provide convenient

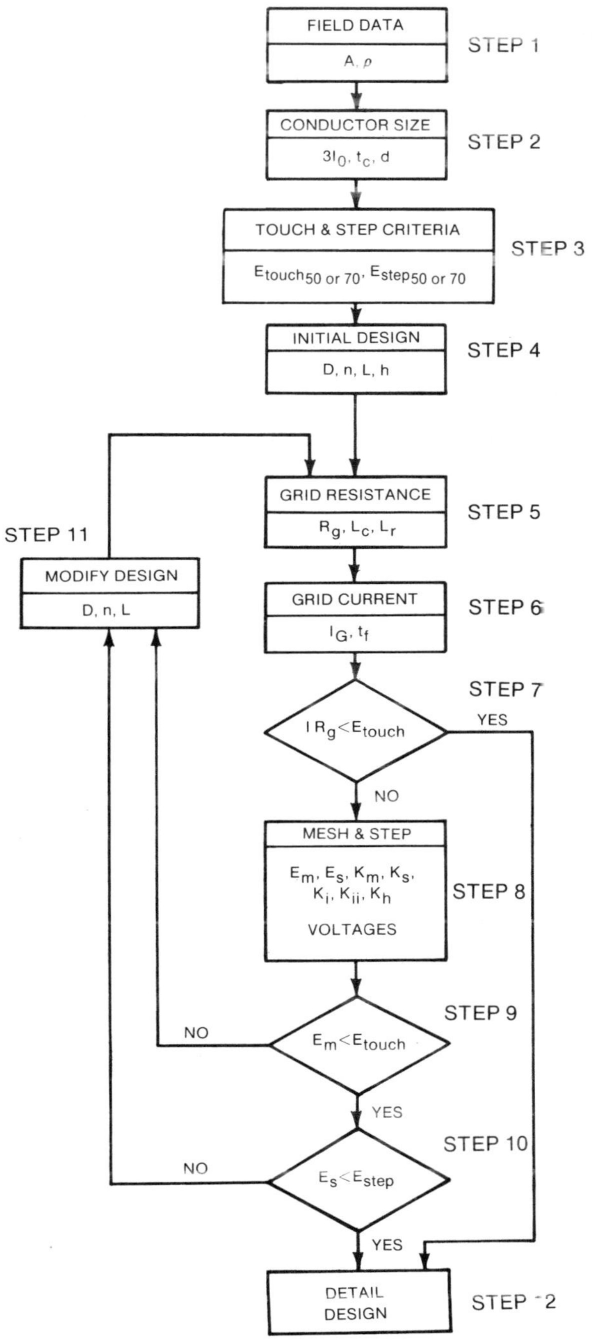

Fig 26
Design Procedure Block Diagram

access for equipment grounds, etc. The initial estimates of conductor spacing and ground rod locations should be based on the current I_G and the area being grounded.

(5) Estimates of the preliminary resistance of the grounding system can be determined by the equations given in 12.2 and 12.3.

For the final design, more accurate estimates of the resistance may be desired, especially when ground rods are used to reach more conductive subsoils. For this application, Eqs 42-44 may be utilized to include the effects of two different soil resistivities in computing the grid resistance and the rodbed resistance. Computer analysis based on modeling the components of the grounding system in detail can compute the resistance with a high degree of accuracy, assuming the soil model is chosen correctly.

(6) The current I_G is determined by the equations given in Section 13. In order to prevent gross over-design of the grounding system, only that portion of the total fault current $3I_0$ that flows through the grid to remote earth (and contributes to the mesh and step voltages and the GPR) should be used in designing the grid. The current I_G should, however, reflect the worst fault type and location, the decrement factor, and any future system expansion.

(7) If the GPR of the preliminary design is below the tolerable touch voltage, no further analysis is necessary. Only additional conductor required to provide access to equipment grounds is necessary.

(8) The calculation of the mesh and step voltages for the grid as designed can be done by the approximate analysis techniques described in 14.5, or by the more accurate computer analysis techniques, as demonstrated in 14.9. Further discussion of the calculations are reserved for those sections.

(9) If the computed mesh voltage is below the tolerable touch voltage, the design may be complete [see Step (10)]. If the computed mesh voltage is greater than the tolerable touch voltage, the preliminary design shall be revised [see Step (11)].

(10) If both the computed touch and step voltages are below the tolerable voltages, the design needs only the refinements required to provide access to equipment grounds. If not, the preliminary design must be revised [see Step (11)].

(11) If either the step or touch tolerable limits are exceeded, revision of the grid design is required. These revisions may include smaller conductor spacings, additional ground rods, etc. More discussion on the revision of the grid design to satisfy the step and touch voltage limits is given in 14.7.

(12) After satisfying the step and touch voltage requirements, additional grid conductors and ground rods may be required. The additional grid conductors may be required if the grid design does not include conductors near equipment to be grounded. Additional ground rods may be required at the base of surge arresters, transformer neutrals, etc.

14.5 Calculation of Maximum Step and Mesh Voltages. Computer algorithms for determining the grid resistance and the mesh and step voltages have been developed in numerous recent references [B2], [B35], [B52], [B57], [B61]. These algorithms required considerable storage capability and may be relatively expensive to execute. In many cases, it is not economically justifiable to use these computer algorithms, or the designer may not have access to a computer with the required capabilities. This section, in conjunction with Appendix A, describes approximate equations for determining the design parameters and establishing the corresponding values of E_m and E_s without the necessity of using a computer. In addition, Appendix B provides curves for a quick estimate or rough check of the calculated values of R_g, E_m, and E_s, or both, based on plotted data, for square grids without ground rods.

Generally,

$$E_m = \rho K_m K_i I_G / L \qquad \text{(Eq 66)}$$

and

$$E_s = \rho K_s K_i I_G / L \qquad \text{(Eq 67)}$$

Thus, the mesh and step voltage values are obtained as a product of geometrical factors (K_m or K_s, respectively), a corrective factor (K_i), which accounts for the increase in current density in the grid extremities, the soil resistivity (ρ), and the average current density per unit of buried conductor (I_G/L).

While the above general Eqs 66 and 67 do not differ from the equations used in the previous editions of the guide, the specific formulas for K_m and K_s have been changed and perform differently than those used in the past. The derivations of the new formulas for K_m and K_s, along with the explanation for the differences between the old and new formulas, are included in Appendix A.

14.5.1 Mesh Voltage (E_m). In Appendix A, Sections 2-5 derive a factor K_m based on the geometry of a ground grid with no ground rods. This K_m is proportional to the mesh voltage E_m, as previously described. The relationship between K_m and E_m depends largely on the current density in the perimeter conductors versus the current density in the inner conductor. To reflect this effect of current density and to correct some of the deficiencies in the equation for K_m in past editions of this guide, the role of K_m has been re-evaluated and two additional weighing terms, K_{ii} and K_h, included in a new equation below, developed by Sverak [B100]:

$$K_m = \frac{1}{2\pi}\left[\ln\left(\frac{D^2}{16hd} + \frac{(D+2h)^2}{8Dd} - \frac{h}{4d}\right) + \frac{K_{ii}}{K_h}\ln\frac{8}{\pi(2n-1)}\right] \qquad \text{(Eq 68)}$$

where

K_{ii} = 1 for grids with ground rods along the perimeter, or for grids with ground rods in the grid corners, as well as both along the perimeter and throughout the grid area

$K_{ii} = \dfrac{1}{(2n)^{2/n}}$ for grids with no ground rods or grids with only a few ground rods, none located in the corners or on the perimeter

$K_h = \sqrt{1 + h/h_o}$

h_o = 1 m (reference depth of grid)

and D, h, n, and d are defined in Table 8.

As explained in Appendix A, a corrective factor K_i is needed to compensate for the fact that the subject mathematical model of N parallel conductors cannot fully account for the effects of a grid geometry, that is, for two sets of parallel conductors that are perpendicular to each other and interconnected at the cross-connection points. (K_i was originally derived as a function that, for a nonsimplified definition of K_m, shown as Eq A26 in Appendix A of this guide, matched the $K_m K_i$ product to the results of Koch's experiment with scale grid models described in Appendix A. This factor is [28]:

$$K_i = 0.656 + 0.172\, n \qquad \text{(Eq 69)}$$

Now a general equation for the mesh voltage E_m can be expressed in terms of ρ, I_G, L, K_m, and K_i:

$$E_m = \frac{\rho\, I_G\, K_m\, K_i}{L} \qquad \text{(Eq 70)}$$

where K_m is determined by Eq 68 and K_i is determined by Eq 69.

If L_c represents the total grid conductor length and L_r represents the total ground rod length, then for grids with ground rods

$$E_m = \frac{\rho\, I_G\, K_m\, K_i}{L_c + 1.15\, L_r} \qquad \text{(Eq 71)}$$

The 1.15 multiplier for L_r in Eq 71 reflects the fact that the current density is higher in the ground rods near the perimeter than in the grid conductors.[29]

For grids with no ground rods, or with only a few rods located within the grid but away from the perimeter

$$E_m = \frac{\rho\, I_G\, K_m\, K_i}{L_c + L_r} \qquad \text{(Eq 72)}$$

14.5.2 Step Voltage (E_s). Section A.7 of Appendix A derives a factor K_s based on the geometry of a ground grid with no ground rods. As with the mesh voltage, this K_s is proportional to the step voltage E_s.

$$E_s = \frac{\rho\, I_G\, K_s\, K_i}{L} \qquad \text{(Eq 73)}$$

[28] Previous editions of this guide defined $K_i = 0.65 + 0.172n$. The correction of 0.65 to 0.656 reflects the obvious fact that for $n = 2$, K_i must be 1.0.

[29] The value of 1.15 is probably too conservative. Indications are that a multiplier of 2.0 or more may be valid for peripheral rods. However, considering that there is a lack of field data and not much information is available on practical experience with grounding systems designed using predominantly peripheral ground rods, judgement should be exercised in the use of Eqs 71 and 72. If only a few, relatively short, ground rods are placed near the center of the grid (that is, for surge arresters, control buildings, etc), the grounding system behaves very much like a grid without ground rods (Eq 72). As more ground rods are placed near the perimeter or the lengths of the ground rods are increased, or both (that is, L_r approaches L_c), the results obtained using Eq 71 become more conservative.

where

$L = L_c + L_r$ for grids with no ground rods or only a few rods in the center away from the perimeter

or

$L = L_c + 1.15L_r$ for grids with ground rods predominantly around the perimeter

For simplification, the maximum step voltage is assumed to occur at a distance equal to the grid depth, h, just outside the perimeter conductor. For the usual burial depth of 0.25 m $< h <$ 2.5 m

$$K_s = \frac{1}{\pi}\left[\frac{1}{2h} + \frac{1}{D+h} + \frac{1}{D}\left(1-0.5^{n-2}\right)\right] \quad \text{(Eq 74)}$$

and for depths smaller than 0.25 m,

$$K_s = \frac{1}{\pi}\left[\frac{1}{2h} + \frac{1}{D+h} + \frac{1}{D} W\right] \quad \text{(Eq 75)}$$

where

$$W = \frac{1}{2} + \frac{1}{3} + \frac{1}{4} \cdots + \frac{1}{n-1}$$

or for $n \geq 6$

$$W \approx \frac{1}{2(n-1)} + \ln(n-1) - 0.423$$

The use of a different equation for K_s, depending on the grid depth h, reflects the fact that the step voltage decreases rapidly with increased depth.

14.6 Estimate of Minimum Buried Conductor Length. A simple equation can be developed to permit a preliminary determination of buried grid conductor necessary to keep the maximum touch voltage within the grounded area below the safe limits established by Eqs 26 and 26a of 6.2. This is done by equating Eq 67 with Eq 26 or 26a of 6.2 as shown below.

For $E_m < E_{\text{touch}50}$, combining Eqs 71 and 26 gives

$$\frac{K_m K_i \rho I_G}{L} < (1000 + 1.5\, C_s(h_s, K)\, \rho_s)\, \frac{0.116}{\sqrt{t_s}} \quad \text{(Eq 76)}$$

Rearranging Eq 76 for L gives

$$L > \frac{K_m K_i \rho I_G \sqrt{t_s}}{(116 + 0.174\, C_s(h_s, K)\, \rho_s)} \quad \text{(Eq 77)}$$

Similarly, for $E_m < E_{\text{touch}70}$, combining and rearranging Eqs 70 and 26 gives

$$L > \frac{K_m K_i \rho I_G \sqrt{t_s}}{(157 + 0.235\, C_s(h_s, K)\, \rho_s)} \quad \text{(Eq 78)}$$

Cases may occur where the conductor length derived from Eqs 77 and 78 is too great to be economically feasible. In such situations, some of the suggestions mentioned in 14.7 may be considered.

Where soil resistivity and total ground current are very low, the length obtained from Eqs 77 or 78 may give a total length of conductor that is too small to properly connect all equipment to be grounded. In this case, more conductor may be required, even though it is not necessary for safety reasons.

14.7 Refinement of Preliminary Design. If calculations based on the preliminary design indicate that dangerous potential differences can exist within the station, the following possible remedies should be studied and applied where appropriate.

(1) Decrease in total grid resistance will decrease the maximum ground grid potential rise and hence the maximum transferred potential. The most effective way to decrease ground grid resistance is by increasing the area occupied by the grid. Deep driven rods or wells may be used if the available area is limited. Decrease in station resistance may or may not decrease appreciably the local gradients, depending on the method used.

(2) Improvement of gradient control. By employing closer spacing of grid conductors, the condition of the continuous plate can be approached more closely. Dangerous potentials within the station can thus be eliminated at a cost. The problem at the perimeter may be more difficult, especially at a small station where earth resistivity is high. However, it is usually possible, by burying the grid perimeter ground conductor outside the fence line, to ensure that the steeper gradients immediately outside this grid perimeter do not contribute to the more dangerous touch contacts. Another effective and economical way to control perimeter gradients is to increase the density of ground rods at the perimeter. This density may be decreased toward the center of the grid. Another approach to controlling perimeter gradients and step potentials is to bury two or more parallel conductors around the perimeter at successively greater depth as distance from the station is increased [B1], [B7], [B67].

(3) Diverting a greater part of the fault current to other paths, for example, by connecting overhead ground wires of transmission lines or by decreasing the tower footing resistances in the vicinity of the substation. In connection with the latter, however, the effect on fault gradients near tower footings should be weighed.

(4) Limiting of short-circuit currents flowing in the ground mat to lower values. If feasible, this will, of course, decrease the total rise in ground mat voltage and all gradients in proportion. Other factors, however, will usually make this impractical. Moreover, if accomplished at the expense of greater fault clearing time, the danger may be increased rather than diminished.

(5) Barring of access to limited areas where it may be impractical to eliminate possibility of excessive potential differences during a fault.

By using one or more of the above methods where necessary, designs can be completed for construction purposes. These should be reasonably liberal, as grounding facilities can usually be installed more cheaply if all go in as part of the general construction job, without the necessity of making additions later.

14.8 Limitations of Simplified Equations for E_m and E_s. Several simplifying assumptions are made in deriving the equations for E_m and E_s, as shown in Appendix A. These assumptions may result in inaccurate results, for some cases, in comparison with the results from more rigorous computer analysis or scale model tests. The inclusion of correction factors into the equations of 14.5 practically eliminates the inaccuracy (within certain ranges for the various parameters) for most practical grid designs.

When using the equations of 14.5, the following limits are recommended for square grids, or for rectangular grids having the same number of conductors in both directions:

$n \leq 25$

$0.25 \text{ m} \leq h \leq 2.5 \text{ m}$

$d < 0.25\, h$

$D > 2.5 \text{ m}$

Although the equations of 14.5 have been tested for n greater than 25 and found to be sufficiently accurate, the tests were not extensive enough to form solid conclusions. Thus, caution should be exercised before exceeding the limits given above.

Furthermore, for equally spaced rectangular grids (that is, with square meshes), the value of n for use on determining the mesh voltage factor K_m and the irregularity factor K_i (using Eqs 68 and 69) should be the geometric mean of the number of conductors in either direction. That is,

$$n = \sqrt{n_A n_B} \qquad \text{for calculating } E_m \qquad \text{(Eq 79)}$$

when n_A and n_B are the number of conductors in each direction. The value of n for use in determining the step voltage factor K_s and the irregularity factor K_i (Eqs 69, 74, and 75) should be the maximum of n_A and n_B,

$$n = \max(n_A, n_B) \text{ for calculating } E_s \qquad \text{(Eq 80)}$$

14.9 Use of Computer Analysis in Grid Design. There are several reasons that may justify the use of more accurate computer algorithms in designing the grounding system. These reasons include:

(1) One or more of the geometric parameters exceed the limits described above

(2) A two-layer soil model is required due to significant variations in soil resistivity

(3) An unsymmetrical grid (that is, L-shaped, with projections, etc) makes it impractical to predetermine the location of the worst touch voltage

(4) Uneven grid conductor or ground rod spacings cannot be analyzed using the approximate methods of 14.5

(5) More flexiblility in determining local danger points may be desired

Several references describe computer algorithms for modeling grounding systems [B2], [B35], [B57]. In general, these algorithms are based on (1) modeling the individual components comprising the grounding system (that is, grid conductors, ground rods, etc), (2) forming a set of equations describing the interaction of these components, (3) solving for the ground-fault current flowing from each component into the earth, and (4) computing the potential at any desired surface point due to all the individual components.

The accuracy of these computer algorithms depends mostly on the size and number of segments each component is broken into for modeling. This determines the accuracy of the current density computed for each conductor segment. Of course, the accuracy also depends on other parameters, such as the soil model and resistivity values, as do the approximate equations of 14.5.

15. Investigations of Transferred Potentials

A serious hazard may result during a fault from the transfer of potentials between the ground-grid areas and outside points, by conductors such as communication and signal circuits, low-voltage neutral wires, conduit, pipes, rails, metallic fences, etc. The danger is usually from contacts of the *touch* type. The importance of the problem results from the very high magnitude of potential difference that is often possible. As mentioned in 6.1, this potential difference may equal or exceed the GPR of the substation. In fact, induced voltages on unshielded communication circuits, static wires, pipes, etc, may result in transferred potentials exceeding the sum of the GPR's of both the faulted substation and the source substation. For any of the cases discussed below, where induced potentials might exist, the metallic paths should be treated as *live* and should be clearly labeled as such.

15.1 Communication Circuits. For communication circuits, schemes have been developed involving protective devices and insulating and neutralizing transformers to safeguard personnel and communications terminal equipment. These will not be discussed here except to emphasize the importance of adequate insulation and isolation from accidental contact of any of these devices, and their wiring, which may reach a high voltage with respect to local ground. This is often overlooked. The introduction of fiber optics to isolate the substation communications terminal from the remote terminal can eliminate the transfer of high potentials. Fiber optics should be considered when potentials cannot easily be controlled by more conventional means.

15.2 Rails. Rails entering the station, when connected either intentionally or otherwise to the ground grid, can theoretically create a hazard at a remote point by transferring there the grid potential rise during a fault. Similarly, if grounded remotely, a hazard can be introduced into the station area. Where considered serious, these hazards can be removed by removable track sections where the rails leave the ground-grid area, or by installing several insulating joints in the rails leaving the grid area. A second set of insulating joints beyond the first would provide against the shunting of a single set by a metal car or the soil itself, and also reduce the somewhat remote hazard of accident from potential differences across a joint itself. The insulating joints must be capable of withstanding the potential difference between remote earth and the potential transferred to the joint. Adequate creepage distance should be ensured to offset any pollution or contamination problems.

15.3 Low-Voltage Neutral Wires. Hazards are possible where low-voltage feeders or secondary circuits, serving points outside the station area, have their neutrals connected to the station ground. When the potential of the station ground grid rises as the result of ground-fault current flow, all, or a large part, of this potential rise may then appear at remote points as a dangerous voltage between this *grounded* neutral wire and the adjacent earth; moreover, where other connections to earth are also provided, the flow of fault current through these may, under unfavorable conditions, create gradient hazards at points remote from the station.

To avoid these difficulties, the low-voltage neutral may be isolated from ground at the station itself, always provided, however, that this does not result in slowing down the clearing time for low-voltage faults to the point where the total hazard is increased rather than diminished. If the low-voltage neutral is isolated from that station ground, it then becomes necessary to avoid hazards at the station due to the introduction, via the neutral wire, of remote earth potential. This implies that this neutral, in and near the station, should be treated as a *live* conductor. It should be insulated from the station ground system by insulation adequate to withstand the maximum ground-grid potential rise, and it should be located so as to minimize the danger of being contacted by personnel.

15.4 Portable Equipment and Tools Supplied from Substation. Similar hazards need to be considered in the case of portable mining, excavating, or material handling equipment, or portable tools, which are supplied electrically from the substation and are used outside of the area of the grid where the mesh potential is held within safe limits. Such loads are often supplied by temporary pole lines or long portable cables. An example is often seen when an addition to an existing station is being constructed.

From what has been said in the previous section, it is apparent that a hazardous transferred potential might appear between equipment and the nearby earth during a fault, if the neutral or grounding wire to the equipment is also connected to the substation ground. In cases such as these, it is common to isolate the supply circuits from the substation ground; to ground the neutrals and equipment to earth at the site of the work; and to make sure that the maximum fault current to the local ground is limited to a low value that will not itself cause gradient hazards.

15.5 Piping. Pipelines should always be tied to the station ground system, preferably at several points, to avoid hazards within the station area. The same is generally true of other conductors, such as cable sheaths or armor, which are more or less in direct contact with the soil. Where these systems are very extensive, they may aid greatly toward reduction of the station ground-grid resistance. To the extent that the pipes carry ground current, the surface equipotential contours will be distorted outward where pipes leave the ground-grid area. The distance the potential will be transferred is determined by the propagation constant, γ, and the characteristic impedance, Z_o, of the pipe. The propagation constant and the characteristic impedance are dependent on the physical parameters of the pipe, the pipe coating, and the soil in which the pipe is buried. EPRI Final Report EL-904 [B77] provides information needed to determine the electrical characteristics of the pipe.

If it is desirable to limit the potential to the substation area, insulated pipe sections may be inserted in the pipeline. The insulated sections should be of sufficient length to avoid shunting by the adjacent soil and be capable of withstanding the potential difference between remote earth and the potential transferred to the joint.

15.6 Auxiliary Buildings. Auxiliary buildings can be treated as part of the substation for grounding purposes, or as separate installations, depending on circumstances. If the buildings and substation are relatively close, and especially if the buildings are linked directly to the substation by such things as water pipes, cable sheaths, phone lines, etc, it is appropriate to treat such buildings and their immediate area as part of the substation. As such, the buildings should be grounded using the same safety criteria as the substation. If the buildings are not as close, and if such conducting links are lacking, it may be decided to treat such buildings as separate units with their own local safety grounds. If served electrically from the substation, they should have their own distribution transformers of a type to provide adequate insulation against transfer of the substation ground-grid potential rise. Secondary neutrals would, in this case, be connected to the local ground at the auxilary buildings only.

16. Investigation of Special Danger Points

16.1 Service Areas. The problems associated with avoiding dangerous step and touch voltage exposure to unauthorized persons outside a substation fence are much the same as those to authorized persons within fenced substation areas [8].

Utilities will often fence a much larger area than initially utilized in a substation and construct a grounding grid only in the utilized area and along the substation fence. The remaining unprotected areas within the fenced area are often used as storage, staging, or general service areas. Temporary grounding is often used in service areas.

A reduced substation grid, which does not include the service area, has both initial cost advantages and future savings resulting from not having the problems associated with "working around" a previously installed total area grid system when future expansion is required into the service area. However, a reduced grid provides less personnel protection compared to a complete substation grid, which includes the service area. Also, because of the smaller area and less conductor length, a service area grid and reduced substation grid will have a higher overall resistance compared to a complete substation grid, which includes the service area.

16.2 Operating Handles. Equipment operating handles deserve special attention because of the higher probability for coincidence of adverse factors. For example, hand operation of equipment, such as a disconnecting switch, requires the presence of the operator near a grounded structure at a point where opening of an energized circuit can sometimes result in an arc to the structure, or perhaps mechanical failure and electrical breakdown of a switch insulator. A large percentage of fatal accidents from voltage gradients are in fact associated with operating handles.

It is relatively easy to protect against these hazards when the operating handle is within a reasonably extensive substation ground-grid area. If the grounding system has been designed conservatively and in line with principles herein described, touch and step voltages near the operating handle, as elsewhere in the station area, should be within safe limits. If conditions are such that accurate calculations are especially difficult, one may wish to add an additional safety factor by using closer mesh or higher resistance surfacing such as crushed rock, or both, in the special danger area.

More difficult is the case of an isolated switch or switches, such as a line sectionalizing switch, where a permanent mat of large size is impractical. Methods used to meet this problem vary greatly and include insulated platforms connected to the handle and ground system, buried plates, grids or ground loops, portable grids connected to the handle and ground system, use of an insulating handle, etc.

It is true that more than one type of satisfactory solution is possible. It has also been shown by calculation and tests that many arrangements that might appear safe at first glance would be extremely dangerous should there ever be an insulation breakdown to the operating mechanism or structure.

One should guard against the fallacy that a cluster of ground rods or even a buried loop will necessarily ensure safety by keeping the surface potential where a man stands at the same potential as the handle. It will not. Current leaving a buried conductor flows in part upward toward the surface. Associated with this upward flow through the resistivity of the ground is a potential drop, which may be a few volts, a few hundred volts, or perhaps many thousands of volts, depending on conditions. An estimate of this voltage between an operating handle and the ground beneath an operator's feet can be made by using Laurent's assumption that the potential difference between the grid and ground at the mesh center in the case of a simple loop is of the order of ρi [B67].

One utility checked, in a model tank, a design in which the operating mechanism was grounded to a cluster of three driven rods. Touch voltage on the order of half the normal line-to-neutral operating voltage was measured. The solution was a portable steel mat that could be rolled up and carried in a line truck and then spread on the ground and bolted to the handle when a switch was to be operated. The three ground rods were retained to ensure fast relay action, since high step potentials will still exist at the boundaries of the mat during a fault.

Another utility using a buried counterpoise in conjuncton with six driven ground rod adopted the use of three ⅛ in. thick perforated aluminum plates (copper and copper-clad steel were avoided to moderate corrosion of transmission tower grillage footings) joined in an L-shaped configuration in cases where high touch potentials were found to be possible. These are laid on the ground around the corner of the structure where the operating handle is mounted. The plates are then covered with one 6 in. layer of crushed rock. The perforations are to prevent pools of water from collecting in the rock, and to help moisture penetrate to the natural soil below the plates. The crushed rock serves several purposes. Since it is of high resistivity, even when wet, essentially all short-circuit current will flow down through the earth rather than up through the rock; the rock will be a high resistance in the path of any short circuit that might flow through the body; it will discourage growth of weeds or sod that would have a low resistance; it will help conserve moisture in the soil beneath; it will discourage theft of the electrode; and it will delineate for the operator the area where he can safely stand. The stranded aluminum counterpoise and the galvanized steel ground rods of the original design have been retained to ensure fast relay action. Also, by occupying a much larger ground area than the plate itself, it decreases the overall resistance, the potential rise of the plate, and to a degree, decreases step potentials at its boundaries.

In a situation of this type, it is normally impractical to protect the man as he steps in or out of the protected area, whether the latter be plate, grid, or insulating platform. Here, however, there is hazard only in the case of local breakdown, as remote line faults will not load the local ground. The chance that a local insulation failure will occur by pure coincidence as the man steps on the platform or grid is doubtless so remote it can be ignored. The best that can be done to protect him if he steps off after a breakdown is to ensure that relays are reasonably fast, to minimize shock duration, and that possible shock voltages are limited to those from step contacts rather than the more lethal touch contacts.

16.3 Fences. Fence grounding is of major importance because the most dangerous *touch* contacts are involved. The outside of the fence is usually accessible to the general public. In addition, the fence may occupy a position on the periphery of the ground-grid area where surface potential gradients are the highest. Past utility practices have been quite varied, but a few facts are clear.

Two different general philosophies of fence grounding have been followed:

(1) Inclusion of the fence within the ground-grid area

(2) Placement of the fence outside the ground-grid area, either with or without close electric coupling between fence and adjacent earth along its length, but with no electric tie between fence and main station grid

Inclusion of the fence within the ground-grid area increases the size of the area and thereby reduces, often substantially, the ground-grid resistance, and hence the maximum ground-grid voltage rise as well. While the fence now takes part fully in this rise, this is not of concern if internal and perimeter gradients of the grid are kept within acceptable limits.

Under the first philosophy, the perimeter conductor of the grid will normally either follow the fence line, or parallel it at a short distance (about 0.5 - 1.5 m or 1.5 - 5 ft) outside. In either case, the perimeter ground conductor and fence should be bonded electrically at frequent intervals.

Placement of the ground conductor directly on the fence line permits the latter to be located on the property line if desired, without obtaining an easement to place a ground conductor on adjacent property. On the other hand, placement of the ground conductor a short distance outside the fence line will decrease the possible touch potential to which a person outside the fence could be subject. Whether or not this difference is important will depend on the circumstances.

Assuming the conditions of a specific idealized example where the perimeter conductor lies 1 m outside the fence, and a man also standing 1 m outside the fence and above this conductor touches the fence, one will encounter a voltage proportional to ρi. This voltage is the vertical difference of potential between the grid conductor where the man is standing and the earth surface immediately above the conductor.

Assume next that the fence is above the conductor, and the man, as before, is 1 m outside. Here, if he touches the fence, he will encounter a potential difference made up of a vertical component as before, plus a horizontal component along 1 m of ground surface, resulting in a total potential difference greater than for the previous arrangement.

Locating the perimeter conductor 1 m outside the fence line instead of directly under the fence increases, only slightly, the touch potential for a man standing inside the fenced area.

If the second philosophy on fences is to be followed, that is, if the fence and its associated grounds are not to be coupled in any way to the main ground (except through the soil), then three factors require consideration:

(1) Is the falling of an energized line on the fence a danger that should be considered?

(2) May hazardous potentials exist at the fence during other types of faults because the fence line crosses the normal equipotential contours?

(3) In practice, can complete metallic isolation of the fence and station ground grid be assured at all times?

In regard to the first point, construction of transmission lines over private fences is commonplace and gives no undue concern. The number of lines crossing a substation fence may be greater, but they are under closer observation; also spans are often shorter and deadended at one or both ends. Hence, this danger can be largely discounted. If one is to design against this danger, then very close coupling of the fence to adjacent ground throughout its length is necessary, and touch potentials inside and out must be reduced to acceptable limits for short-circuit currents of essentially the same maximum value as for the station. Isolation from the main station grid, which would otherwise dissipate a part of this current, will make this more difficult.

In regard to the second point, fences will seldom follow exactly the normal equipotential lines that result from fault current flowing in the station grid. If coupling of the fence to ground is left to chance, the fence might, under some circumstances, assume the potential of a point on the ground where the coupling was relatively good, and attain a high voltage in relation to the adjacent ground surface at other points. Voltage and currents involved here are less than would result from the fall of an energized line but may be sufficient in all but the simplest cases to warrant close coupling of the fence to ground, preferably by a buried conductor along its full length and frequently bonded to it.

In regard to the third point, a disadvantage of isolating the fence from the main grid (aside from sacrificing improvement of station grid resistance) is the chance of inadvertent electric connection between the grid and the fence areas. Telephone, signal, or distribution circuits from the station to a gate house, water pipes, etc, could transfer main grid potentials and introduce dangerous local potential differences during faults. If the fence itself is not tightly coupled to the nearby ground by its own adequate ground system, then any such inadvertent connections to the main grid could create a hazard along its entire length under fault conditions. This hazard could only be partially negated by utilizing broken sections or insulated joints at regular intervals.

16.4 Cable Sheath Grounding. Metallic cable sheaths, unless effectively grounded, may attain dangerous voltages to earth. These voltages may result from insulation failure, charges due to electrostatic induction, and flow of currents in the sheath, or from the voltage rise during faults discharging to the station ground system to which the sheaths are connected. All grounding connections should be made to the shield in such a way as to provide a permanent low-resistance bond.

The wire or strap used to connect the cable shield ground connection to the permanent ground shall be sized to carry the available fault current (Section 9).

Sheath currents on single-conductor cables can be reduced by grounding one end of the sheaths only, when the cable length is not excessive. For long cables, normal cable practices are indicated, using bonding transformers, cross bonding, insulated joints, and so forth. Another alternative is grounding through primary cells or similar devices for cathodic protection.

The sheaths of shielded control cables should be grounded at both ends to eliminate induced potentials. If the control cable sheath is grounded at widely separated points, large potential gradients in the ground grid during faults may cause excessive sheath currents to flow. One solution is to run a separate conductor in parallel with the control cable connected to the two sheath ground points. The current will then be diverted away from the sheath. This separate conductor (usually bare copper) is typically routed along the top of the inside wall of the cable trench.

Nonshielded cables are subject to transient induced voltage magnitudes of 190% or more than the induced voltages on shielded cables [B74]. Induced voltages in nonshielded cables can be reduced by as much as 60% by grounding both ends of an unused wire. The effects of fault currents on the conditions to be encountered with any of these grounding arrangements can only be determined by careful analysis of each specific case.

16.5 GIS Bus Extensions. A number of unique problems are encountered in the grounding of gas-insulated substations (GIS) as compared with conventional substations. The grounded metal enclosure of GIS equipment can be a source of dangerous touch voltages during fault conditions. (Section 8 provides techniques for evaluating touch voltages in GIS.)

16.6 Surge Arresters. Surge arresters must always be provided with a reliable low-resistance ground connection. They should be connected as close as possible to the terminals of the apparatus to be protected and have as short and direct a path to earth as practicable. While many utilities provide separate ground leads from arresters mounted on metal structures, other utilities use the arrester mounting structures as the surge arrester ground path because the large cross section of the steel members provides a lower resistance path than a copper cable of the usual size. In these cases it is important to ensure adequate electric connection from the structure to both arrester ground lead and ground grid; and also to be sure that the steel cross-sectional area is adequate for conductivity, and that no high resistance is introduced into joints from paint films, rust, etc. Aluminum structures, where used, can obviously provide a ground path of high conductivity and current carrying capacity. ANSI/IEEE C57.12.00-1980 [8] requires copper-faced ground pads near the base for grounding the transformer tank. The standard also provides for a ground pad at the top of the tank for fastening a copper ground lead from the lightning arrester when it is mounted on the transformer. The tank plate between the welded-on ground pads provides a higher conductivity path than a separate cable, and when this arrangement is used, it makes a reliable low resistance ground connection.

16.7 Note on Separate and Common Grounds. The practice of having separate grounds within a substation area is rarely used for the following reasons:

(1) Higher resistances for both safety (separate) and system grounds are produced than would be the case for a single uniform ground system

(2) In the event of insulation failures in the station, high currents could still flow in the safety ground

(3) Because of a high degree of coupling between separate electrodes in the same area, the safety objective (of keeping the GPR of the safety grounds low for line faults) would not be accomplished

(4) Often dangerous potentials would be possible between nearby grounded points because decoupling of the separate grounds is possible, at least to some extent

17. Notes on the Construction of a Grounding System

Following the ground-grid design, the grounding plan is usually drawn on the fence and foundation plan drawing. The ground grid is then ready for construction.

The method, or combination of methods chosen, will depend on a number of factors, such as size of a grid, type of soil, size of conductor, depth of burial, availability of equipment, cost of labor, and any physical or safety restrictions due to nearby existing structures or energized equipment.

There are two commonly employed methods to install the ground grid. These are: (1) the trench method and (2) the cable plowing method. Both of these methods employ machines. Where these machines are not employed due to lack of space to move them or small size of the job site, the ground grid is installed by hand digging.

17.1 Ground-Grid Construction — Trench Method. The flags are staked on the perimeter along two sides to identify the spacing between parallel conductors. These markers also serve as a guide for the trenching machine. The trenches are dug using a trenching machine usually along the side having the larger number of parallel conductors. These trenches are dug to the specified depth (usually about 0.5 m or 1.5 ft). Conductors are installed in these ditches and ground rods are driven and connected to the conductors. Pigtails for equipment grounds may also be placed at this time. These initial ditches are then backfilled with dirt up to the location of the cross connections.

The next step is to dig cross conductor ditches (often to a shallower depth), once again using markers as a guide. Care must be taken when digging these ditches to avoid snagging the conductor laid in the backfilled ditches at cross points. The conductors are installed in the ditches and any remaining ground rods are driven and connected to the conductors. Remaining pigtails are also connected to these conductors. Cross type connections are made between perpendicular conductor runs. The ditches are then backfilled with dirt.

An alternative method consists of confining the work to a small section of the total yard and completing this section entirely before moving on to a new area. In this event, the trenches are all dug at the same depth prior to any conductor being placed. Installation of conductors and ground rods are the same as in the previous method.

17.2 Ground-Grid Construction — Conductor Plowing Method. Another procedure for the installation of ground conductors, which may prove economical and quick when conditions are favorable and proper equipment is available, is to plow the conductors in. A special narrow plow is used, which may be either attached to, or drawn by, a tractor or four-wheel drive truck, if there is sufficient maneuvering room. The plow may also be drawn by a winch placed at the edge of the yard. The conductor may be laid on the ground in front of the plow, or a reel of conductor may be mounted on the tractor or truck, or on a sled pulled ahead of the plow. The conductor is then fed into the ground along the blade of the plow to the bottom of the cut. Another method is to attach the end of the conductor to the bottom of the plow blade, and pull it along the bottom of the cut as the plow progresses. In this case, care should be taken to ensure that the conductor does not work its way upward through the loosened soil.

The cross conductors are plowed in at slightly less depth to avoid damage to previously laid conductors. The points of crossing, or points where ground rods are to be installed, are then uncovered, and connections are made as described in 17.3.

With adequate equipment, and the absence of heavy rock, this method is suitable for all of the conductor sizes and burial depths normally used. The reader is referred to IEEE Std 590-1977 [10] for discussion on the subject.

17.3 Installation of Joints, Pigtails, and Ground Rods. Once the conductors are placed in their trenches, the required connections are then made. Generally, the points of crossing require a cross type connection, while tee connections are used for taps to a straight conductor run located along the perimeter. Types of connections are many and varied and of course depend upon the joint, material being joined, and the standard practice of the utility concerned. Mechanical joints utilizing a clamping force or a wedge action as well as brazed or exothermic joints have all been used, as discussed in Section 9.

Exothermic joints utilize a method of welding copper to copper or copper to steel in which an outside source of heat or power is not required. Powdered metals (copper oxide and aluminum) are poured from a container into a graphic crucible and ignited by means of a flint gun. The reduction of the copper oxide by the aluminum (exothermic reaction) produces molten copper and aluminum slag. The molten copper flows over the conductors in the confines of the graphite mold, melting them and welding them together.

Mechanical joints rely on physical forces to provide adequate pressure to ensure an electrically sound connection. The necessary force is achievable through the use of hydraulics or mechanics for compression type fittings, impact or screw action for wedge type fittings, or mechanically tightened bolts and nuts for clamp type bolted fittings. Some mechanical joint fittings are designed to incorporate a force limiting feature during the installation procedure so as to provide the correct force necessary. These limiting features are of course peculiar to the particular connector used, but include such items as shear bolts designed to limit torque and precision dies to limit compression.

Pigtails are left at appropriate locations for grounding connections to structures or equipment. These pigtails may be the same cable size as the underground grid or a different size depending on the number of grounds per device, the magnitude of the ground fault current, and the design practices of the utility concerned. The pigtails are then readily accessible after backfilling for the above grade connections. The above grade connections are made by the usual methods previously mentioned.

Also prior to backfilling, the installation of the ground rods is usually accomplished. This may be done using a hydraulic hammer, air hammer, or other mechanical device. Considerable developmental work has been done in this area and presently an air hammer installed on a truck complete with a ground rod guiding device is available for efficient and easy ground rod driving.

Joining of two ground rods is done either by welding or by using a threaded coupler. Either method has been proven adequate. The connection between ground rod and grid conductor can be made using any method described previously in this section.

17.4 Construction Sequence Consideration for Ground-Grid Installation. A ground grid is normally installed after the yard is graded, foundations are poured, and deeper underground pipes and conduits are installed and backfilled. The security fence may be installed before or after the ground-grid installation. In cases where deeper underground pipes and conduits are not installed before ground-grid installation, an attempt should be made to coordinate the trenching procedured in a logical manner.

17.5 Safety Considerations During Subsequent Excavations. It has been shown in previous chapters of this guide that the insulating value of a layer of clean crushed rock or gravel is an aid to safety under ground fault conditions. Therefore, when an excavation is necessary after a rock surfacing has been applied, care should be taken to avoid mixing the lower resistivity soil from the excavation with the surrounding rock surfacing material.

During subsequent excavations there are more chances to snag the ground conductor. In such a case a check should be made to determine if there is a break in the conductor and joints. A break in the conductor or joints, or both, must be immediately repaired.

18. Field Measurements of a Constructed Grounding System

18.1 Measurements of Grounding System Impedance. As has already been indicated, only approximate results can usually be expected from a precalculation of station ground impedance. A careful measurement of the impedance of the installation as constructed is therefore desirable.

It is not inferred that extreme precision is always possible in measurement, but results should be more dependable than calculated values and adequate for their purposes, provided care is used to select methods that will avoid gross errors.

In this section only general methods are discussed. For more detailed information refer to ANSI/IEEE Std 81-1983 [3] (several important points of this guide have been used here, where applicable). While in this section the ohmic value is referred to as resistance, it should be remembered that there is a reactive component that should be taken into consideration when the ohmic value of the ground under test is less than ½ Ω, and the area is relatively large. This reactive component has little effect on grounds with an impedance higher than 1 Ω.

18.1.1 Two-Point Method (Ammeter-Voltage Method). This method measures the total resistance of the unkown and an auxiliary ground. Since the resistance of the auxiliary ground is presumed to be negligible in comparison with the resistance of the unknown ground, the measured value in ohms is then expressed as the resistance of the unknown ground.

This method is subject to large errors for low ohmic value driven grounds, but it may be useful if a "go or no go" type of test is all that is needed.

18.1.2 Three-Point Method. This method involves the use of two test electrodes with their resistances designated as r_2 and r_3, and with the electrode to be measured designated as r_1. The resistance between each pair of electrodes is measured and designated as r_{12}, r_{13}, and r_{23}, where $r_{12} = r_1 + r_2$, etc. Solving the simultaneous equations, it follows that:

$$r_1 = \frac{(r_{12}) - (r_{23}) + (r_{13})}{2} \qquad \text{(Eq 81)}$$

Therefore, by measuring the resistance of each pair of ground electrodes in series and substituting these values in Eq 81, the value of r_1 can be determined.

If the two test electrodes have substantially higher resistance than the electrode under test, the errors on the individual measurements will be greatly magnified in the final results. In addition, this method can give erroneous values, such as zero or negative resistance, if the electrodes are not separated by a sufficiently large distance. Consequently, when measuring a single driven electrode, the spacing between the three ground electrodes should be 10 m or more. For larger areas the minimum spacing should be on the order of the dimensions of the ground system. Consequently, this method then becomes awkward for the large substations.

18.1.3 Ratio Method. This method compares the resistance of the electrode under test to that of a known resistance, generally by the same electrode configuration as in the fall-of-potential method described in 18.1.5. Being a comparison method, the ohmic readings are independent of the test current magnitude provided that the test current is high enough to give adequate sensitivity.

18.1.4 Staged-Fault Tests. It may be necessary to stage a high-current test where specific information is desired for a particular grounding design. This test would also give quantities from which the ground impedance could readily be determined.

This type of test would require the use of an oscillograph that would record the voltage between selected points. However, the magnitude of the voltage may be quite large and require a potential transformer to step down the voltage to a manageable level. The maximum voltage and potential transformer ratio should be determined in advance of the staged-fault test so that no test equipment is overstressed. The fall-of-potential test can be used to determine the expected voltage of a staged-fault test.

Another important consideration is the calibration of the oscillograph circuit, which is composed of a potential transformer with a possible high resistance in the primary. This resistance is composed of the remote potential ground in series with a long lead. A satisfactory calibration of the deflection of the oscillograph element may be made by inserting a measured voltage in the primary circuit in series with the lead and the remote potential ground as used during the test.

The location of the actual points to be measured is, of course, dependent on the information desired; but in all cases due allowance should be made for coupling between test circuits.

18.1.5 Fall-of-Potential Method. This method has several variations and is applicable to all types of ground impedance measurements. (See Fig 27.) The basic measurement normally used consists essentially of passing a current through the station ground via a ground electrode *C* remote from the station, and measuring the voltage between the station ground and the remote ground at *P*. The term "remote" implies very large electrode spacings since the electrodes are considered to be placed in locations where the earth current density approaches zero.

For measuring resistance, the current source is connected between the station ground mat *E* and a current electrode located at a distance of several hundred feet from the station. The potential-measuring circuit is then connected between the station mat *E* and a potential electrode *P*, with measurements being made at various locations of the electrode outside the station. This potential electrode may be moved toward the current electrode in equal increments of distance, starting near the station, and the resistance readings obtained at the various locations may be plotted against distance from the station. The resulting graph should

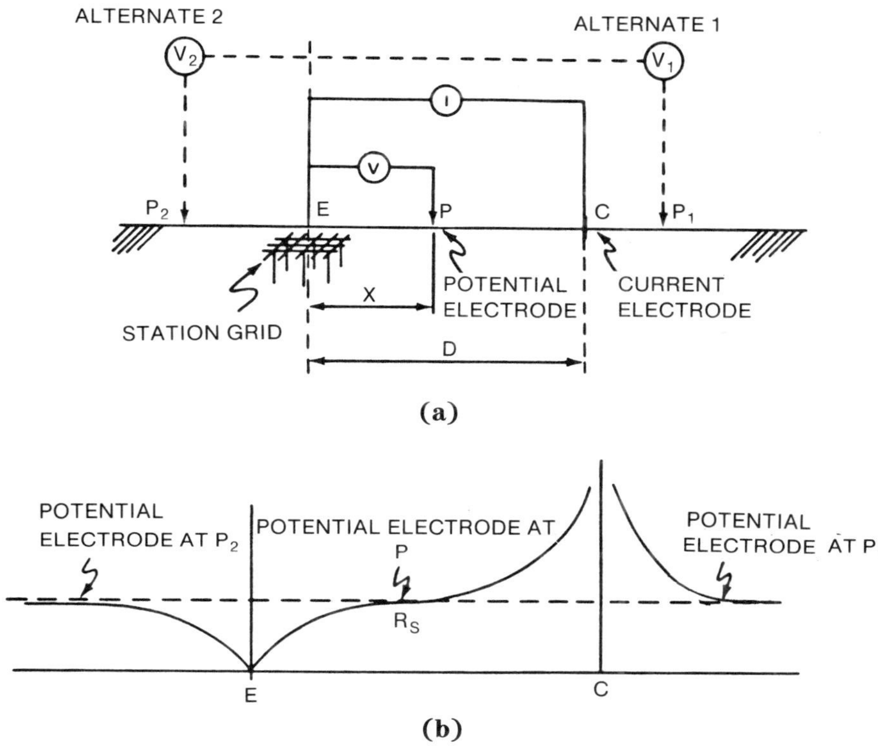

Fig 27
(a) Fall-of-Potential Method and
(b) Earth Surface Potentials for Various Spacings "X"

resemble curve *EPC* of Fig 27(b). From *E* to *P*, the voltage per ampere of test current rises, but the voltage gradient decreases, becoming practically zero at *P*. Continuing toward *C*, the effect of current converging on the current test probe becomes apparent and a rising voltage gradient is observed as the current probe is approached. The horizontal portion of the graph represents a zone where the effect of current convergence toward the ends of the current circuit is negligible. This zone is outside the "area of influence" of both the station and test grounds; R_S, the resistance measured on the flat portion, is the station ground resistance.

In order to obtain a flat portion of the curve, it is necessary that the current electrode be effectively outside the "influence" of the ground to be tested. This influence is sometimes called "extent" of station ground and may be considered as the distance beyond which there is a negligible effect on the measured rise of ground voltage caused by ground current. Theoretically, the influence extends to infinity; but practically, there is a limit because the influence varies inversely as some power of the distance from the ground to be tested. This influence is determined and allowed for during the test on ground grids or deep-driven ground rods of 1 Ω or less. In the case of small areas such as single-rod grounds, tower footings (not connected to overhead wires or counterpoises), the influence can be rendered negligible by keeping spacings on the order of 50 m, which is practical and easy to achieve on site.

For large grounding grids the spacings required may not be practical or even possible, especially since the transmission line overhead ground wires and feeder neutrals connected to station ground effectively extend the area of influence. Consequently, the flat portion of the curve will not be obtained and other methods of interpretation must be used. These methods are discussed in ANSI/IEEE Std 81-1983 [3].

It should be noted that placement of the potential probe P at the opposite side with respect to electrode C (that is, at P_2) will always result in a measured apparent impedance smaller than the true resistance. In addition, when P is located on the same side as electrode C (that is, at P_1), there is a particular location that gives the true resistance.

The primary advantage of the fall-of-potential method is that the potential and current electrodes may have substantially higher resistance than the ground system being tested without significantly affecting the accuracy of the measurements.

18.2 Field Survey of Potential Contours and Touch and Step Voltages. The best assurance that a station is safe would come from actual field tests of step and touch voltages with a heavy current load on the ground mat. Because of the expense, few utilities are likely to make these tests as a regular routine practice. If, however, gross discrepancies between calculated and measured resistance or known anomalies in the ground resistivities throw doubt on the calculated step and touch voltages, then such tests may be considered. This is especially true when the computed values are close to tolerable limits, and further improvement of the ground to provide a larger safety factor would be difficult or costly.

In such situations, it may be worthwhile to load the ground mat with a test current (preferably in the order of about 100 A) and actually take measurements of potential gradients at selected locations throughout the station and around its perimeter. A recent EPRI project [B2] included such a field test [B57]. This project included comparisons of the field test results with a computer solution. The method of measurement was found to be quite feasible and gives good results.

The basic method for such gradient measurements involves passing a test current through the station ground via a remote current electrode, as in station ground resistance measurements, and measuring the resulting touch and step voltages. To obtain the potentials existing under actual fault conditions, the test values are multiplied by the ratio of actual ground-fault current to test current.

Since the potentials of interest are those existing at the surface of the earth, the potential probe used is of a type that makes a surface contact.

The relatively high contact resistances involved generally rule out the use of instruments designed for ground resistance measurements since they operate over a limited range of potential probe resistance. To use a voltmeter-ammeter method, it is usually necessary to have a high-impedance voltmeter, and use test currents high enough to overcome the effects of residual ground currents.

Several methods of measuring and recording potentials may be used. Using a high-impedance voltmeter, profiles and contours of open-circuit contact voltages may be plotted for the entire station. By assuming suitably conservative values of body-and-foot-to-ground resistances, and safe body current, the maximum safe value of open-circuit contact voltage can be determined and hazardous touch and step potentials can be located on the potential map.

Langer [B66] and Bodier [B11] have described measurement techniques in which the effect of actual contact and body resistances are simulated. The operator wears rubber gloves and rubber-soled boots equipped with metallic-mesh contact surfaces. Voltages between these metal contact surfaces are measured by a vacuum tube voltmeter shunted by a resistance equal to an assumed value of body resistance and current is measured by a milliammeter. The ratio of shock current to total ground current is thus determined.

By including foot-to-earth contact resistances as a part of the test procedure, the effect of variations in surface conductivity is taken into account. Thus, the additional safety factor provided by surface coverings of crushed stone, pavement, etc, is included in the test results.

Additional information on making field measurements of potentials can be found in ANSI/IEEE Std 81-1983 [3].

18.3 Assessment of Field Measurements for Safe Design. With the figure for measured resistance available, the maximum ground mat potential rise can be recalculated. If substantially different from that based on the computed resistance, the precautions taken against transferred potentials may need review.

The measured resistance does not provide a direct means of rechecking the computed step and touch potentials, as these are derived from the resistivity. However, if the difference between the computed and measured station grid resistance is very large, the resistance or resistivity figures may come under suspicion, the latter being, in general, less reliable. Each case will have to be judged on its merits to determine whether the discrepancy is such as to warrant further investigation or additional measurement of the resistivities, employment of larger safety factors, or direct measurement of danger potentials or shock currents as described above.

18.4 Periodic Checks of Installed Grounding System. Some utilities recheck station ground resistance at monthly intervals for a few years to determine seasonal variation and effect of soil stabilization after completion of construction. It is also well advised to review the ground system from time to time for possible changes in system conditions that may affect the maximum value of ground current, as well as extensions to the station itself that may affect the maximum current, the station ground resistance, or local potential differences.

19. Physical Scale Models

It often is difficult to draw valid conclusions concerning a general grounding problem solely from actual field data. The lack of consistent results caused by the inability to control the test, such as weather conditions, and other variables affecting the condition of the soil, and difficulties in data collecting, all hamper the ability to run and duplicate tests. Since it is helpful to have verification of theoretical assumptions or computer techniques, or both, scale models have been used to bridge the gap. Models can be used to determine the resistance and potential profiles of ground-grid arrangements.

The early scale model tests used water to represent uniform soil. Using small models in large tanks gave consistent results and enabled various models and conditions to be tested and the effects of different parameters to be observed.

In the late 1960's, a two-layer laboratory model was developed at Ecole Polytechnique to verify computer techniques [B76]. This method used concrete blocks to represent the lower layer of soil. A technique was later developed by Ohio State University that used agar, a gelatin-like substance frequently used in biological studies, to simulate the lower levels of soil. EPRI Project 1494-3 [B97] describes the methodology for performing such model tests. In this project, accurate uniform soil and two-layer soil models were used to study the effects of many parameters on resistance and surface potentials.

The results of model tests have shown that scale models can be effectively used for parametric studies for grounding grid design and for verifying computer simulations of ground-grid parameters [B102].

20. Bibliography

[B1] AIEE Working Group on Substation Grounding Practices. Application Guide on Methods of Substation Grounding. *AIEE Transactions*, vol 73, part III-A, 1954, pp 271 - 278.

[B2] *Analysis Techniques for Power Substation Grounding Systems*. EPRI Final Report EL-2682, vols 1 and 2, Oct 1982.

[B3] ARMSTRONG, H. R. Grounding Electrode Characteristics from Model Tests. *AIEE Transactions*, vol 72, part III, 1953, pp 1301 - 1306.

[B4] ARMSTRONG, H. R. and SIMPKIN, L. J. Grounding Electrode Potential Gradients from Model Tests. *IEEE Transactions on Power Apparatus and Systems*, Oct 1960, pp 618 - 623.

[B5] BELLASCHI, P. L. Lightning Currents in Field and Laboratory. *AIEE Transactions*, vol 54, 1935, pp 837 - 843.

[B6] BIEGELMEIER, U. G. Die Bedeatung der Z-Schwelle des Herzkammerfilimmerns fur die Festlegung von Beruhrunggsspanungs greuzeu bei den Schutzma Bradhmer gegen elektrische Unfalle. *E&M*, vol 93, no 1, 1976, pp 1 - 8.

[B7] BIEGELMEIER, U. G. and ROTTER, K. Elektrische Widerstrande und Strome in Menschlicken Korper. *E&M*, vol 89, 1971, pp 104 - 109.

[B8] BIEGELMEIER, U. G. and LEE, W. R. New Considerations on the Threshold of Ventricular Fibrillation for AC Shocks at 50 - 60 Hz, *Proceedings of the IEEE*, vol 127, 1980, pp 103 - 110.

[B9] BLATTNER, C. J. Study of Driven Ground Rods and Four Point Soil Resistivity Tests. *IEEE Transactions on Power Apparatus and Systems*, vol PAS-101, no 8, Aug 1982, pp 2837 - 2850.

[B10] BODIER, G. La Secrites des Personnes et la Question des Mises a la Terre dans les Postes de Distribution, *Bulletin de la Société Francaise des Electriciens*, ser 6th, vol VII, no 74, Oct 1947, pp 545 - 562.

[B11] BODIER, M. G. Systematic Investigation of Potential Gradients In and Around a Transformation Substation. *Bulletin de la Société Francaise des Electriciens*, July 1951.

[B12] BOGAJEWSKI, W., DAWALIBI, F., GERVAIS, Y., and MUKHEDKAR, D. *Effects of Sustained Ground Fault Current on Concrete Poles*. Paper 81 WM 202-1, presented at the IEEE-PES Winter Meeting, Atlanta, GA, Feb 1981.

[B13] CARSON, J. R. Wave Propagation in Overhead Wires with Ground Return. *Bell Systems Technical Journal*, vol 5, 1926.

[B14] CLEM, J. E. Reactance of Transmission Lines With Ground Return. *IEEE Transactions on Power Apparatus and Systems*, vol PAS-50, Sept 1931, pp 901-918.

[B15] *Computation of Zero Sequence Impedances of Power Lines and Cables*. Engineering Report no 37, Joint Subcommittee on Development and Research of EEI and Bell Telephone Systems, July 22, 1936.

[B16] Conversion Subcommittee of AIEE Substations Committee. Recommended Grounding Practices for Single Polarity DC Structures. Paper no 57-719, *AIEE Transactions*, vol 76, part III, 1957.

[B17] CRAWFORD, E. and GRIFFITH, M. A Closer Look at "The Facts of Life" in Ground Mat Design. *IEEE Transactions on Industry Applications*, vol IA-15, no 3, May/June 1979, pp 241-250.

[B18] *Currents in Earth Wires at the Top of Pylons and Their Effect on the Potentials of the Pylons and Stations in the Neighborhood of a Point of Short-Circuit to Earth*. Contribution no 78, CCITT, Study Group V, Nov 1963.

[B19] DALZIEL, C. F. A Study of the Hazards of Impulse Currents. *AIEE Transactions*, vol 72, part III, Oct 1953, pp 1032-1043.

[B20] DALZIEL, C. F. Dangerous Electric Currents. *AIEE Transactions*, vol 65, 1946, pp 579-585, 1123-1124.

[B21] DALZIEL, C. F. Effect of Wave Form on Let-Go Currents. *AIEE Transactions*, vol 62, 1943, pp 739-744.

[B22] DALZIEL, C. F. Electric Shock Hazard. *IEEE Spectrum*, Feb 1972, pp 41-50.

[B23] DALZIEL, C. F. Threshold 60-Cycle Fibrillating Currents. *AIEE Transactions*, vol 79, part III, 1960, pp 667-673.

[B24] DALZIEL, C. F. and LEE, R. W. Reevaluation of Lethal Electric Currents. *IEEE Transactions on Industry and General Applications*, vol IGA-4, no 5, Oct 1968, pp 467-476.

[B25] DALZIEL, C. F. Temporary Paralysis Following Freezing to a Wire. *AIEE Transactions*, vol 79, part III, 1960, pp 174-175.

[B26] DALZIEL, C. F., LAGEN, J. B., and THURSTON, J. L. Electric Shock. *AIEE Transactions*, vol 60, 1941, pp 1073-1079.

[B27] DALZIEL, C. F. and LEE, W. R. Lethal Electric Currents. *IEEE Spectrum*, Feb 1969, pp 44-50.

[B28] DALZIEL, C. F. and MANSFIELD, T. H. Effect of Frequency on Perception Currents. *AIEE Transactions*, vol 69, 1950, pp 1161-1168.

[B29] DALZIEL, C. F. and MASSOGLIA, F. P. Let-Go Currents and Voltages. *AIEE Transactions*, vol 75, part II, 1956, pp 49-56.

[B30] DALZIEL, C. F., OGDEN, E., and ABBOTT, C. E. Effect of Frequency on Let-Go Currents. *AIEE Transactions*, vol 62, Dec 1943, pp 745-750 and 1000.

[B31] DAWALIBI, F. and MUKHEDKAR, D. *Ground Fault Current Distribution in Power Systems — The Necessary Link*. Paper A77 754-5, presented at the IEEE Summer Power Meeting, Mexico City, 1977.

[B32] DAWALIBI, F. Ground Fault Current Distribution Between Soil and Neutral Conductors. *IEEE Transactions on Power Apparatus and Systems*, vol PAS-99, no 2, Mar/Apr 1980, pp 452-461.

[B33] DAWALIBI, F., BOUCHARD, M., and MUKHEDKAR, D. Survey on Power System Grounding Design Practices. *IEEE Transactions on Power Apparatus and Systems*, vol PAS-99, no 4, July/Aug 1980, pp 1396-1405.

[B34] DAWALIBI, F. and MUKHEDKAR, D. Influence of Ground Rods on Grounding Systems. *IEEE Transactions on Power Apparatus and Systems*, vol PAS-98, no 6, Nov/Dec 1979, pp 2089-2098.

[B35] DAWALIBI, F. and MUKHEDKAR, D. Optimum Design of Substation Grounding in Two-Layer Earth Structure; Part I — Analytical Study, Part II — Comparison Between Theoretical and Experimental Results, and Part III — Study of Grounding Grids Performance and New Electrodes Configuration. *IEEE Transactions on Power Apparatus and Systems*, vol PAS-94, no 2, Mar/Apr 1975, pp 252-261, 262-266, 267-272.

[B36] DAWALIBI, F. and MUKHEDKAR, D. Parametric Analysis of Grounding Systems. *IEEE Transactions on Power Apparatus and Systems*, vol PAS-98, no 5, Sept/Oct 1979, pp 1659-1668.

[B37] DAWALIBI, F., MUKHEDKAR, D., and BENSTED, D. Soil Effects on Ground Fault Currents. *IEEE Transactions on Power Apparatus and Systems*, vol PAS-100, July 1982, p 3442.

[B38] DERI, A., TEVAN, G., SEMLYEN, A., and CASTANHEIRA, A. The Complex Ground Return Plane — A Simplified Model for Homogeneous and Multilayer Earth Return. *IEEE Transactions on Power Apparatus and Systems*, vol PAS-100, no 8, Aug 1981, pp 3686-3693.

[B39] DESIENO, C. F., MARCHENKO, P. P., and NASSELL, G. S. General Equations for Fault Currents in Transmission Line Ground Wires. *IEEE Transactions on Power Apparatus and Systems*, vol PAS-89, Nov/Dec 1970, pp 1891-1900.

[B40] DICK, E. P., FUJINMOTO, N., FORD, G. L., and HARVEY, S. Transient Ground Potential Rise in Gas-Insulated Substations — Problem Identification and Mitigation. *IEEE Transactions on Power Apparatus and Systems*, vol PAS-101, no 10, Oct 1982, pp 3610-3619.

[B41] DICK, W. K. and HOLLIDAY, H. R. Impulse and Alternating Current Tests on Grounding Electrodes in Soil Environment. *IEEE Transactions on Power Apparatus and Systems*, vol PAS-97, no 1, Jan/Feb 1978, pp 102-108.

[B42] DODDS, T. H. and SPINDLE, H. E. Grounding of Gas Insulated Substations. *IEEE Transactions on Power Apparatus and Systems*, vol PAS-94, no 4, July/Aug 1975, pp 1183-1191.

[B43] ELEK, A. Hazards of Electric Shock at Stations During Fault and Method of Reduction. *Ontario Hydro Research News*, vol 10, no 1, 1958, pp 1-6.

[B44] ENDRENYI, J. Fault Current Analysis for Substation Grounding Design. *Ontario Hydro Research Quarterly*, 2nd Quarter, 1967.

[B45] ENDRENYI, J. Analysis of Transmission Tower Potentials During Ground Faults. *IEEE Transactions on Power Apparatus and Systems*, vol PAS-86, Oct 1967, pp 1274-1283.

[B46] *EPRI Transmission Line Reference Book 345 kV and Above*. Electric Power Research Council, Palo Alto, CA, 1975.

[B47] EWY, K. A. and SMOLLECK, H. A. *A Graphical Explanation of the Resistance and Surface-Potential Calculations for Grounding Systems in Two-Layer Earth*. Paper 83 SM 430-6, presented at the 1983 IEEE-PES Summer Power Meeting, Los Angeles, CA.

[B48] FAGAN, E. J. and LEE, R. H. The Use of Concrete-Enclosed Reinforcing Rods as Grounding Electrodes. *IEEE Transactions on Industry and General Applications*, vol IGA-6, no 4, July/Aug 1970, pp 337-348.

[B49] FERRIS, L. P., KING, B. G., SPENCE, P. W., and WILLIAMS, H. Effect of Electric Shock on the Heart. *AIEE Transactions*, vol 55, May 1936, pp 498-515, 1263.

[B50] FORD, G. L. and GEDDES, L. A. Transient Ground Potential Rise in Gas-Insulated Substations — Assessment of Shock Hazard. *IEEE Transactions on Power Apparatus and Systems*, vol PAS-101, no 10, Oct 1982, pp 3620-3629.

[B51] GARRETT, D. L. *Determination of Maximum Ground Fault Current Through Substation Grounding System Considering Effects of Static Wires and Feeder Neutrals*, Proceedings of Southeastern Electric Exchange, Atlanta, GA, 1981.

[B52] GARRETT, D. L. and HOLLEY, H. J. Calculation of Substation Ground Resistance Using Matrix Techniques. *IEEE Transactions on Power Apparatus and Systems*, vol PAS-99, no 5, Sept/Oct 1980, pp 2008-2011.

[B53] GEDDES, L. A. and BAKER, L. E. Response to Passage of Electric Current Through the Body. *Journal of Association for the Advancement of Medical Instruments*, vol 2, Feb 1971, pp 13-18.

[B54] GEIGES, K. S. Electric Shock Hazard Analysis. *AIEE Transactions*, vol 75, part III, 1956, pp 1329-1331.

[B55] GRAYBILL, H. W., KOEHLER, H. C., NADKARNI, J. D., and NICHOLAS, J. H. Termination of High-Pressure Oil Cables in Gas-Insulated Minisubstation Equipment. *IEEE Transactions on Power Apparatus and Systems*, vol PAS-93, no 5, Sept/Oct 1974, pp 1669-1674.

[B56] GROSS, E. T. B., *et al.* Grounding Grids for High Voltage Stations. *AIEE Transactions*, vol 72, Aug 1953, pp 799-810.

[B57] HEPPE, R. J. Computation of Potential at Surface Above an Energized Grid or Other Electrode, Allowing for Nonuniform Current Distribution. *IEEE Transactions on Power Apparatus and Systems*, vol PAS-98, no 6, Nov/Dec 1979, pp 1978-1989.

[B58] HEPPE, R. J. Step Potentials and Body Currents for Near Grounds in Two-Layer Earth. *IEEE Transactions on Power Apparatus and Systems*, vol PAS-98, no 1, Jan/Feb 1979, pp 45-59.

[B59] IEEE Working Group on Electrostatic Effects of Transmission Lines, General Systems Subcommittee. Electrostatic Effects of Overhead Transmission Lines, Part 1 — Hazards and Effects, *IEEE Transactions on Power Apparatus and Systems*, vol PAS-91, Mar/Apr 1972, pp 422-426.

[B60] JONES, W. R. Bentonite Rods Assure Ground Rod Installation in Problem Soils. *IEEE Transactions on Power Apparatus*, vol PAS-99, no 4, July/Aug 1980, pp 1343-1346.

[B61] JOY, E. B., MELIOPOULOS, A. P., and WEBB, R. P. *Touch and Step Calculations for Substation Grounding Systems*. Paper A-79-052-2, presented at IEEE/PES Winter Power Meeting, New York, NY, Feb 1979.

[B62] KERCEL, S. W. Design of Switchyard Grounding Systems Using Multiple Grids. *IEEE Transactions on Power Apparatus and Systems*, vol PAS-100, no 3, Mar 1981, pp 1341-1350.

[B63] KINYON, A. L. Earth Resistivity Measurements for Grounding Grids. *AIEE Transactions*, vol 20, Dec 1961, pp 795-800.

[B64] KISELEV. Research into Electric Shock. *Electrical Review*, vol 31, Dec 1965.

[B65] KOUWENHOVEN, W. B., *et al.* AC Shocks of Varying Parameters Affecting the Heart. *AIEE Transactions*, vol 78, part 1, 1959, pp 163-169.

[B66] LANGER, H. Messungen von Erderspannungen in einem 220 kV Umspanwerk. *Electrotechnische Zietschrift*, vol 75, no 4, Feb 1954, pp 97-105 (English translation available in AIEE No. 80-1961, Appendix V, pp 91-102).

[B67] LAURENT, P. G. Les Bases Generales de la Technique des Mises a la Terre dans les Installations Electriques. *Bulletin de la Societé Francaise des Electriciens*, vol 1, ser 7, July 1951, pp 368-402 (English translation available in Appendix I).

[B68] LEE, W. R. Death from Electrical Shock. *Proceedings of the IEEE*, vol 113, no 1, Jan 1966, pp 144-148.

[B69] LOUCKS, W. W. A New Approach to Substation Grounding. *Electrical News and Engineering*, May 15, 1954.

[B70] MAHONAR, V. N. and NAGAR, R. P. Design of Steel Earthing Grids in India. *IEEE Transactions of Power Apparatus and Systems*, vol PAS-98, no 6, Nov/Dec 1979, pp 2126-2134.

[B71] *Manual on Ground Resistence Testing*. Publication no 25-J, James G. Biddle Co, 1952.

[B72) MELIOPOULOS, A. P., PAPALEXOPOULOS, A., and WEBB, R. P. *Current Division in Substation Grounding System*. Proceedings of the 1982 Protective Relaying Conference, Georgia Institute of Technology, Atlanta, GA, May 1982.

[B73] MILLER, HART, and BROWN. Stray Current and Galvanic Corrosion of Reinforced Steel in Concrete. *Material Performance*, vol 15, no 5, May 1976, pp 20-27.

[B74] MITANI, H. Magnitude and Frequency of Transient Induced Voltage in Low-Voltage Control Circuits of Power Stations and Substations. *IEEE Transactions on Power Apparatus and Systems*, vol PAS-99, no 5, Sept/Oct 1980, pp 1871-1878.

[B75] MORGAN, V. T. Rating of Conductors for Short-Design Currents. *Proceedings of the IEEE*, vol 118, no 3-4, Mar/Apr 1971, pp 555-570.

[B76] MUKHEDKAR, D., GERVAIS, Y., and DEJEAN, J. P. Modeling a Grounding Electrode. *IEEE Transactions on Power Apparatus and Systems*, vol PAS-92, no 1, Jan 1973, pp 295-297.

[B77] *Mutual Design Considerations for Overhead AC Transmission Lines and Gas Transmission Pipelines*. EPRI Final Report EL-904, vols 1 and 2, Sept 1978.

[B78] NAHMAN, J. M. and SALAMON, D. D. Analytical Expressions for the Resistance of Grounding Grids in Nununiform Soil. *IEEE Transactions on Power Apparatus and Systems*, vol PAS-103, pp 880-885, Apr 1984.

[B79] NAHMAN, J. M. and SALAMON, D. D. *Analytical Expressions for the Resistance of Rodbeds and of Combined Grounding Systems in Nonuniform Soil*. IEEE Paper 86 WM126-7, presented at the IEEE-PES Winter Meeting, New York, Feb 2-7, 1986.

[B80] NIEMAN, J. Unstellung von Hochstspannungs-Erdungsalagen Aufden Betrieb Mit Starr Geerdetem Sternpunkt. *Electrotechnische Zeitschrift*, vol 73, no 10, May 1952, pp 333-337.

[B81] NUNNALLY, H. N., WEBB, R. P., JOY, E. B., and MELIOPOULOS, A. P. Computer Simulation for Determining Step and Touch Potentials Resulting from Faults or Open Neutrals, in URD Cable. *IEEE Transactions on Power Apparatus and Systems*, vol PAS-98, no 3, May/June 1979, pp 1130-1136.

[B82] OLLENDORFF, F. *Erdstrome* (Ground Currents). Stuttgart, Germany: Springer-Verlag, 1928; Burkhauser-Verlag-Baselund, 1969.

[B83] OSYPKA, P. Quantitative Investigation of Current Strength and Routing in AC Electrocution Accident Involving Human Beings and Animals. *Technische Hochschule Braunschweig*, Brunswick, West Germany, 1966/SLA Translation Center TT-6611470.

[B84] PALMER, L. S. Examples of Geotechnical Surveys. *Proceedings of the IEE*, Paper 2791-M, vol 106, June 1959, pp 231 - 244.

[B85] PATEL, S. G. *Field Measurements at Texas Valley Substation*. Presented at the Workshop on High Voltage Power System Grounding, Georgia Institute of Technology, Atlanta, GA, May 12 - 14, 1982.

[B86] ROSA, E. B., McCOLLUM, B., and PETERS, O. S. *Electrolysis in Concrete*. Department of Commerce, Technical Paper of Bureau of Standards, no 18, Mar 1913, pp 1 - 137.

[B87] RUDENBERG, R. Distribution of Short-Circuit Currents in Ground. *Elektrotechnische Zeitschrift*, vol 31, 1921.

[B88] RÜDENBERG, R. Basic Considerations Concerning Systems. *Electrotechnische Zeitschrift*, vols 11 and 12, 1926.

[B89] RÜDENBERG, R. Grounding Principles and Practices-1, Fundamental Considerations on Grounding Currents. *Electrical Engineering*, vol 64, no 1, Jan 1945, pp 1 - 13.

[90] "Safe Substation Grounding, Part I," IEEE Committee Report, J. G. Sverak, Chairman, *IEEE Transactions on Power Apparatus and Systems*, vol PAS-100, no 9, pp 4281 - 4290, Sept 1981.*

[B91] "Safe Substation Grounding, Part II," IEEE Committee Report, J. G. Sverak, Chariman, *IEEE Transactions on Power Apparatus and Systems*, vol PAS-101, no 10, pp 4006 - 4023, Oct 1982.

[B92] SCHWARZ, S. J. Analytical Expression for Resistance of Grounding Systems. *AIEE Transactions*, vol 73, part III-B, 1954, pp 1011 - 1016.

[B93] SEBO, S. A. Zero Sequence Current Distribution Along Transmission Lines. *IEEE Transactions on Power Apparatus and Systems*, vol PAS-88, June 1969, pp 910 - 919.

[B94] SEMLYEN, A. Ground Return Parameters of Transmission Lines as Asymptotic Analysis for Very High Frequencies. *IEEE Transactions on Power Apparatus and Systems*, vol PAS-100, no 3, Mar 1981 pp 1031 - 1038.

*References [B90] and [B91] contain the early (preballot) versions of Sections 1 - 6 and 7 - 12 of this guide. As such, neither the equations nor the text of these references should be used in lieu of the corresponding material of the guide.

[B95] SMITH, G. E. *Resistivity of Some Surface Coverings Used in Substation Yards as a Function of Resistivity of the Wetting Agent and as a Function of Time After Wetting.* Internal Report of Carolina Power and Light Co, Apr 20, 1978.

[B96] STAUFFACHER, E. R. Short Time Current Carrying Capacity of Copper Wire. *General Electric Review*, vol 31, no 6, June 1928, pp 326 - 327.

[B97] *Substation Grounding Scale Model Tests.* EPRI Interim Report EL-3099, May 1983.

[B98] SUNDE, E. D. *Earth Conduction Effects in Transmission Systems.* New York: McMillan, 1968.

[B99] SVERAK, J. G. Optimized Grounding Grid Design Using Variable Spacing Technique. *IEEE Transactions on Power Apparatus and Systems*, vol PAS-95, no 1, Jan/Feb 1976, pp 362 - 374.

[B100] SVERAK, J. G. Simplified Analysis of Electrical Gradients Above a Ground Grid; Part I — How Good is the Present IEEE Method? *IEEE Transactions on Power Apparatus and Systems*, vol PAS-103, no 1, Jan 1984, pp 7 - 25.

[B101] SVERAK, J. G. Sizing of Ground Conductors Against Fusing. *IEEE Transactions on Power Apparatus and Systems*, vol PAS-100, no 1, Jan 1981, pp 51 - 59.

[B102] SVERAK, J. G., BOORAEM, C. H., and KASTEN, D. G. *Post-Design Analysis and Scale Model Tests for a Two Grid Earthing System Serving the 345 kV GIS Facilities at Seabrook Power Plant.* Paper 410-06, Proceedings of the CIGRE Symposium on High Currents in Power Systems Under Normal, Emergency and Fault Conditions, Brussels, , June 3 - 5 1985.

[B103] TAGG, G. F. *Earth Resistances.* New York: Pitman, 1964.

[B104] TAYLOR, H. G. The Earth-Loading Capacity of Earth Electrodes. *IEE Journal*, vol 77, Oct 1935, p 542.

[B105] THAPAR, B. and GROSS, E. T. B. Grounding Grids for High Voltage Stations — Part IV - Resistance of Grounding Grids in Nonuniform Soil. *IEEE Transactions on Power Apparatus and Systems*, vol PAS-82, Oct 1963, pp 782 - 788.

[B106] TOWNE, H. M. Lightning Arrester Grounds — Parts I, II, and III. *General Electric Review*, vol 35, Mar - May, 1932, pp 173 - 280.

[B107] *Use of Aluminum for Substation Buses.* Working Group 57.1. IEEE Paper 62-280.

[B108] VERMA, R., MERAND, A., and BARBEAU, P. Design of Low Resistence Grounding System for Hydro-Electric Plant Located on Highly Resistive Soils. *IEEE Transactions on Power Apparatus and Systems*, vol PAS-97, no 5, Sept/Oct 1978, pp 1760 - 1768.

[B109] VERMA, R. and MUKHEDKAR, D. Ground Fault Current Distribution in Substation, Towers and Ground Wire. *IEEE Transactions on Power Apparatus and Systems*, vol PAS-98, May/June 1979, pp 724-730.

[B110] WENNER, F. A Method of Measuring Earth Resistances. *Bulletin of the Bureau of Standards*, Rep no 258, vol 12, no 3, Feb 1916, pp 469-482.

[B111] YU, L. Determination of Induced Currents and Voltages in Earth Wires During Faults. *Proc. IEE*, vol 120, no 6, June 1973, pp 689-692.

[B112] ZABORSZKY, J. Efficiency of Grounding Grids with Nonuniform Soil. *AIEE Transactions*, vol 74, Dec 1955, pp 1230-1233.

Appendixes

(These Appendixes are not a part of ANSI/IEEE Std 80-1986, IEEE Guide for Safety in AC Substation Grounding.)

Appendix A
Mathematical Analysis of Gradient Problem

(This Appendix is based on J. G. Sverak, "Simplified Analysis of Electrical Gradients Above a Ground Grid I — How Good Is the Present Method?," *IEEE Transactions on Power Apparatus and Systems*, vol PAS-103, no 1, pp 7-25, January 1984.)

In the previous editions of this guide, the following equation was provided for determining the value of a mesh voltage (in volts) on the earth's surface above the center of a corner mesh — assuming an equally spaced rectangular grid, which is buried in depth h in a homogeneous soil of uniform resistivity. This grid may consist of N parallel conductors spaced D apart, and of an undetermined number of cross connections. All grid wires are assumed to be of diameter d. The spacing of parallel conductors D, as well as d and h, are in meters.

$$E_{\text{mesh}} = \sigma \frac{I}{L} K_i K_m \qquad \text{(Eq A1)}$$

where

σ = average soil resistivity in Ω-m
I = maximum rms current flowing between ground grid and earth
L = total length of buried conductors, including cross connections, and (optionally) the combined length of ground rods in m
K_i = corrective factor for current irregularity
K_m = mesh factor defined for N parallel conductors,

$$K_m = \frac{1}{2\pi} \ln \frac{D^2}{16hd} + \frac{1}{\pi} \ln \left(\frac{3}{4}\right)\left(\frac{5}{6}\right)\left(\frac{7}{8}\right) \cdots \text{etc} \qquad \text{(Eq A2)}$$

the number of factors in parenthesis in the second term above being two less than the number of parallel conductors in the basic grid, excluding cross connections

For step voltage calculations, the simplified formula was

$$E_{\text{step}} = \sigma K_s K_i I / L \qquad \text{(Eq A3)}$$

where

$$K_s = \frac{1}{\pi}\left[\frac{1}{2h} + \frac{1}{D+h} + \frac{1}{2D} + \frac{1}{3D} \cdots \text{etc}\right]$$

the total number of terms within the brackets being equal to the number of the parallel conductors in the basic grid, excluding cross connections

(Eq A4)

The corrective factor for current irregularity K_i has the form of a simple linear function of the number of parallel conductors in one direction, that is, neglecting any cross connections:

$$K_i = 0.172\,N + C;\; C = 0.65 \qquad \text{(Eq A5)}$$

NOTE: Exactly, $C = 0.656$ since K_i must be equal to 1 for $N = 2$.

This semiempirical factor was derived as a matching function that adjusted the K_m value calculated by means of Eq A26 of the full fundamental model described herein, to those results of Koch's experiment, which are presented in Appendix J of this guide.

In summary, Koch measured the distribution of surface potentials for a number of square grid models placed in an electrolytic tank. Each model was made of a copper wire 0.2 mm in diameter, with each grid arranged in a square having 120 mm sides, set on the water surface. The individual grids studied and the calculated products of $K_i \cdot K_m$ coefficients for the corner mesh of each grid are shown in Fig A1.

In view of these experimental conditions the applicability of Eq A1 was originally limited only to designs that satisfied the following criteria. Conductor spacing D, depth of burial h, and conductor diameter d of an equally spaced rectangular grid must be so chosen as to assure: $D >> h >> d$; for D, h, and d in m.[30]

Table A1 shows a recalculation of Koch's model data for h = 0.1 mm and the grids A (N=2), B (N=3), C (N=5), and D (N=9), as well as for additional subdivisions up to N=30. Furthermore, in order to illustrate the behavior of the simplified formula for larger depths, the adjacent columns provide corresponding values for h increased to 0.5 mm and to 1 m, respectively. The resulting products of $K_i \cdot K_m$ are printed as follows:

Description	Column	K_m of Equation
Simplified K_m formula $\cdot K_i$	I	A2
Abbreviated K_m model $\cdot K_i$	II	A6
Full K_m model $\cdot K_i$	III	A26

[30] Since Koch plotted the mesh center voltages as percentages of the total voltage rise (IR of the grid model), the figures reported by him had each been subtracted from 100%, and then each of the respective percentages multiplied by IR. The individual grid resistances were calculated by Schwarz's formula and checked (agreement within about 6%) by using that of Laurent. Nevertheless, it should be noted that Koch's models do not constitute a truly close scaled-down grid design for practical use. For instance, if a copper conductor of AWG 2/0 size is assumed to be used in the actual design, d = 10.5 mm. Consequently, for a 64 mesh grid, the actual size would be

$$120 \text{ mm} \cdot (10.5 \text{ mm} / 0.2\text{mm}) = 6300 \text{ mm} = 6.3 \text{ m}, D = 0.79 \text{ m, and } h = 0.005 \text{ m}$$

In his paper, Koch gave a 13.8 m size, which was based on the assumption of a 30 · 3 mm ground strap. [Since not the cross section, but the circumference of a conductor determines its performance (per unit of length) in discharging current into the ground, the paper shows that an equivalent d is 23 mm for the given strap dimensions.]

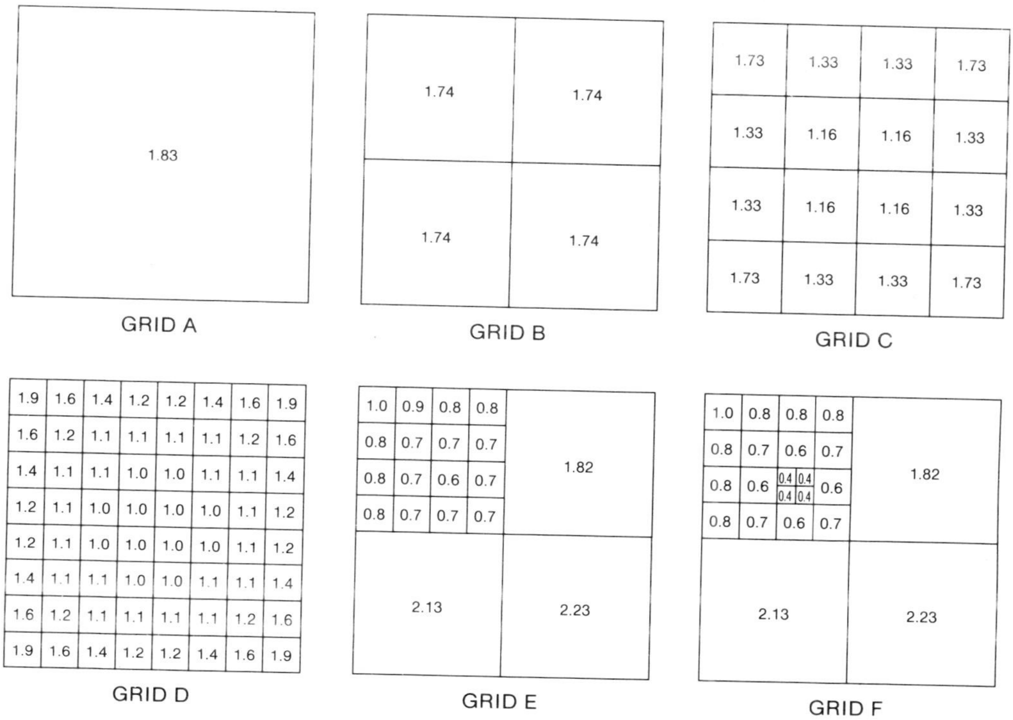

(FIGURES IN DIAGRAMS ABOVE THE PRODUCTS OF COEFFICIENTS $K_M \times K_I$ DETERMINED FROM KOCH EXPERIMENTAL DATA)

GRID	A	B	C	D	E	F
MAXIMUM VALUE $K_M \times K_I$ AS RECORDED ABOVE	1.83	1.74	1.73	1.90	2.23	2.23
COEFFICIENT K_M COMPUTED BY METHOD OF APPENDIX A	1.82	1.50	1.18	0.86	1.50	1.50
COEFF. $K_I = \frac{K_M \times K_I}{KM}$	1.00	1.16	1.47	2.21	1.49	1.49

Fig A1
Products of Coefficients $K_M \cdot K_I$ Determined from Koch Experimental Data

However, in practical use the simplified formula Eq A1 has been found to produce results that are 10–40% low for denser grids — and becoming negative for high N's, if the spacing D is approaching the order of the depth parameter, typically, for D less than 5 m and h greater than 0.25 m.

Part of the difficulty has undoubtedly resulted from a certain misconception as to what the limits imposed by Eq A1 are. Referring to Table A1, were these limits stated, say, as $D > 20h > 10d$, many a problem might have been averted in the past.

However, the root of most application problems with Eq A1 is the fact that the simplification process used led to a highly asymmetrical model, Fig A2.

Specifically, in Eq A1 the factor K_m is not based on a full mathematical model, which is described herein, Section 1, Eqs A6–A26, but referring to the 1961-1976 editions, K_m was derived from an equation that took into account only one conductor for calculating the vertical component of the mesh voltage (E_y), but N conductors for the horizontal component (E_x). This equation is shown below, in a notation that is consistent with that used in the following sections.

$$E_{\text{mesh}} = E_y + E_x = \sigma\, i\, K_i\, K_{mo} \qquad \text{(Eq A6)}$$

where

E_y = vertical component of the mesh voltage
E_x = horizontal component of the mesh voltage

and

$$K_{mo} = K_{my}(1,1) + K_{mx}(1,N)$$

$$K_{my}(1,1) = \frac{1}{2\pi} \ln\left[\frac{4h^2}{4hd - d^2}\right]$$

$$K_{mx}(1,N) = \frac{1}{\pi} \sum_{k=0}^{N-1} \ln\left[\frac{4h^2 + (2k-1)^2 D^2}{4h^2 + 4(kD)^2}\right]$$

To remedy this situation, the new simplified equations developed in this appendix are based on a symmetrical model, Fig A3. Here, the first two peripheral conductors that form the "corner mesh" are placed in exact depth h, while for the remaining conductors, the depth is approximated with a conservative error bias.

The following Sections 1.1-1.6 document this development in terms of the geometry of the basic mathematical model shown in Fig A4. Applicability limits for the equations based on the symmetrical model of Fig A3 are

$N < 25$; $D > 2.5$m; $0.25 \text{ m} < h < 2.5$m; $d < 0.25\, h$; 2.5:1 maximum length-to-width ratio for a rectangular grid

New applicability limits have also been imposed on the old simplified formula for step voltage calculations, Eq A3.

Since Eq A3 has been found to produce values too high for values of D approaching the order of h, and h equal or greater than 0.5 m, in this standard the use of above K_s is limited to cases of $D > 10h$ and $h < 0.25$ m; shown as Eq A76, and a new formula has been developed for $h \geq 0.25$ m.

Table A1
$K_M K_I$ for Koch's Model

		DEPTH = 0.10 (mm)			0.50 (mm)			1.00 (mm)		
N	D	{I}	{II}	{III}	{I}	{II}	{III}	{I}	{II}	{III}
2	120.00	1.71	1.82	1.82	1.45	1.47	1.47	1.34	1.35	1.35
3	60.00	1.63	1.76	1.76	1.33	1.35	1.35	1.20	1.21	1.21
4	40.00	1.62	1.77	1.77	1.28	1.30	1.30	1.13	1.14	1.14
5	30.00	1.62	1.79	1.79	1.24	1.26	1.26	1.07	1.08	1.08
6	24.00	1.63	1.82	1.82	1.20	1.23	1.23	1.01	1.03	1.03
7	20.00	1.64	1.84	1.84	1.16	1.20	1.20	0.96	0.98	0.98
8	17.14	1.64	1.87	1.87	1.12	1.16	1.16	0.90	0.92	0.92
9	15.00	1.64	1.89	1.89	1.08	1.12	1.12	0.84	0.87	0.86
10	13.33	1.64	1.90	1.90	1.03	1.07	1.07	0.77	0.80	0.80
11	12.00	1.63	1.91	1.91	0.98	1.03	1.03	0.70	0.74	0.74
12	10.91	1.62	1.92	1.92	0.92	0.97	0.97	0.62	0.67	0.67
13	10.00	1.60	1.92	1.92	0.86	0.92	0.92	0.54	0.60	0.59
14	9.23	1.58	1.92	1.92	0.80	0.86	0.86	0.46	0.52	0.52
15	8.57	1.56	1.91	1.91	0.73	0.79	0.79	0.37	0.45	0.44
16	8.00	1.53	1.90	1.90	0.65	0.73	0.72	0.28	0.37	0.36
17	7.50	1.49	1.89	1.89	0.58	0.66	0.65	0.18	0.28	0.27
18	7.06	1.46	1.87	1.87	0.50	0.58	0.58	0.08	0.20	0.18
19	6.67	1.42	1.85	1.85	0.41	0.50	0.50	-0.02	0.11	0.09
20	6.32	1.38	1.83	1.83	0.33	0.42	0.42	-0.13	0.02	0.00
21	6.00	1.33	1.80	1.80	0.24	0.34	0.34	-0.24	-0.07	-0.09
22	5.71	1.28	1.77	1.77	0.14	0.26	0.25	-0.35	-0.16	-0.19
23	5.45	1.23	1.74	1.74	0.05	0.17	0.16	-0.46	-0.26	-0.29
24	5.22	1.17	1.70	1.70	-0.05	0.08	0.07	-0.58	-0.35	-0.39
25	5.00	1.11	1.66	1.66	-0.16	-0.02	-0.02	-0.70	-0.45	-0.49
26	4.80	1.05	1.62	1.62	-0.26	-0.11	-0.12	-0.83	-0.55	-0.59
27	4.62	0.99	1.58	1.58	-0.37	-0.21	-0.22	-0.95	-0.65	-0.69
28	4.44	0.92	1.53	1.53	-0.48	-0.31	-0.32	-1.08	-0.75	-0.80
29	4.29	0.85	1.48	1.48	-0.59	-0.41	-0.42	-1.21	-0.85	-0.92
30	4.14	0.78	1.43	1.43	-0.71	-0.51	-0.53	-1.35	-0.95	-1.01

A1. Analysis of Gradient Problem — Basic Mathematical Model

Generally, in a boundless homogeneous medium of resistivity σ, the voltage difference between two points X_1 and X_2 at a respective distance r_1 and r_2 from a line source dissipating current i per unit length is

$$E_{12} = \frac{\sigma i}{2\pi} \int_{r_1}^{r_2} (1/r)\,dr = \frac{\sigma i}{2\pi} \ln (r_2/r_1) \qquad \text{(Eq A7)}$$

If the line source is buried in shallow depth h below a flat ground surface, and the points X_1 and X_2 are both placed on the surface, the voltage difference between them can be calculated as if it is caused by wire (1) in depth h, and by a mirror image of the wire (1*) placed symmetrically in a distance $-h$ above the ground plane; assuming now, again, the medium σ as filling the entire space; Fig A5.

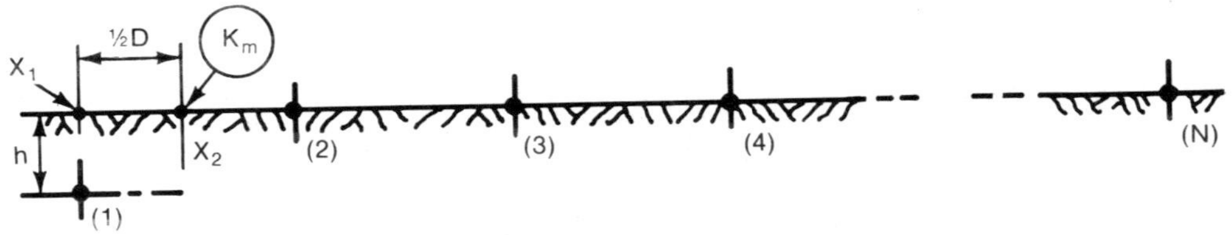

Fig A2
Asymmetrical Model of Old Simplified Definition of K_M per Eq A1

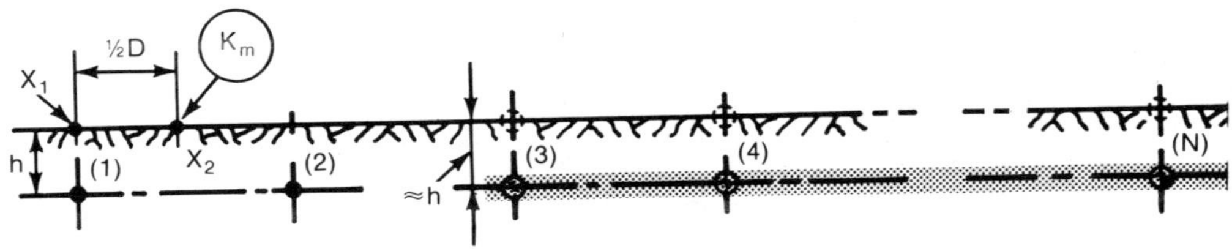

Fig A3
Symmetrical Model of New Simplified Definition of K_M per Eq A64

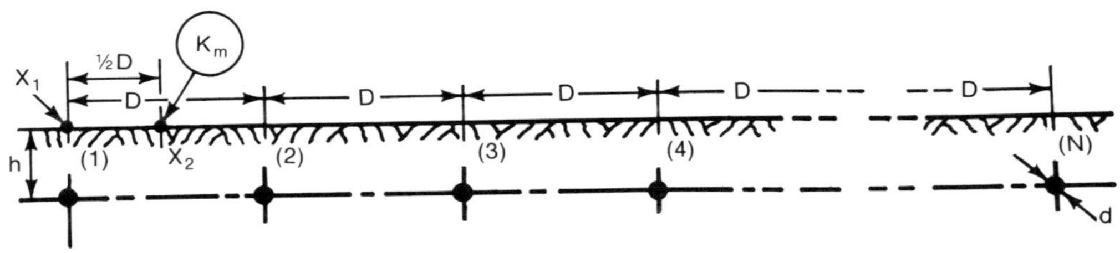

Fig A4
Full Model of Nonsimplified Definition of K_M E_S

From the geometry of Fig A5, it is apparent that at any point $X(x,0)$ on the surface (here $X = X_i$, $i = 1,2$), the surface current density per unit area δ_s due to current i flowing both from the wire (1) and the wire image (1*) is, as a vector in space,

$$\delta_s = \frac{i}{2\pi r}(\cos\alpha + j\sin\alpha) + \frac{i}{2\pi r^*}(\cos\alpha - j\sin\alpha) \qquad \text{(Eq A8)}$$

where

α = angle between the direction of the current vectors δ and δ^*, Fig A5, and the horizontal plane

Since

abst (r) = abst (r^*) = $(x^2 + h^2)^{1/2}$, and $\cos\alpha = x/r$, $\sin\alpha = y/r$, it follows that Eq A8 can be rewritten as

$$\delta_s = \frac{i}{2\pi r}(2\cos\alpha) = \frac{i}{\pi}\,\frac{x}{x^2+h^2}$$

Consequently, the voltage difference between points X_1 and X_2, as a scalar, is

$$E_x(1) = \sigma\int_{X_1}^{X_2}\delta_s\,ds = \frac{\sigma i}{\pi}\int_{x_1}^{x_2}\frac{x\,dx}{x^2+h^2} = \frac{1}{2\pi}\,\sigma i\ln\left[\frac{x_2^{\,2}+h^2}{x_1^{\,2}+h^2}\right] \qquad \text{(Eq A9)}$$

Consider now a set of N equally spaced parallel wires and their images, as shown in Fig A6. Here the distance between two given points on the earth's surface, $0(0,0)$ and $X(-\frac{1}{2}D,0)$ is $\frac{1}{2}D$, and the distance between any two line sources is D. With such a geometry, the difference in the surface potentials from 0 to X, produced by kth wire and its image, can be expressed as

$$E_x(k) = \frac{\sigma i}{\pi}\int_{0}^{-\frac{1}{2}D}\frac{x(k)\,dx}{x(k)^2+h^2}\ ;\ x(k) = (k-1)D + x;\ k = 1,2,\cdots \qquad \text{(Eq A10)}$$

If the assumption is made that the electrical field of an individual wire is not affected by the presence of other wires, then the effect of N line sources and N images on the resulting voltage between 0 and X is a sum of the individual contributions determined by superposition, wire by wire:

$$E_x = \sum_{1}^{N} E_x(k) = \sigma i\,K_{mx}(1,N) = \sigma i\sum_{1}^{N}\frac{1}{2\pi}\ln\left[\frac{[(k-1\frac{1}{2})D]^2+h^2}{[(k-1)D]^2+h^2}\right] \qquad \text{(Eq A11)}$$

Alternatively, the factor $K_{mx}(1,N)$, which represents this effect of N wires on the surface voltage along the x-axis (from the position above the first wire toward the point above the centerline between the first and the second wire), can be stated as

$$K_{mx}(1,N) = \frac{1}{2\pi}\sum_{0}^{N-1}\ln\left[\frac{4h^2+(2k-1)^2D^2}{4h^2+4(kD)^2}\right] \qquad \text{(Eq A12)}$$

Although the potential difference between the 0 and X has been obtained by Eq A1, their potential with respect to a remote ground remains unknown and has to be determined. If all N wires and their images are assumed to be at potential E_0 during a ground fault, so that

$$E_0 = R_0 I_0 \qquad \text{(Eq A13)}$$

where

- R = the grounding system resistance
- I_0 = the total current flowing into ground, that is, $I_0 = i \cdot L$, L being the total length of buried wires

the voltage at point 0, produced by the first conductor and its image, is

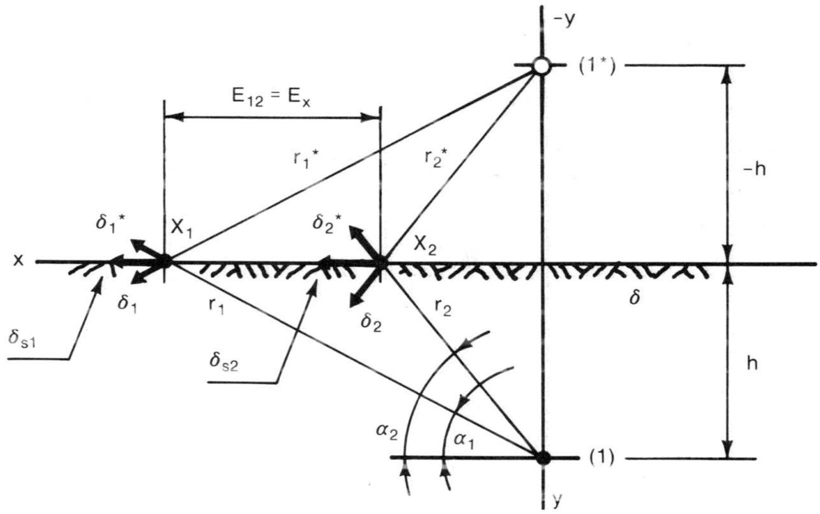

Fig A5
Geometry for Derivation of E_x (1) per Eq A9

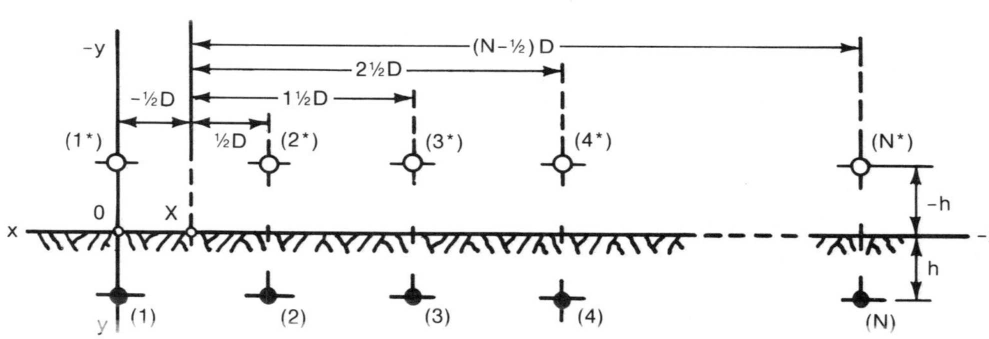

Fig A6
Distances Between the Centerline of Corner Mesh and Parallel Conductors and Their Images

$$V_0(1) = E_0 - E_y(1) \qquad \text{(Eq A14)}$$

Here, using Eq A6 and integrating from the surface of both the real and the image line source to the point 0 on the air-earth boundary, $E_y(1)$ is calculated as follows:

$$E_y(1) = \frac{1}{2\pi} \sigma i \left[\int_{½d}^{h} \frac{1}{y} \, dy + \int_{2h-½d}^{h} \frac{1}{y^*} \, dy^* \right] \qquad \text{(Eq A15)}$$

which can be expressed under one integral as

$$E_y(1) = \frac{\sigma i}{2\pi} \int_{\frac{1}{2}d}^{h} \left(\frac{1}{y} - \frac{1}{2h-y}\right) dy = \frac{\sigma i}{2\pi} \left[\ln\left(\frac{2h}{d}\right) + \ln\left(\frac{2h}{4h-d}\right)\right],$$

or

$$E_y(1) = \frac{\sigma i}{2\pi} \ln \left[\frac{4h^2}{4hd - d^2}\right] \qquad \text{(Eq A16)}$$

Similarly, as before, the voltage at point 0 produced by any other wire than the first one generally is

$$V_0(k) = E_0 - E_y(k); \; k = 2, 3, \cdots, N \qquad \text{(Eq A17)}$$

Using Eq A6 again, once for the real source and once for its image, the general form of $E_y(k)$ can be written as

$$E_y(k) = C \int \frac{1}{r_k} \, dr + C \int \frac{1}{r_k^*} \, dr^* \qquad \text{(Eq A18)}$$

Based on the geometry of Fig A7, it can be seen that for a vertical voltage drop the following substitutions hold for all k's, including $k = 1$:

$$r_k = \sqrt{y^2 + (k-1)^2 D^2}\,; \; dy = dr \sin \beta_k = y \, dy / r_k\,; \; C = \frac{\sigma i}{2\pi} \qquad \text{(Eq A19)}$$

$$r_k^* = \sqrt{(2h-y)^2 + (k-1)^2 D^2}\,; \; dy = -dr^* \sin \beta_k^* = (y-2h) \, dy / r_k^* \qquad \text{(Eq A20)}$$

so that the integral solution for $E_y(k)$ becomes

$$E_y(k) = \frac{\sigma i}{2\pi} \int_{\frac{1}{2}d}^{h} \frac{y \, dy}{y^2 + (k-1)^2 D^2} + \frac{\sigma i}{2\pi} \int_{\frac{1}{2}d}^{h} \frac{(y-2h) \, dy}{(2h-y)^2 + (k-1)^2 D^2} \qquad \text{(Eq A21)}$$

$$= \frac{\sigma i}{4\pi} \ln \left\{ \left[\frac{(k-1)^2 D^2 + h^2}{(k-1)^2 D^2 + \frac{1}{4} d^2}\right] \cdot \left[\frac{(k-1)^2 D^2 + h^2}{(k-1)^2 D^2 + (2h - \frac{1}{2} d)^2}\right] \right\}$$

By superposition, the voltage at point 0 due to the effect of N real and N imaginary line sources is

$$V_0 = E_0 - \sum_{1}^{N} E_y(k) = E_0 - \sigma i \, K_{my}(1,N) = E_0 - E_y\,; \; E_y = \sum_{1}^{N} E_y(k) \qquad \text{(Eq A22)}$$

where

$$K_{my}(1,N) = \frac{1}{4\pi} \sum_{0}^{N-1} \ln \left[\frac{(kD)^2 + h^2}{(kD)^2 + \frac{1}{4} d^2} \cdot \frac{(kD)^2 + h^2}{(kD)^2 + (2h - \frac{1}{2} d)^2}\right] \qquad \text{(Eq A23)}$$

Now, the voltage at point X can also be determined; referring to Eq A11, it follows that

$$V_X = E_0 - (E_x + E_y) = E_0 - \sigma i \, (K_{mx}(1,N) + K_{my}(1,N)) \qquad \text{(Eq A24)}$$

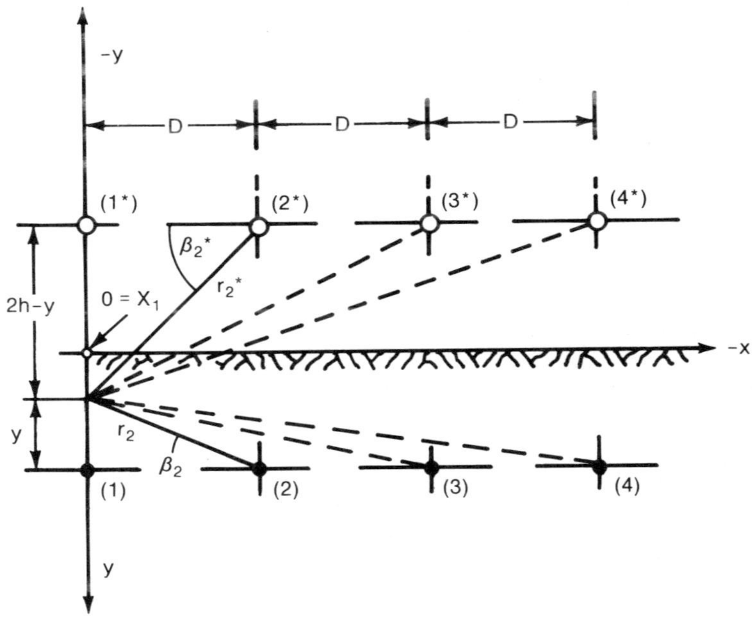

Fig A7
Geometry for Derivation of E_y (k) per Eq A21

Since the potential of all grounded structures is E_0, the touch voltage E_t between the structures and any point on the earth's surface that is along the center line drawn between the first and second peripheral conductor of a set of N parallel conductors is

$$V_t = E_0 - V_X = E_x + E_y \tag{Eq A25}$$

With the understanding that in terms of this N-conductor model there are no true corners, V_t can be viewed as a *corner mesh* voltage of the N conductor set, $E_{\text{mesh}_0(N)}$. Thus

$$E_{\text{mesh}_0(N)} = E_x + E_y = \sigma i \, K_m \tag{Eq A26}$$

where

K_m = a mesh voltage factor; $K_m = K_{mx}(1,N) + K_{my}(1,N)$

The subindex "0" symbolizes the fact that the assumption of a uniform current density nullifies any end effects.

A2. Derivation of Mesh Factor K_m for $N = 2$ and $h \geq 4d$

Based on the analysis of the basic model in Section A1, for N = 2 the mesh voltage factor can be expressed in terms of its x-components and y-components, wire by wire, as

$$K_m = K_{mx}(1,2) + K_{my}(1,2) = K_{mx}(1) + K_{mx}(2) + K_{my}(1) + K_{my}(2) \quad \text{(Eq A27)}$$

where

$$K_{mx}(1) = \frac{1}{2\pi} \ln \left[(4h^2 + D^2)/(4h^2)\right] \quad \text{(Eq A28)}$$

$$K_{mx}(2) = \frac{1}{2\pi} \ln \left[(4h^2 + D^2)/(4h^2 + 4D^2)\right] \quad \text{(Eq A29)}$$

$$K_{my}(1) = \frac{1}{2\pi} \ln \left[(4h^2)/(4hd - d^2)\right] \quad \text{(Eq A30)}$$

$$K_{my}(2) = \frac{1}{4\pi} \ln \left[(h^2 + D^2)/(D^2 + \tfrac{1}{4}d^2)\right] + \frac{1}{4\pi} \ln \left[(h^2 + D^2)/(D^2 + (2h - \tfrac{1}{2}d)^2)\right] \quad \text{(Eq A31)}$$

Combining Eqs A28 and A30 and simplifying for small d and $h \geq 4d$ by neglecting all $\tfrac{1}{2}d$ and $\tfrac{1}{4}d^2$ terms, one gets

$$K_{mx}(1) + K_{my}(1) \approx \frac{1}{2\pi} \ln \left[(1 + D^2/4h^2)(4h^2/4hd - 0)\right] \approx \frac{1}{2\pi} \ln \left[(h/d) + (D^2/4hd)\right] \quad \text{(Eq A32)}$$

and similarly, for Eqs A29 and A31, using a square of Eq A31,

$$K_{mx}(2) + K_{my}(2) \approx \frac{1}{4\pi} \ln \left\{\left[(4h^2 + D^2)/(4h^2 + 4D^2)\right]^2\right\} + \frac{1}{4\pi} \ln \left\{(h^2 + D^2)^2/\left[(D^2 + 0)(D^2 + (2h - 0)^2\right]\right\} \quad \text{(Eq A32A)}$$

which, after some manipulation and multiplying of arguments, gives

$$K_{mx}(2) + K_{my}(2) \approx \frac{1}{2\pi} \ln (\tfrac{1}{4} + h/2D) \quad \text{(Eq A33)}$$

The detailed simplification procedure of obtaining Eq A33 is as follows. Neglecting d^2 in Eq A31 and using a square of Eq A29, these two equations can be added by a multiplication of arguments under logarithm:

$$K_{mx}(2) + K_{my}(2) = \frac{1}{4\pi} \ln \left[\frac{(4h^2 + D^2)(4h^2 + D^2)}{(4h^2 + 4D^2)(4h^2 + 4D^2)} \frac{(h^2 + D^2)}{(D^2 + 0)} \frac{(h^2 + D^2)}{D^2 + (2h - 0)^2}\right]$$

$$= \frac{1}{4\pi} \ln \left[\frac{(4h^2 + D^2)^2}{16(h^2 + D^2)^2} \frac{(h^2 + D^2)^2}{D^2(D^2 + 4h^2)}\right] = \frac{1}{4\pi} \ln \left[\frac{4h^2 + D^2}{16D^2}\right]$$

$$= \frac{1}{4\pi} \ln \left[\frac{h^2}{4D^2} + \frac{1}{16}\right] = \frac{1}{4\pi} \ln \left[(h/2D)^2 + (\tfrac{1}{4})^2\right] \quad \text{(Eq A34)}$$

Now, deliberately, the following simplification is made:

$$\frac{1}{4\pi} \ln \left[(h/2D)^2 + (1/4)^2\right] \approx 1/2 \frac{1}{2\pi} \ln \left[(h/2D + 1/4)^2\right] \qquad \text{(Eq A35)}$$

Of course, a detailed analysis of Eq A35 reveals that

$$\frac{h^2}{4D^2} + \frac{1}{16} \neq \frac{h^2}{4D^2} + \frac{h}{4D} + \frac{1}{16}$$

because an extra term $h/4D$ is present in the simplified Eq A33, shown on the right side of Eq A35. However, this added term plays a very useful role — for small values of h and large D it is negligible, while for dense spacings and the depth of grid burial approaching D, it slightly increases the combined value of the argument under the logarithm. This effect tends to rectify the conceptual deficiency of the basic mathematical model in this respect. Hence, Eq A31 can be viewed as a simplified equation with a correctly applied error bias.

Denoting formally the simplified expressions Eqs A32 and A33 as $K'_{mxy}(1)$ and $K'_{mxy}(2)$, it holds that

$$K'_{mxy}(1) + K'_{mxy}(2) = \frac{1}{2\pi} \ln \left[(h/d)(1 + D^2/4h^2)1/4 + h/2D)\right] \qquad \text{(Eq A36)}$$

Since the order in which the individual members of Eq A27 are summed does not matter, obviously

$$K'_{mx}(1,2) + K'_{my}(1,2) = K'_{mxy}(1) + K'_{mxy}(2) \qquad \text{(Eq A37)}$$

and the mesh factor for $N = 2$ and $h \geq 4d$, is approximately

$$K_m(N{=}2) \approx \frac{1}{2\pi} \ln \left[\frac{D^2}{16hd} + \frac{(D+2h)^2}{8Dd} - \frac{h}{4d}\right] \qquad \text{(Eq A38)}$$

A3. Effect of Other Conductors for $N > 2$; $K_{mx}(3,N)$ and $K'_{mx}(3,N)$ Series

For $N > 2$, the calculation of K_m has to take into account the presence of $N - 2$ conductors outside the *corner mesh*, formed by the first two conductors in Fig A6.

Generally, starting with the third conductor in Fig A6, the effect of conductors 3, 4, 5, ..., N on the surface voltage V_X at point X can be expressed in terms of Eq A24 as a factor $K_m(3,N)$ in the following transcription:

$$V_X = E_0 - \sigma i \left[K_m(1,2) + K_m(3,N)\right] \qquad \text{(Eq A39)}$$

where, in addition to the already defined symbols,

$K_m(1,2)$ is identical to K_m for $N = 2$, defined by Eq A38

and

$$K_m(3,N) = K_{mx}(3,N) + K_{my}(3,N) \qquad \text{(Eq A40)}$$

The expressions $K_{mx}(3,N)$ and $K_{my}(3,N)$ can be calculated by means of two earlier equations, Eq A12 for $K_{mx}(1,N)$ and Eq A23 for $K_{my}(1,N)$, if the starting value of k in the summations is changed from $k=0$ to $k=2$.

However, for the latter y-component, the following simplifying assumption can reasonably be made. As one can observe in Fig A7, if $D > h$ (typically $D \geq 2.5$ m, and $h \leq 1$ m), then any point along the vertical line connecting (1) and (1*) is nearly equidistant from any outside conductor and its corresponding image, that is from (3) to (3*), (4) and (4*), etc. For this reason, these N - 2 conductors will not make any significant contribution to the vertical current components produced by the first two wires that are represented in Eq A38. Therefore, it can be assumed that $K_{my}(3,N) \approx 0$, and, for a nonzero h,

$$K_m(3,N) \approx K_{mx}(3,N) \qquad \text{(Eq A41)}$$

where

$$K_{mx}(3,N) = \frac{1}{2\pi} \sum_{k=2}^{N-1} \ln \left[\frac{(2k-1)^2 D^2 + 4h^2}{4k^2 D^2 + 4h^2} \right]$$

A further simplification can be made for h approaching zero. For convenience, let

$$K'_{mx}(3,N) = \lim_{h \to 0} \left[K_{mx}(3,N) \right] \qquad \text{(Eq A42)}$$

Then, the effect of N - 2 outside conductors can be expressed by means of the following numerical series, obtained by a limiting operation as is shown below.

$$2\pi K'_{mx}(3,N) \doteq \lim_{h \to 0} \sum_{2}^{N-1} \ln \left[\frac{(2k-1)^2 D^2 + 4h^2}{4k^2 D^2 + 4h^2} \right] = \ln \prod_{2}^{N-1} \left[\frac{(2k-1)D}{2kd} \right]^2 ; \qquad \text{(Eq A43)}$$

and so

$$K'_{mx}(3,N) \doteq \frac{1}{\pi} \ln \left[(3/4)(5/6)(7/8) \cdots (2N-3)/(2N-2) \right] \qquad \text{(Eq A44)}$$

A4. Numerical Approximation of $K'_{mx}(3,N)$ Series for $h \to 0$

For N parallel conductors representing a grounding grid, the mesh voltage factor K_m can be viewed as consisting of three components:

$$K_m = K'_{my}(1,2) + K'_{mx}(1,2) + K'_{mx}(3,N) \qquad \text{(Eq A45)}$$

where in particular, for a zero burial depth,

$$K'_{mx}(3,N) = (1/\pi) \ln \left[(3/4)(5/6)(7/8) \cdots \right] \qquad \text{(Eq A46)}$$

reflects the beneficial effect of $(N-2)$ parallel conductors outside the *corner mesh* on lowering the voltage difference between the voltage on grounded metal and that existing on the earth's surface, above a center line between the first two peripheral conductors forming the first mesh. Thus,

$$K'_{mx}(3,N) = 1/\pi \sum_{k=1}^{n=N-2} \ln\ (a_k) = (1/\pi)\ln(S_N) \qquad \text{(Eq A47)}$$

where

$$S_N = \prod_{K=3}^{N} \left[\frac{2K-3}{2K-2}\right] \qquad \text{(Eq A48)}$$

Since

$$\lim_{n\to\infty} (a_n) \Bigg| = 1, \text{ and } 0.75 \le a_k < a_{k+1} < 1, \text{ for } k = 1,2,\ldots$$

$K_{mx}(3,N)$ is subtractive; the higher N, the higher negative value of the logarithm of S_N results.

Consequently, in order to approximate the series by a simple function, and to keep errors sufficiently small and on the conservative side, we will seek such a functional S_N^*, which will satisfy the following conditions:

$$S_N^* = \sup\ (S_N) \text{ for any } N \in (3 \le N < \infty) \qquad \text{(Eq A49)}$$

$$\lim_{N\to\infty} (S_N^* - S_N) \Bigg| = +0 \qquad \text{(Eq A50)}$$

and, for any countable finite series, such a small positive error margin σ, $\sigma \ge 0$, which will be acceptable. To prevent runaway errors for high N, this last requirement represents the need for a convergence, that is, for any small positive number $\gamma \le \sigma$, there always exists a positive integer I, such that

$$(S_N^* - S_K) \le \gamma \text{ for } K \in (3, N+I) \qquad \text{(Eq A51)}$$

Consider now the following two double factorials:

$$(2n-1)!! = (2n-1)(2n-3)(2n-5)\cdots 5 \cdot 3 \cdot 1$$

$$(2n)!! = (2n)(2n-2)(2n-4)\cdots 6 \cdot 4 \cdot 2$$

Let

$$S_n = \left[\frac{(2n-1)!!}{(2n)!!}\right] \qquad \text{(Eq A52)}$$

Comparing Eqs A48 and A52, it can be seen that for $n = N - 1$,

$$S_n = \left[\frac{(2N-3)!!}{(2N-2)!!}\right] = (1/2)\ S_N \qquad \text{(Eq A53)}$$

As shown in mathematical handbooks, $\pi/2$ can be represented by the following infinite series:

$$1/2\,\pi = (2)(2/3)(4/3)(4/5)(6/5)(6/7)(8/7)(8/9)\cdots \qquad \text{(Eq A54)}$$

A closer analysis of Eq A54 reveals that a very useful relationship exists between the number π and S_n, as defined in Eq A52:

$$\lim_{n \to \infty} \left\{ \left[\frac{(2n - 1)!!}{(2n)!!} \right]^2 (2n + 1) \right\} = 2/\pi \qquad \text{(Eq A55)}$$

Using Eq A53, it can be seen that

$$2/\pi = \lim_{n \to \infty} \left[(2n + 1)(S_n)^2\right] = \lim_{N \to \infty} \left[(2N - 1)(S_N/2)^2\right] \qquad \text{(Eq A56)}$$

On this basis, for any countable finite series S_N, $N \epsilon \, (3 \leq N \leq \infty)$, the following inequality holds:

$$(2N - 1)(S_N)^2 \geq 8/\pi, \quad \text{or } S_N \geq (8/\pi(2N - 1))^{1/2} \qquad \text{(Eq A57)}$$

Therefore, the sought functional S_N^*, is

$$S_N^* = \sqrt{\frac{8}{\pi(2N - 1)}} \qquad \text{(Eq A58)}$$

Proof: Expressing S_n by means of a gamma function $\Gamma(x)$ for the particular x; $x = n+1$, and $x = n+½$ values,

$$\Gamma(n + 1) = n! \quad \text{and} \quad \Gamma(n + ½) = \frac{(2n)!\sqrt{\pi}}{n!\,(2)^{2n}}$$

Thus,

$$S_n = \frac{(2n - 1)!!}{(2n)!!} = \frac{2^n\,\Gamma(n + ½)}{\sqrt{\pi}} \cdot \frac{1}{2^n\,\Gamma(n + 1)} = \frac{(2n)!}{2^{2n}\,(n!)^2} \qquad \text{(Eq A59)}$$

Substitution of Eq A59 into Eq A56 yields

$$2/\pi = \lim_{n \to \infty} \left\{ (2n+1) \left[\frac{(2n)!}{(2^{2n}\,(n!)^2} \right]^2 \right\} \qquad \text{(Eq A60)}$$

Now, consider two integers: μ and v, $(1 < \mu < v < n < \infty)$; say, $\mu = 5$, $v = 7$, and let $n = 25$. Then

$S_{(5)} > S_{(7) > S_{n=25}} > 2/\pi$

should hold.

Provide

$11\,(10!/(2^{10}(5!)^2)^2 > 15\,(14!/2^{14}(7!)^2)^2 > 51\,(50!/(2^{50}(25!)^2)^2 > 2/\pi$

Calculated,

$0.66618 > 0.65818 > 0.64289 > 2/\pi \approx 0.63662$

or, more vividly,

$2/0.9556\,\pi > 2/0.9672\,\pi > 2/0.9902\,\pi > 2/\pi$

Hence, $K'_{mx}(3,N)$ can be approximated with reasonable accuracy, as

$$K^*_{mx}(3,N) = \frac{1}{\pi}\ \ln\left(\sqrt{\frac{8}{2N-1)\pi}}\ \right) = (1/2\,\pi)\ \ln\,(8/(\pi(2N-1))) \qquad \text{(Eq A61)}$$

A5. Correction of $K'_{mx}(3,N)$ Series for $0 < h \le 2.5$ m

In comparison with the relatively simple expression for $K'_{mx}(3,N)$ series, Eq A60, the calculation of $K'_{mx}(3,N)$ series for nonzero depth h of a grid burial is tedious.

A natural question thus arises: instead of using Eq A41 to calculate $K_{mx}(3,N)$ directly, would it be possible to use Eq A61 in conjunction with some convenient correction factor that would make the $K'_{mx}(3,N)$ series behave more like that of Eq A41?

For the usual depth range $0 < h \le 2.5$ m, a suitable corrective factor is

$$H_m(h,h_0) = 1\ /\ \sqrt{1 + h/h_0}\ ;\ h_0 = 1 \text{ m} \qquad \text{(Eq A62)}$$

Therefore, combining Eqs A61 and A62, a much briefer expression can eventually be used to approximate K_{mx} (3, N):

$$K_{mx}(3,N) \approx H_m(h,h_0)\ K^*_{mx}(3,N)$$

$$\approx \frac{1}{2\pi\ \sqrt{1 + h/h_0}}\ \ln\left[\frac{8}{\pi\ (2N-1)}\right] \qquad \text{(Eq A63)}$$

A6. Development of Simplified Equations for E_{mesh}

A kernel of a simplified formula for K_m can be developed now, using the previously developed expressions A38 and A63, to obtain a formulation for $N > 2$ and nonzero depth h.

$$K_m = \frac{1}{2\pi}\left[\ln\left(\frac{D^2}{16hd} + \frac{(D+2h)^2}{8Dd} - \frac{h}{4d}\right) + \frac{1}{\sqrt{1+h/h_0}}\ \ln\left(\frac{8}{\pi(2N-1)}\right)\right] \qquad \text{(Eq A64)}$$

However, in order to use this expression for calculating E_{mesh}, in conjunction with the irregularity factor K_i, it is necessary to accommodate the fact that K_i — being a simple multiplier — increases not only the current in the peripheral conductors, but also in all remaining ones. Since the grid resistance R_g and the total current are fixed values for any given design, and $i = I/L$, K_i actually upsets Ohm's law:

$$\frac{I_0}{R_g} = \frac{iL}{R_g} \neq K_i\ \frac{iL}{R_g} \quad \text{if } K_i > 1 \qquad \text{(Eq A65)}$$

Obviously, the only way to compensate for this problem is to use a second corrective factor K_{ii} for the average current characterizing the majority of conductors external to the corner meshes. This factor has to be less than 1:

$$\frac{iL}{R_g} = \frac{K_i L_1 + K_{ii} L_2}{R_g} i; \text{ for } L_1 + L_2 = L \text{ and } K_{ii} < 1 \qquad \text{(Eq A66)}$$

Furthermore, since the use of ground rods may have a different effect on the current distribution within the grid (if the rods are placed predominantly along the perimeter, or spread evenly within the grid area), it is useful to establish the equation for calculating E_{mesh} in the following form, Eq A67, leading to two versions of the same formula, stated below as Eqs A68 and A69, respectively.

$$E_{\text{mesh}} = \sigma I_0 K_i \left[\frac{1}{2\pi L} K_{mxy}(1,2) + \frac{1}{\pi L} K_{ii} H_m(h,h_0) K^*_{mx}(3,N) \right] \qquad \text{(Eq A67)}$$

where

K_{ii} = corrective factor for inner area currents
L_c = total length of horizontal grid conductors
L_r = total length of ground rods
L = $L_c + 1.15\, L_r$ for grids with rods along perimeter and $K_{ii} = 1$
L = $L_c + L_r$ for grids with rods evenly spread over A, and $K_{ii} = (2N)^{-(2/N)}$

$$E_{\text{mesh}} = \frac{\sigma K_i I_0}{2\pi(L_c + 1.15 L_r)} \left[\ln\left(\frac{D^2}{16hd} + \frac{(D+2h)^2}{8Dd} - \frac{h}{4d} \right) + \frac{1}{\sqrt{1+h/h_0}} \ln \frac{8}{\pi(2N-1)} \right] \qquad \text{(Eq A68)}$$

$$E_{\text{mesh}} = \frac{\sigma K_i I_0}{(L_c + L_r) 2\pi} \left[\ln\left(\frac{D^2}{16hd} + \frac{(D+2h)^2}{8Dd} - \frac{h}{4d} \right) + \frac{1}{(2N)^{2/N}\sqrt{1+h/h_0}} \ln \frac{8}{\pi(2N-1)} \right] \qquad \text{(Eq A69)}$$

Figure A8 illustrates the result of computer calculations for three cases of an identical square grid having 16 meshes: (1) without any ground rods, (2) with 24 rods spread evenly within the grid area, and (3) with 24 rods placed along the perimeter. The grid design data are as follows: soil resistivity 100 Ω/m, 40 · 40 m grid area, total length of grid conductors 400 m, each rod 4.1 m long, depth of burial 0.5 m, and diameter of both horizontal and vertical electrodes 0.02 m.[31]

The basic idea behind two different definitions of K_{ii} is as follows. For Eq A68 there is little or no difference between the magnitudes of an "average" effective current flowing in the conductors forming the corner meshes and those in inner areas of the grid. In addition, the use of a 1.15 multiplier for the rod length L_r reflects the theoretical premise of a relatively higher efficiency of peripheral ground rods in emitting current into the earth, in comparison with an equivalent per unit length of grid conductor, documented in the literature.

[31] Computer program MALT, Version 2, Revision 6, of Safe Engineering Services LTD, Montreal, Canada, was used for this experiment.

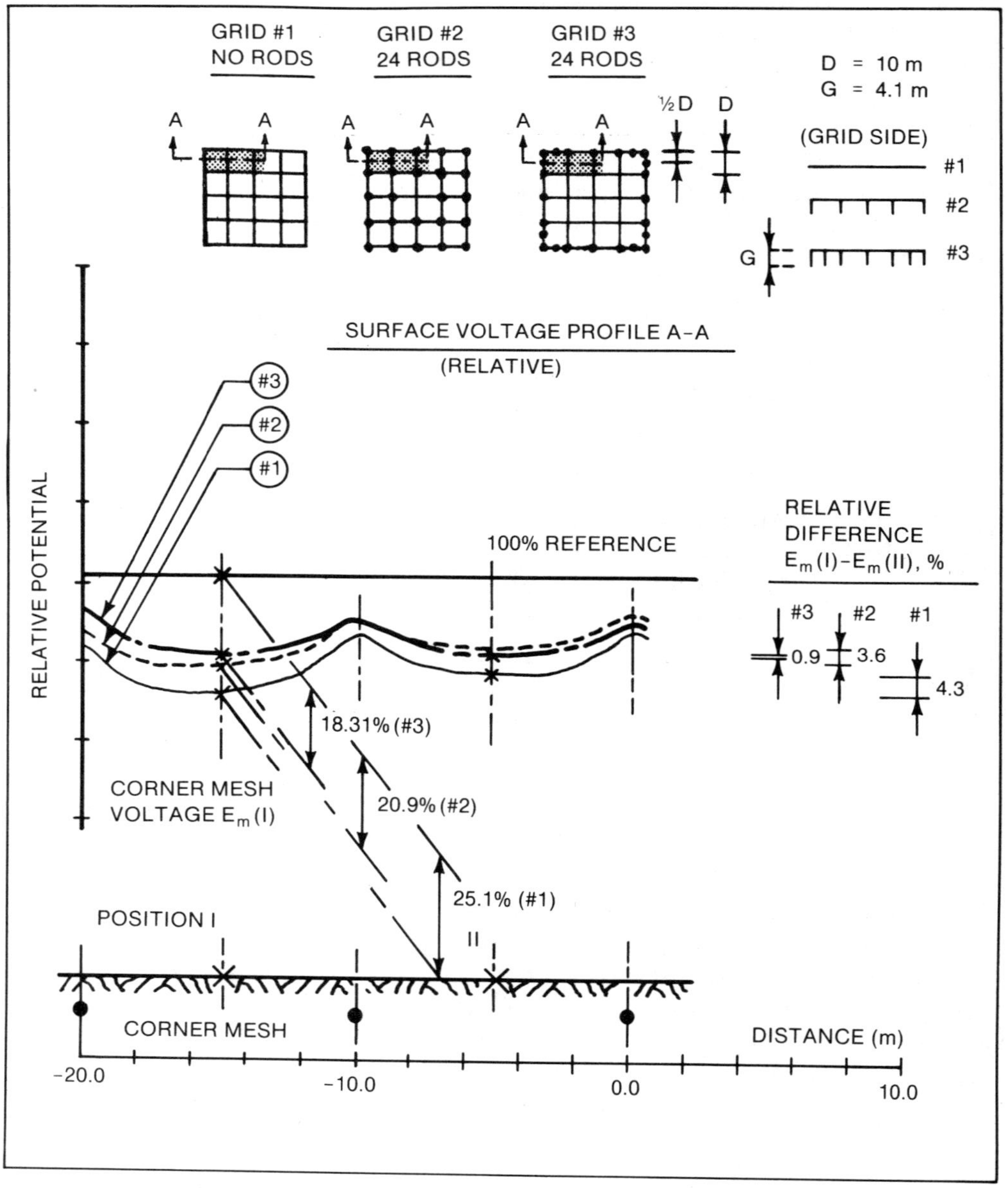

Fig A8
Comparison of Surface Potential Profiles of Three Identical Grids
(1) Grids Without Ground Rods
(2) Grids With 24 Evenly Distributed Ground Rods
(3) Grids With 24 Peripheral Ground Rods

For Eq A69 the assumption is that with the increasing N, the resulting relative increase of the electrically *flat* inner area of the grid and of the portions of conductor length carrying lower current somewhat reduces the overall effect of $N - 2$ outside wires of the subject N conductor model on diminishing the corner mesh voltage. Figure A9 illustrates how the chosen semiempirical factor K_{ii} of Eq A69 modifies the K_i curve for the outside wires.

A7. Step Voltage Calculations

By definition, the step voltage is a voltage difference between two points on the ground surface, X_1 and X_2, which are separated by 1 m distance. Strictly speaking, in order to determine the maximum step voltage near a grounding grid, one should proceed in the following two steps: (1) find the point (that is, its coordinate) x_0 of the maximum gradient $G(x_0)$, and (2) search all possible pairs of points X_1 and X_2 within $a \pm 1$ m distance from x_0, to find the pair that would produce the largest voltage difference, that is

$$E_{\text{step}} = \max \int_{x_1}^{x_2} G\ (x)\, dx \quad \text{for } x_1 + 1 \geq x_0 \geq x_2 - 1 \text{ and } \operatorname{abst}(x_1 - x_2) = 1 \text{ m}; \qquad \text{(Eq A70)}$$

where

x_0 is the coordinate of the maximum gradient point: $G(x_0) = \max G(x)$

Comparing Eq A70 with Eqs A9 and A10, it can be seen that for one conductor in depth h, the gradient at point X on the ground surface is

Fig A9
Effect of K_{ii} Upon K_i As a Function of a Number of Parallel Conductors N

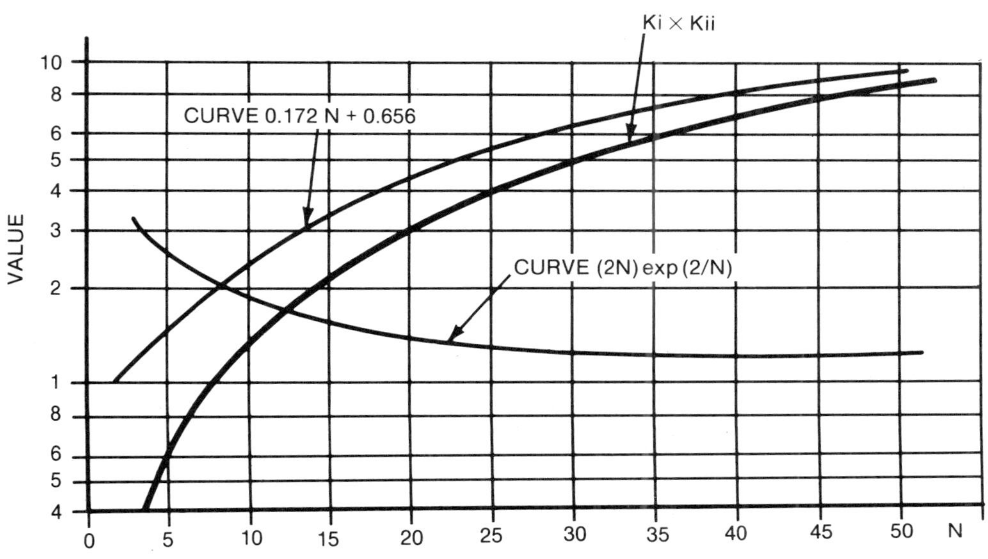

$$G(x) = \sigma\, \delta_s(x,h) = \sigma\, \frac{i}{\pi}\left[\frac{x}{x^2 + h^2}\right] \quad \text{(Eq A71)}$$

and for N conductors,

$$G(x) = \sigma\, \frac{i}{\pi} \sum_{k=1}^{N} \left[\frac{x(k)}{x(k)^2 + h^2}\right] ;\; x(k) = (k-1)\,D + x \quad \text{(Eq A72)}$$

However, in view of the fact that the step voltages are inherently less dangerous than the touch or mesh voltages resulting from the same current, several convenient simplifications can be made.

First, assuming that $D + x > h$ for $D > 2.5$ m, and $0 < h < 2.5$ m, as long as x is at least $\approx$ ½ h or more, then $(D + x)^2 >> h^2$, and Eq A72 approximated from the second term by a simpler series is

$$G(x) = \sigma\, \frac{\mathrm{i}}{\pi}\left[\frac{x}{x^2 + h^2}\right] + \sum_{k=2}^{N}\left[\frac{1}{(k-1)D + x}\right] \quad \text{(Eq A73)}$$

The error involved is conservative, making the gradient higher. The point of maximum gradient can now be determined by taking the derivative of Eq A73, putting it equal to zero, and solving for x:

$$x^2 = h^2 - \left[\frac{1}{(D+x)^2} + \frac{1}{(2D+x)^2} + \cdots\right](x^2 + h^2)^2 \quad \text{(Eq A74)}$$

Second, the subtractive second term in Eq A74 can be neglected and the maximum gradient assumed to occur where $x = h$. Since the neglected term, representing the effect of $N - 1$ wires beyond the first one, would bring the point of maximum gradient slightly closer toward the point on the surface directly above the first conductor, this assumption somewhat compensates for the error of Eq A73.

Third, assume that the gradient determined by solving Eq A73 for $x = x_0$, with $x_0 \approx h$, remains constant within a ± ½ m distance, so that the maximum step voltage is the numerical value of maximum V/m:

$$E_{\text{step}} = \sigma\, \frac{i}{\pi}\left[\frac{1}{2h} + \frac{1}{D+h} + \frac{1}{2D+h} + \cdots + \frac{1}{(N-1)D + h}\right] K_i \quad \text{(Eq A75)}$$

Fourth, for $D \geq 10\,h$, or more strictly, for $0 \leq h \leq 0.25$ m, h can be neglected in all the terms of the above series, starting with the third term. The error is again on the conservative side, roughly in the order of 10–15%.

$$E_{\text{step}} = \sigma\, \frac{i}{\pi}\left[\frac{1}{2h} + \frac{1}{D+h} + \frac{1}{2D} + \frac{1}{3D} + \cdots + \frac{1}{(N-1)D}\right] K_i \quad \text{(Eq A76)}$$

Although there is no simple formula for the sum of a finite harmonic series of m members, that is $1 + ½ + ¼ + \cdots + \frac{1}{m}$, the solution can be approximated[32] by

[32] J. J. Tuma, Engineering Mathematics Handbook, 2nd Ed. New York, NY: McGraw-Hill, 1979.

$$\sum_{k=1}^{m} \frac{1}{k} \approx x(m) + c \quad \text{for } m \geq 5 \tag{Eq A77}$$

where

C = 0.577215665

and $x(m)$ is a rather complicated function of m. For the purpose of simplifying Eq A76, the following approximation can be made for $m \geq 5$ and k = 2, 3, $\cdots$ m:

$$\sum_{k=2}^{m} \frac{1}{k} \approx \frac{1}{2m} + \ln m - 0.423 \tag{Eq A78}$$

Consequently, for $n \geq 6$, Eq A76 becomes

$$E_{\text{step}} = \frac{\delta i}{\pi} \left[\frac{1}{2h} + \frac{1}{D+h} + \frac{1}{D} \left(\frac{1}{2(N-1)} + \ln (N-1) - 0.423 \right) \right] \tag{Eq A79}$$

Fifth, for the usual range of h, 0.25 m $\leq h \leq$ 2.5 m, the cumulative error in the simplified terms representing the effect of the third, fourth, $\cdots$, Nth wire can be reduced by using a more quickly diminishing geometric series:

$$E_{\text{step}} = \sigma \frac{i}{\pi} \left[\frac{1}{2h} + \frac{1}{D+h} + \frac{1}{2D} + \frac{1}{4D} + \cdots + \frac{1}{(2)^{N-2} D} \right] K_i \tag{Eq A80}$$

Using a sum of the geometric series and retaining the first two terms that contain h, Eq A80 may be simply written as

$$E_{\text{step}} = \sigma \frac{i}{\pi} \left[\frac{1}{2h} + \frac{1}{D+h} + \frac{1}{D} \left[1 - (1/2)^{N-2} \right] \right] K_i \tag{Eq A81}$$

A8. Grid Resistance Formula

For calculating the resistance of R_g of a grounding grid, the earlier editions of this guide provided the following formula suggested by Laurent.

$$R_g = \sigma \left(\frac{1}{4r} + \frac{1}{L} \right) \quad \text{for } h = 0; \; r = \sqrt{A/\pi} \tag{Eq A82}$$

Since for an infinite L, Eq A82 becomes an expression for the resistance of a metallic plate at zero depth, in order to obtain a correction for nonzero depths in the usual range of $0 < h < 2.5$ m, the above formula can be combined with another semiempirical formula for the resistance of a plate, given by Laurent in [B67].

$$R_p = \sigma \frac{1}{8r} \left(1 + \frac{r}{2.5h + r} \right) \quad \text{for nonzero } h \tag{Eq A83}$$

A simplified equation that combines Eqs A82 and A83 is

$$R_g\ (h) = \sigma \left[\frac{1}{L} + \frac{1}{\sqrt{20/A}} \left(1 + \frac{1}{1 + h\sqrt{20/A}} \right) \right] \qquad \text{(Eq A84)}$$

where

h = grid burial depth in m
A = area of the grid in m^2
L = total length of the conductor in m

A9. Fortran Routines for K_m, K_s, and R_g

Subroutine for K_m. A subroutine for calculating K_m by means of a full IEEE model of NC parallel conductors of diameter CDIA in-depth DPTH, spaced SPAC apart, is listed below. This routine has a built-in correction factor CII for mitigating the error due to neglected distortions of the gradient field by the presence of NC conductors, which are simultaneously dissipating current into the ground.

In the main program K_m is obtained as the value of CMF, by calling subroutine FULNEW, as shown below. All parameters are in meters. The value of P_i = 3.1415, etc, has to be declared in the main program.

Function for R_g.

```
      CALL FULNEW(CMF,2,CDIA,DPTH,SPAC,NC,2,PI)
```

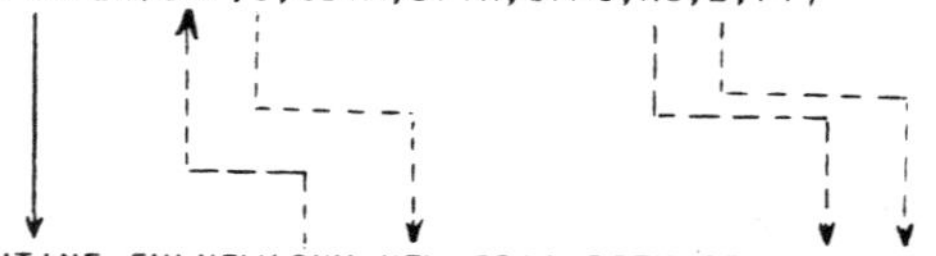

```
      SUBROUTINE FULNEW(CKM,KEY,CDIA,DPTH,SPAC,NX,NY,PI)
C     ------------------------------------------------
      TERM(A,B,C,D)=(A*A+B*B)/(C*C+D*D)
      RA=CDIA/2.
      RB=2.*DPTH-RA
      DO 7 I=1,2
      GOTO(1,2),I
    1 N1=1
      N2=2
      GOTO 3
    2 N1=3
      N2=NX
    3 T1=1.
      T2=1.
      T3=1.
      DO 4 M=N1,N2
      DK=SPAC*FLOAT(M-1)
      DM=SPAC*FLOAT(2*M-3)/2.
      T1=T1*TERM(DPTH,DM,DPTH,DK)
      IF(M.GT.NY)GOTO 4
      T2=T2*TERM(DPTH,DK,RA,DK)
      T3=T3*TERM(DPTH,DK,RB,DK)
    4 CONTINUE
      GOTO(5,6),I
    5 EX2=2.*ALOG(T1)
      EY2=ALOG(T2)+ALOG(T3)
      GOTO 7
    6 CII=1.
      IF(KEY.EQ.2)CII=FLOAT(2*NX)**(-2./NX)
      EXF=EX2+CII*2.*ALOG(T1)/SQRT(1.+DPTH)
      EYF=EY2+CII*(ALOG(T2)+ALOG(T3))
      GOTO 8
    7 CONTINUE
    8 CKM=(EXF+EYF)/(4.*PI)
    9 RETURN
      END
```

(a) By entering "1" as the value of KEY in the calling statement, and using 2 for NY (number of calculated E_y components), the value of K_m, which is returned as CKM, corresponds to the condition of ground rods placed predominantly along a grid perimeter.

(b) Entering 2 for KEY and again using 2 for NY yields K_m, which is consistent with conditions of no ground rods or ground rods evenly spread over the grid area.

Functions for K_s. For a depth below 0.25 m:

```
      FUNCTION STEPF(N,S,H,PI)
      ------------------------
      M=N-1
      C=1./(2.*H)
      DO 1 I=1,M
    1 C=C+1./(FLOAT(I)*S+H)
      STEPF=C/PI
      RETURN
      END
```

For a depth over 0.25 m up to 2.5 m:

```
      FUNCTION STEPN(N,S,H,PI)
      ------------------------
      M=N-2
      A=1./(2.*H)
      B=1./(S+H)
      C=1.-(.5)**M
      STEPN=(A+B+C/S)/PI
      RETURN
      END
```

Function for R_g.

```
      FUNCTION RG20(RHO,AREA,CTL,H)
C     -----------------------------
      A=1./CTL
      B=1./SQRT(20.*AREA)
      C=1.+H*SQRT(20./AREA)
      RG20=RHO*(A+B*(1.+1./C))
      RETURN
      END
```

Appendix B
Graphical Analysis of Square Ground Grids Without Ground Rods in Uniform Soil

(This Appendix is taken from EPRI Final Report 2682, Vol 1, Analysis of Substation Grounding Systems, developed by the Georgia Institute of Technology.)

The graphical analysis summarized here is described in Joy *et al.* and is available in full in the EPRI Final Report EL 2682, Vol 1. For the purposes of this guide, the number of graphs has been greatly reduced to include variations of only those parameters that have the most impact on the grid resistance, touch voltage, and step voltage. The developed graphs apply only to square grids without ground rods and with uniform conductor spacings in both directions. Though the EPRI report gives interpolation guidelines for extending the use of graphs to rectangular grids with rectangular meshes, these are not included here.[33]

The following explanations and example illustrate the use of the graphical data for the calculations of grid resistance R_g, ground potential rise GPR, mesh voltage E_m, and corner step voltage E_s.

corner mesh voltage. The corner mesh voltage E_m is calculated by multiplying the GPR by the corner mesh voltage percentage obtained from Fig B1. The corner mesh voltage is thus

$$E_m = \text{GPR} \cdot \frac{\text{graphed value of corner mesh voltage percentage}}{100} \quad \text{(Eq B1)}$$

Figure B1 gives the corner mesh voltage percentage of GPR for a grid depth of 0.5 m and conductor diameter of 0.01 m. The grid depth and conductor diameter have been found to have negligible effect on E_m for grid depths between 0.25 and 1.0 m and for conductor diameters between 2.5 and 10.0 mm.

corner step voltage. The corner step voltage E_s is calculated using the GPR and the corner step voltage percentage obtained from Figs B3-B5.

$$E_s = \text{GPR} \cdot \frac{\text{graphed value of corner step voltage percentage}}{100} \quad \text{(Eq B2)}$$

[33] For rectangular grids, the numerical proceduresof Appendix A are more straightforward and include the effects of ground rods. For comparison of both procedures, refer to the discussion part of E. B. Joy *et al.*, Graphical Data for Grounding Grid Analysis, IEEE Transactions PAS, vol PAS-102, no 9, pp 3046-3048, September 1983.

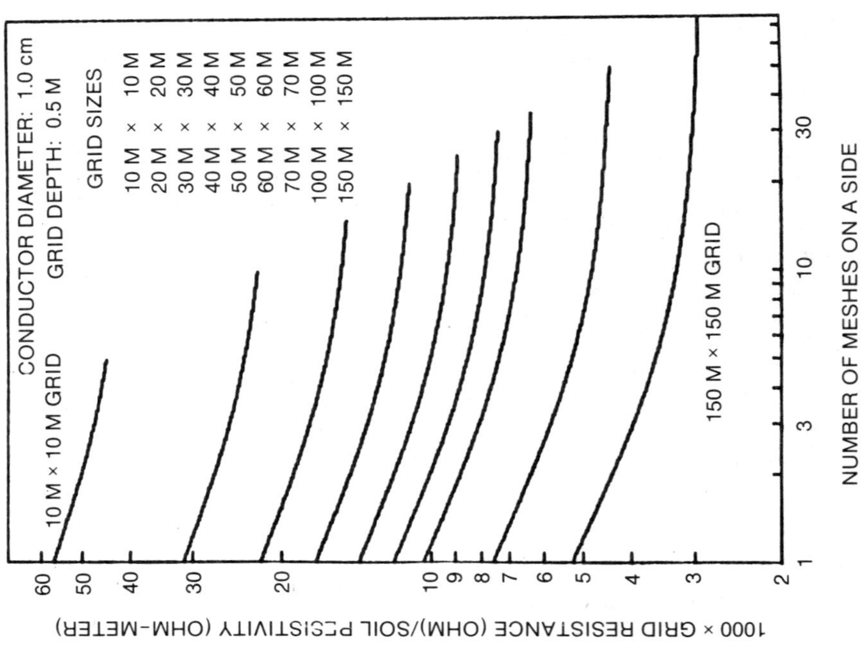

Fig B1
Corner Mesh Voltage for a Square Grid in Depth 0.5 m and 1 cm Conductor Diameter

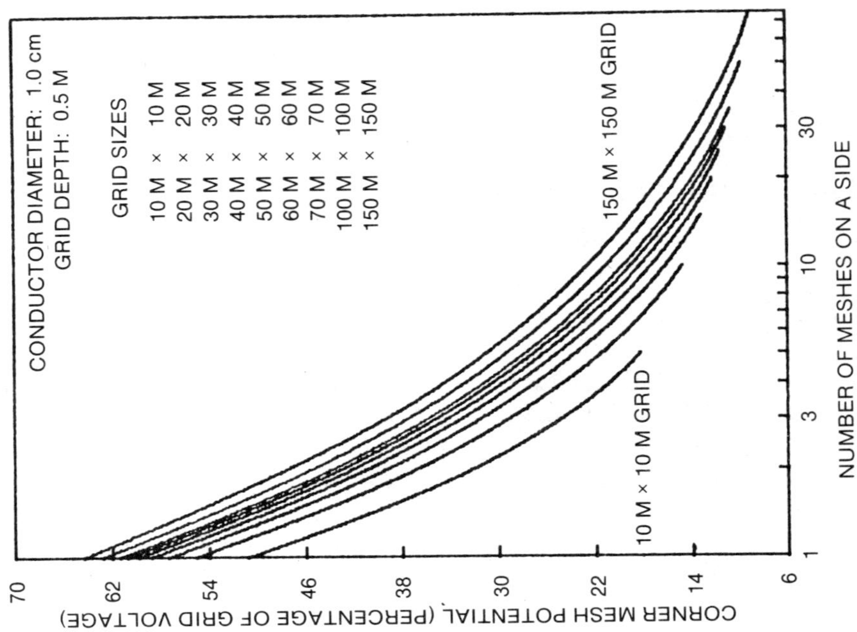

Fig B2
Grid Resistance for a Square Grid in Depth 0.5 m and 1 cm Conductor Diameter

grid resistance. The grid resistance is given by

$$\text{grid resistance} = \frac{\text{soil resistivity } (\Omega\text{-m}) \cdot \text{graphed value}}{1000} \; \Omega \qquad \text{(Eq B3)}$$

The graph of grid resistance (Fig B2) is for a grid depth of 0.5 m and conductor diameter of 10 mm. The grid depth and conductor diameter have been found to have negliglble effect for grid depths between 0.25 and 0.5 m and for conductor diameters between 2.5 and 10 mm.

ground potential rise. The ground potential rise (GPR) is calculated based on the current I_G injected into the grid and the grid resistance R_g:

$$\text{GPR} = R_g * I_G \qquad \text{(Eq B4)}$$

Figures B3 - B5 display the corner step voltage percentage for three grid depths, since the grid depth has a significant effect on the step voltage. The conductor

Fig B3
Corner Step Voltage for a Square Grid in Depth 0.25 m and 1 cm Conductor Diameter

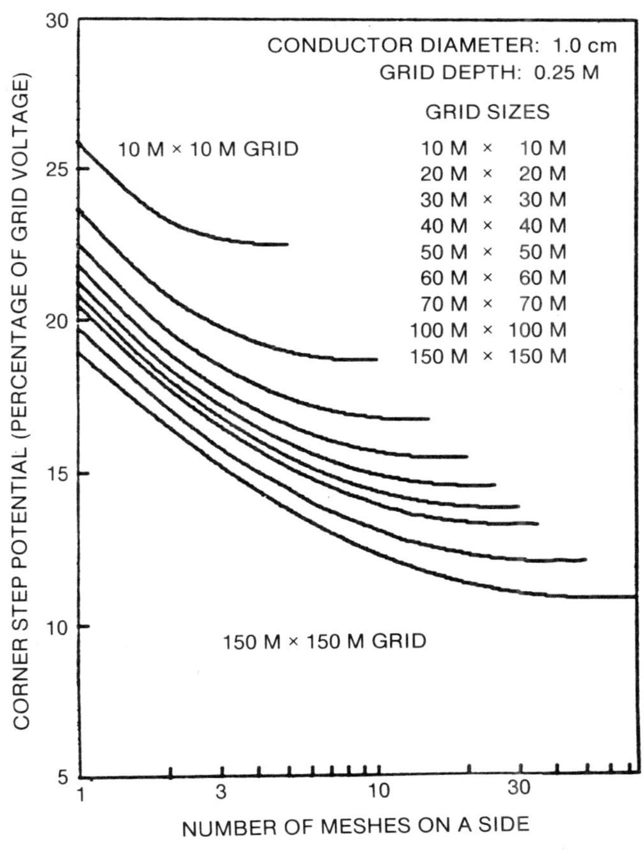

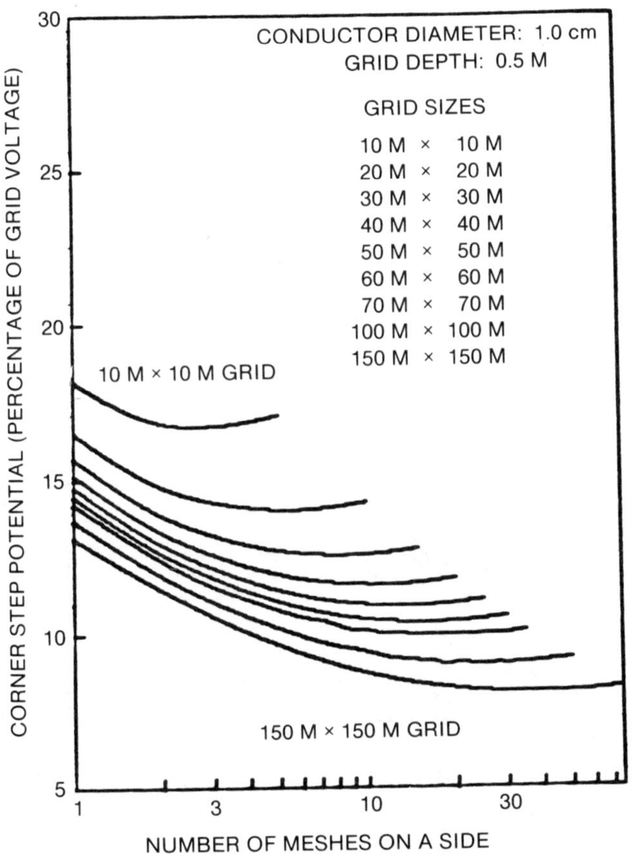

Fig B4
Step Voltage Outside a Square Grid in Depth of 1 m and 1 cm Conductor Diameter

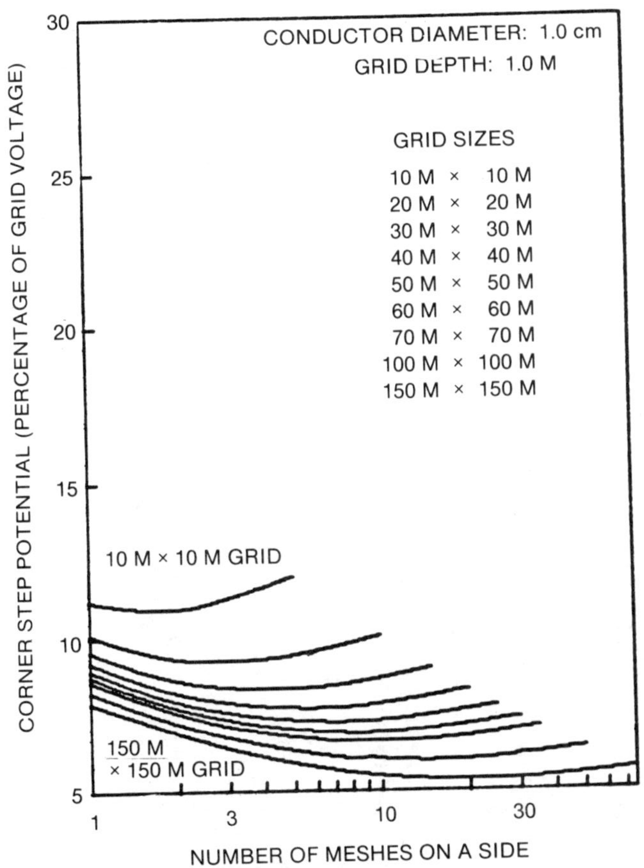

Fig B5
Corner Mesh Voltage for a Square Grid in Depth 1.0 m and 1 cm Conductor Diameter

diameter has been found to have negligible effect on the step voltage for conductor diameters between 2.5 and 10.0 mm.

Example: Find the resistance to remote earth, ground potential rise, corner mesh voltage, and the corner step voltage of a 50 · 50 m ground grid in a uniform soil with resistivity of 200 Ω-m. The grid has 100 equally spaced meshes (10 meshes on a side) and the grid is at a depth of 0.5 m and the grid conductors have a diameter of 10 mm. The fault current into the grid is 1000 A.

Figure B2 shows the grid resistance R_g for the case of a 10 mm conductor diameter and 0.5 m grid depth. The fifth curve from the top of Fig B2 is the curve for the 50 · 50 m grid. The point on this curve, for the value of 10 meshes on a side, is approximately 9.4. Thus, the grid resistance is

$$\text{grid resistance} = \frac{200 \cdot 9.4}{1000} = 1.88\ \Omega \qquad \text{(Eq B5)}$$

The grid voltage is calculated as

$$\text{GPR} = (1.88) * (1000) = 1880\ \text{V} \qquad \text{(Eq B6)}$$

The corner mesh voltage percentage for the grid is given in Fig B1. For the case of 10 meshes on a side, the corner mesh voltage percentage is approximately 18%. Thus, the corner mesh voltage is

$$V_m = (1880) * \frac{18}{100} = 338\ \text{V} \qquad \text{(Eq B7)}$$

The corner step voltage percentage for the grid is given in Fig B4. Again, for the case of 10 meshes on a side, the step voltage is approximately 11%. Thus the step voltage is

$$V_s = (1880) * \frac{11}{100} = 207\ \text{V} \qquad \text{(Eq B8)}$$

Appendix C
Sample Calculations

This appendix illustrates the application of equations, tables, and graphs for designing a substation grounding system. The specific objectives are as follows:

(1) To demonstrate the use of graphical means of Appendix B and tabulated data of Section 9 for expedient assessment of a rough design of a substation grounding grid in uniform soil

(2) To show the application of principal equations of the guide and Appendix A for several refinements of the design concept toward a satisfactory final design solution

(3) With the use of a computer, to illustrate such design conditions for which the use of simplified calculations of this guide would not be appropriate for a safe design, as only some of the equations may be used with caution

In view of these objectives, the following series of Examples 1-3 neither represents, nor is intended to be, the best or most efficient way to design a grounding system. The purpose of additional Exhibits 1 and 2 is solely tutorial.

For the series of Examples 1-3, the following design data are given:

Fault duration t_f	= 0.5 s
Fault impedance Z_1	= 4.0 + j10.0 Ω
Fault impedance Z_0	= 10.0 + j40.0 Ω
Current division factor S_f	= 0.6
Line-to-line voltage at worst-fault location	= 115 000 V
Soil resistivity ρ	= 400 Ω-m
Crushed rock resistivity (wet) ρ_s	= 2500 Ω-m
Thickness of crushed rock surfacing h_s	= 0.1 m (4 in)
Depth of grid burial h	= 0.5 m
Available grounding area A	= 63 · 84 m

The crushed rock resistivity is assumed to be a conservative estimate based on actual measurements of typical rock samples. The fault impedances and current division factor S_f are determined for the worst-fault type and location, including any conceivable system additions over the next 25 years. Thus, no additional safety factor for system growth is added. In addition, it is assumed that the substation will not be cleared by circuit breakers with an automatic reclosure scheme. Thus, the fault duration and shock duration are equal.

C1. Square Grid Without Ground Rods — Example 1

Using the step-by-step procedure as described in 14.4 and illustrated in Fig 26, the following design evaluations can be made.

Step 1: Field Data. Although the substation grounding grid is to be located within a rectangle of 63 · 84 m (5292 m^2), for the initial design assessment it may be expedient to assume that a square 70 · 70 m grid with no ground rods will be used. Consequently, the area occupied by such a grid is

$$A = 4900 \text{ m}^2$$

The average value of 400 Ω-m, determined from the soil resistivity measurements, is assumed to be that of uniform soil.

Step 2: Conductor Size. Ignoring the station resistance, the symmetrical ground fault current $I_f \approx 3\,I_0$, is

$$3I_0 = \frac{(115000/\sqrt{3}\,)\ (3)}{2\ (4.0 + \mathrm{j}10.0) + (10.0 + \mathrm{j}40.0)}$$

and, hence

$$|3I_0| = 3180 \text{ A}$$

Since the fault duration is 0.5 s, according to Table 6 of 13.4, the decrement factor $D_f = 1.00$, and the rms asymmetrical fault current is also 3180 A. For a conservative design, this current magnitude will be used to determine the minimum diameter of ground conductors.

Assuming the use of copper wires, the nomogram in Fig 14 can be used to obtain the required cross-sectional area on per unit basis. Reading on a 97% curve for hard-drawn copper, the value is approximately 2.55 mm^2/kA at a 0.5 s time, based on the melting temperature of 1084 °C.

Denoting the cross-sectional area as A_c,

$$A_c \approx 3.18 \text{ kA} \cdot 2.55 \text{ mm}^2/\text{kA} \approx 8.1 \text{ mm}^2$$

or, reading on the cmils/A scale,

$$A_c \approx 3180 \text{ A} \cdot 5 \text{ cmils/A} \approx 15\,900 \text{ cmils}$$

Since $A_c = \pi d^2/4$ the conductor diameter is less than 3.5 mm, or 0.0035 m.

Based on this computation, a copper wire as small as size AWG No 8 could eventually be used, but due to the mechanical strength and ruggedness requirements, a larger conductor with diameter d = 0.01 m (0.414 in) is usually preferred.

Consequently, at this stage, the designer may opt to check if, alternatively, the use of a less conductive (30%) copper-clad steel wire and the imposition of a more conservative maximum temperature limit of 700 °C will still permit the use of a conductor with diameter d = 0.01 m.

Using Table 1 of Section 9, in conjunction with Eq 33,

$$\text{TCAP} = 3.846,\ K_0 = 245,\ \alpha_r = 0.00378,\ \rho_r = 5.862$$

Next, applying T_m = 700 °C, T_a = 40 °C, I = 3.18 kA, and t_c = 0.5 s,

$$A_c^t = 3.18 \sqrt{\frac{\dfrac{(0.5)\,(0.00378)\,(5.862)\cdot 10^4}{3.846}}{\ln\,[1 + (700-40)/(245+40)]}} = 15.59 \text{ mm}^2$$

In this case, $d_{min} = \sqrt{(15.59)\,(4/\pi)} = 4.46$ mm, that is $\approx$ 0.005 m, which is less than d = 0.01 m desired. Hence, a 30% copper-clad steel wire of AWG 2/0 size is a viable alternative for grid wires, even if the conservative maximum temperature limit of 700 °C is imposed.

Step 3: Touch and Step Criteria. For a 0.102 m (4 in) layer of crushed rock surfacing, with resistivity of 2500 Ω-m, and for an earth with resistivity of 400 Ω-m, reflection factor K is

$$K = \frac{400 - 2500}{400 + 2500} = -0.72$$

Figure 8 of 5.4 of this guide indicates for $K = -0.72$ the resistivity of crushed rock is to be derated by a reduction factor $C_s \approx 0.63$.

Assuming that for the particular station the location of grounded facilities within the fenced property[34] is such that the person's weight can be expected to be at least 70 kg, Eq 23 and 26 may be used to compute the tolerable step and touch voltages as follows:

$$E_{step70} = [1000 + 6\,(0.63)\,2500]\,0.157/\sqrt{0.5}$$
$$= 2320 \text{ V, and}$$

$$E_{touch70} = [1000 + 1.5\,(0.63)\,2500]\,0.157/\sqrt{0.5}$$
$$= 746 \text{ V}$$

Step 4: Initial Design. Assume a preliminary layout of 70 · 70 m grid with equally spaced conductors, as shown in Fig C1: spacing D = 7 m, grid burial depth h = 0.5 m, and no ground rods. The total length of buried conductors, L, is 2 · 11 · 70 m = 1540 m.

Step 5: Determination of grid resistance. Using the graphical analysis technique of Appendix B, for the 70 · 70 m grid with 10 meshes on a side, a curve in Fig B1 of Appendix B indicates that the grid resistance R_g will be, approximately,

$$R_g = (400)\,(6.7)/1000 = 2.68, \text{ say } 2.7\ \Omega$$

(If computed by Eq 40 of the guide, for L = 1540 m, and grid area A = 4900 m^2 the obtained resistance value would be 2.78 Ω.)

Step 6: Maximum Grid Current I_G. Per procedure and definitions of 13.1, the maximum grid current I_G is determined by Eq 58 and the value of decrement

[34] That is, not accessible to the general public.

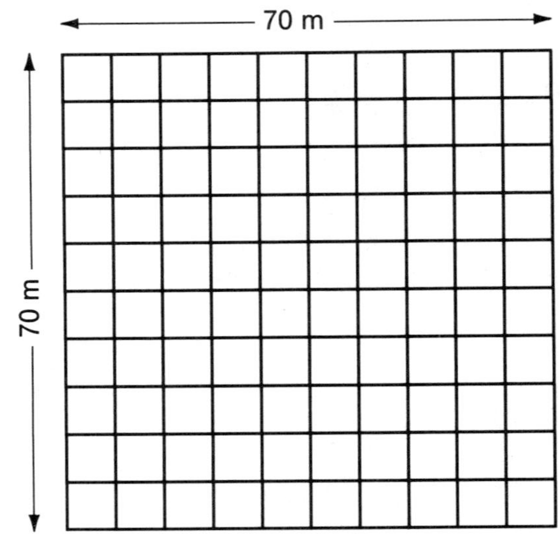

Fig C1
Square Grid Without Ground Rods

factor D_f. Referring to Step 2, for $D_f = 1.0$, and the given current division factor $S_f = 0.6$,

I_G = (0.6) (3180) (1) = 1908 A

Step 7: Now it is necessary to compare the product of I_G and R_g, GPR, to the tolerable touch voltage, $E_{\text{touch}_{70}}$. Calculated,

GPR = (1908) (2.7) = 5152 V

far exceeds 746 V, determined in Step 3 as the safe value of $E_{\text{touch}_{70}}$. Therefore, further design evaluations are necessary.

Step 8: Mesh Voltage. Using the graphical analysis again, Fig B2 of Appendix B gives approximately the corner mesh voltage as 20% of GPR. Thus,

$E_m \approx (5152)\ (0.2) \approx 1030$ V

(If calculated by Eq 72, E_m is 1125 V, using $K_i = 2.55$, $K_{ii} = 0.57$, and $K_m = 0.89$.)

Step 9: Because the estimated mesh voltage is considerably higher than the tolerable touch voltage (that is, 1030 V versus 746 V), the grid design must be modified.

C2. Square Grid With Ground Rods — Example 2

In the previous Example 1, Step 10 of the design procedure has not been reached due to the failure to meet the criterion of Step 9. Generally, there are two approaches to modifying the grid design to meet the tolerable touch voltage requirements: (1) reduce the GPR to a value below the tolerable touch voltage or to a value low enough to result in a value of E_m below the tolerable touch voltage, and (2) modify the design in such a way as to decrease the value of E_m. Both of these approaches are related in that they both involve changing any or all of the following: grid conductor spacing, total conductor length, grid depth, addition of ground rods, etc. In this example, the preliminary design will be modified to include 20 ground rods, each 7.5 m (24.6 ft) long, around the perimeter of the grid, as shown in Fig C2 below.

NOTE: Because the graphical means of Appendix B cannot account for the effects of ground rods, only the principal equations of the guide based on Appendix A will be used in the remaining calculations.

Step 5. The use of Eq 40 for $L = 1540\text{ m} + 20 \cdot 7.5\text{ m} = 1690\text{ m}$, and $A = 4900\text{ m}^2$ yields the following value of grid resistance R_g:

$$R_g = (400)\left[\frac{1}{1690} + \frac{1}{\sqrt{(20)(4900)}}\left(1 + \frac{1}{1 + 0.5\sqrt{20/4900}}\right)\right] = 2.75\ \Omega$$

Steps 6 and 7. The revised GPR is (1908) (2.75) = 5247 V >> 746 V.

Step 8. Using Eq 68 to compute K_m, and Eq 69 to determine K_i, gives

Fig C2
Square Grid With Twenty 7.5 m Rods

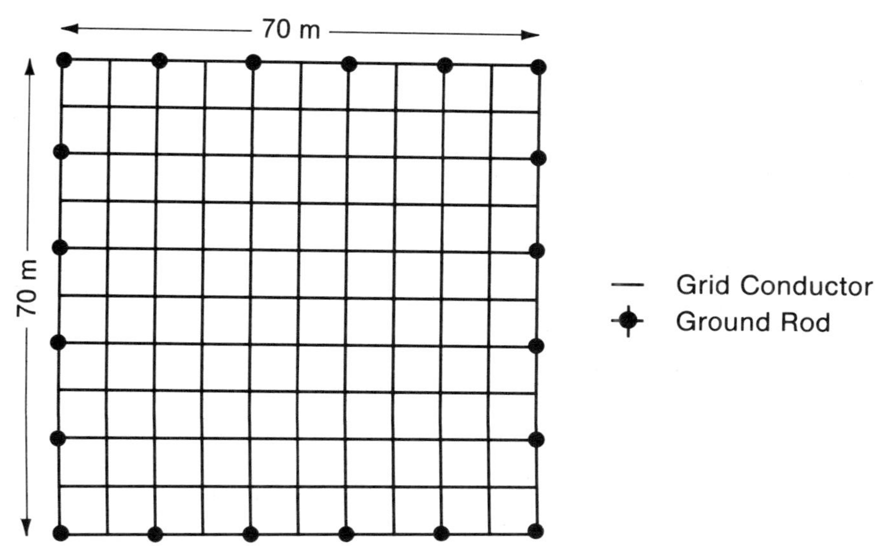

$$K_m = \frac{1}{2\pi} \left\{ \ln \left[\frac{7^2}{(16)(0.5)(0.01)} + \frac{[7 + (2)(0.5)]^2}{(8)(7)(0.01)} - \frac{0.5}{(4)(0.01)} \right] + \frac{1}{\sqrt{1 + 0.5}} \ln \frac{8}{\pi[(2)(11) - 1]} \right\} = 0.77,$$

for $K_{ii} = 1$ (peripheral ground rods)

and

$$K_i = 0.656 + 0.172\,(11) = 2.55$$

Applying Eq 71 to obtain E_m,

$$E_m = (400)\,\frac{(1908)(0.77)(2.55)}{1540 + 1.15\,(150)} = 874 \text{ V}$$

Step 9. Although the calculated corner mesh voltage is lower than that of Example 1, it still is higher than the tolerable touch voltage (874 V versus 746 V). Obviously, the use of longer ground rods and greater number of rods will eventually enable us to meet the criterion value. The needed increase of buried length ΔL can be estimated by using the ratio of 874 V – 746 V, as follows:

$$\Delta L = \left(\frac{874}{746} - 1 \right) (1540 + 72.5) = 293.8 \text{ m, say } 300 \text{ m}$$

Therefore, an increase of individual rod length from 7.5 – 15 m, and installation of 36 rods instead of 20, would produce the desired effect:

$$E_m = (400)\,\frac{(1908)(0.77)(2.55)}{1540 + 1.15\,(540)} = 693 \text{ V} < 746 \text{ V},$$

provided that the majority of 36 rods is placed along the grid perimeter.

C3. Rectangular Grid With Ground Rods — Example 3

In this example the preliminary grid design will be reconciled in terms of the actual shape of the grounding area. Realizing that the full grounding area is only about 8% larger than that used in the previous calculations, most of the conclusions from Example 2 can be used for arriving at a suitable final design solution.

Choosing, again, spacing $D = 7$ m, for a rectangular $63 \cdot 84$ m grid, the grid wire pattern is $10 \cdot 13$, and the grid conductor combined length is $13 \cdot 63 \text{ m} + 10 \cdot 84 \text{ m} = 1659$ m. Assuming the use of 38 ground rods, each 10 m long, the total buried length is

$$L = 1659 \text{ m} + (38)(10 \text{ m}) = 2039 \text{ m}.$$

Figure C3 illustrates the corresponding design arrangement.

Step 5. Using again Eq 40, but for $L = 2039$ m, and $A = 5292 \text{ m}^3$, gives:

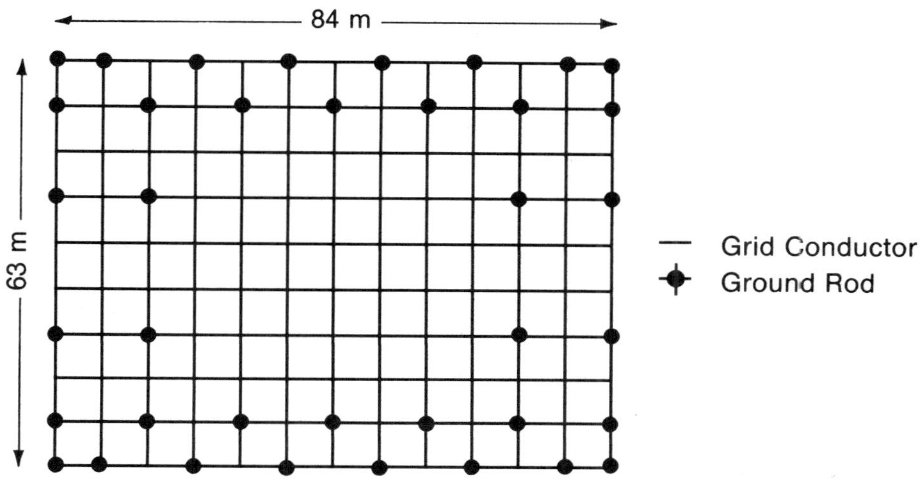

Fig C3
Rectangular Grid With Thirty-Eight 10 m Ground Rods

$$R_G = 400 \left[\frac{1}{2039} + \frac{1}{\sqrt{(20)\,(5292)}} \left(1 + \frac{1}{1 + 0.5\sqrt{20/5292}} \right) \right] = 2.62\ \Omega$$

Steps 6 and 7. Using $A\ I_G = 1908$ A as before, and $R_G = 2.62\ \Omega$,

GPR = (1908) (2.62) = 4998.96 $\approx$ 5000 V > 746 V

Step 8. For the particular design arrangement shown in Fig C3, Eq 71 can again be used to estimate the corner mesh voltage. However, since the grid is rectangular, the value of n to be used in the mesh voltage computation must be determined. First, to see if K_m and K_i need to be recalculated:

$$n = \sqrt{(13)\,(10)} = 11.4 \approx 11$$

Since this geometric mean value of n rounded to the nearest integer is identical to the value of n for the grid of Example 2, and since D, d, and h are also the same, the values of K_m and K_i from Example 2 can be directly applied to calculate E_m. Hence,

$$E_m = \frac{(400)\,(1908)\,(0.77)\,(2.55)}{1659 + (1.15)\,(380)} = 714.95, \text{ say } 715\ \text{V}$$

Step 9: Touch Voltage Criterion. This calculated mesh voltage is lower than 746 V of the $E_{\text{touch}_{70}}$ limit.

Step 10: Step Voltage Criterion. Because the step voltage has not been calculated yet, Eqs 69, 73, and 74 are used to compute K_i, E_s, and K_s, respectively. The value of n to be used in the step voltage computation is different from that used in Step 8, being the maximum of the number of parallel conductors in either direction. Thus, $n = 13$. Using Eqs 69 and 74,

$$K_i = 0.656 + (13)(0.172) = 2.89$$

and

$$K_s = \frac{1}{\pi}\left[\frac{1}{2(0.5)} + \frac{1}{7+0.5} + \frac{1}{7}\left(1 - (0.5)^{13-2}\right)\right] = 0.41$$

Applying the obtained K_i and K_s values in Eq 73, the step voltage is

$$E_s = \frac{(400)(1908)(0.41)(2.89)}{1659 + 380} = 433 \text{ V}$$

Step 11. The computed E_s is well below the tolerable step voltage, determined in Step 3 of Example 1, that is, 443 V $\ll$ 2320 V.

Step 12. A safe design has been obtained. At this point, all equipment pigtails, additional ground rods for surge arresters, etc, should be added to complete the grid design details.

C4. Equally Spaced Grid With Ground Rods in Two-Layer Soil — Exhibit 1

The SGA package of grounding computer programs, described in EPRI's Final Report EL-2682, Vol 2, was used to model an equally spaced grid in two-layer soil.

Fig C4
Equally Spaced Square Grid With Nine Rods in Two-Layer Soil

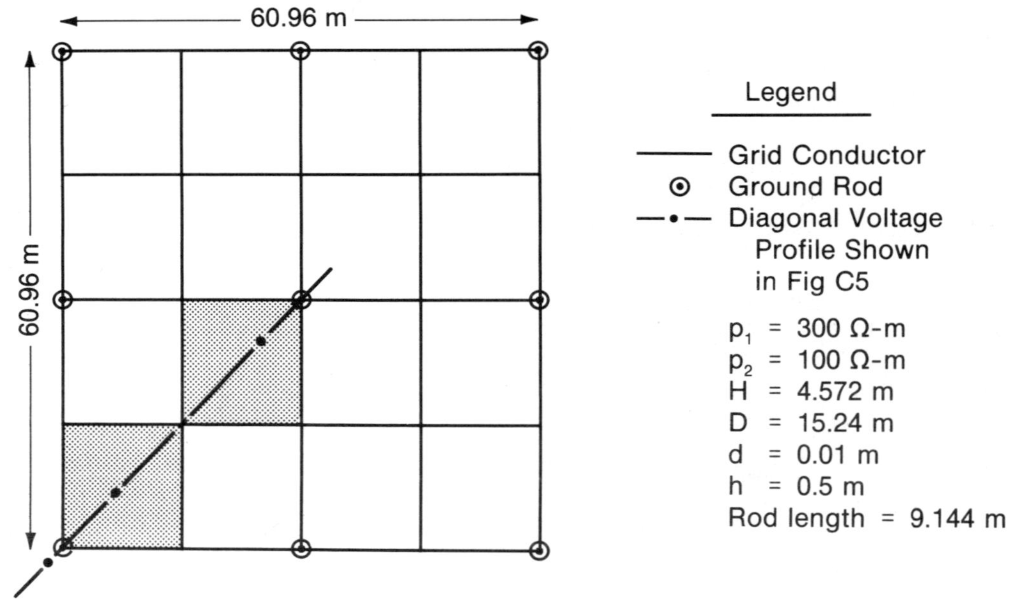

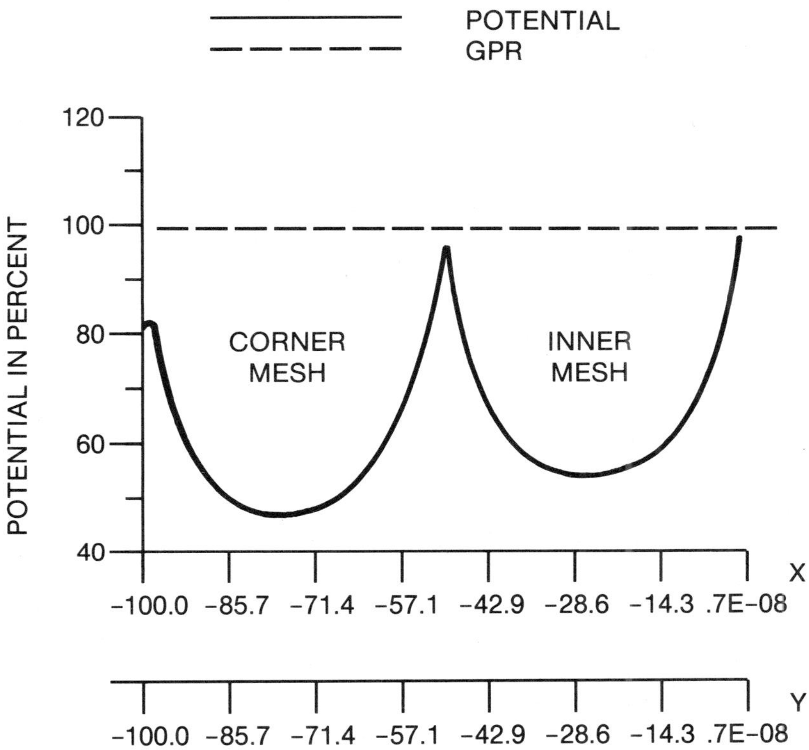

Fig C5
Diagonal Voltage Profile for a Grid of Fig C4 in Two-Layer Soil

As shown in Fig C4, the 61 · 61 m (200 · 200 ft) grid consisted of four meshes per side, and had nine ground rods, each 9.144 m (30 ft) long. The diameter of ground rods was 0.0127 m (0.5 in). The grid consisted of four meshes per side, formed by wires of a 0.01 m diameter, buried 0.5 m below the earth's surface. The depth of the upper layer of a 300 Ω-m soil 4.47 m (15 ft); the lower soil had resistivity of 100 ohm/m.

The computer-calculated values of resistance, corner mesh voltage, and maximum step voltage, were as follows:

$$R_g = 1.353\ \Omega,\ E_m = 49.66\%\ \text{of GPR, and}\ E_s = 18.33\%\ \text{of GPR}$$

As it can be determined from Fig C5, the mesh voltage coordinates were X = -75.00 ft, and Y = -75.00 ft, that is, the center of the corner mesh. The maximum step voltage (not shown) was calculated outside the grid, between the grid corner (X, Y = -100 ft) and the point at X, Y = 102.12 ft, that is, approximately over 1 m distance in a diagonal direction away from the grid corner.

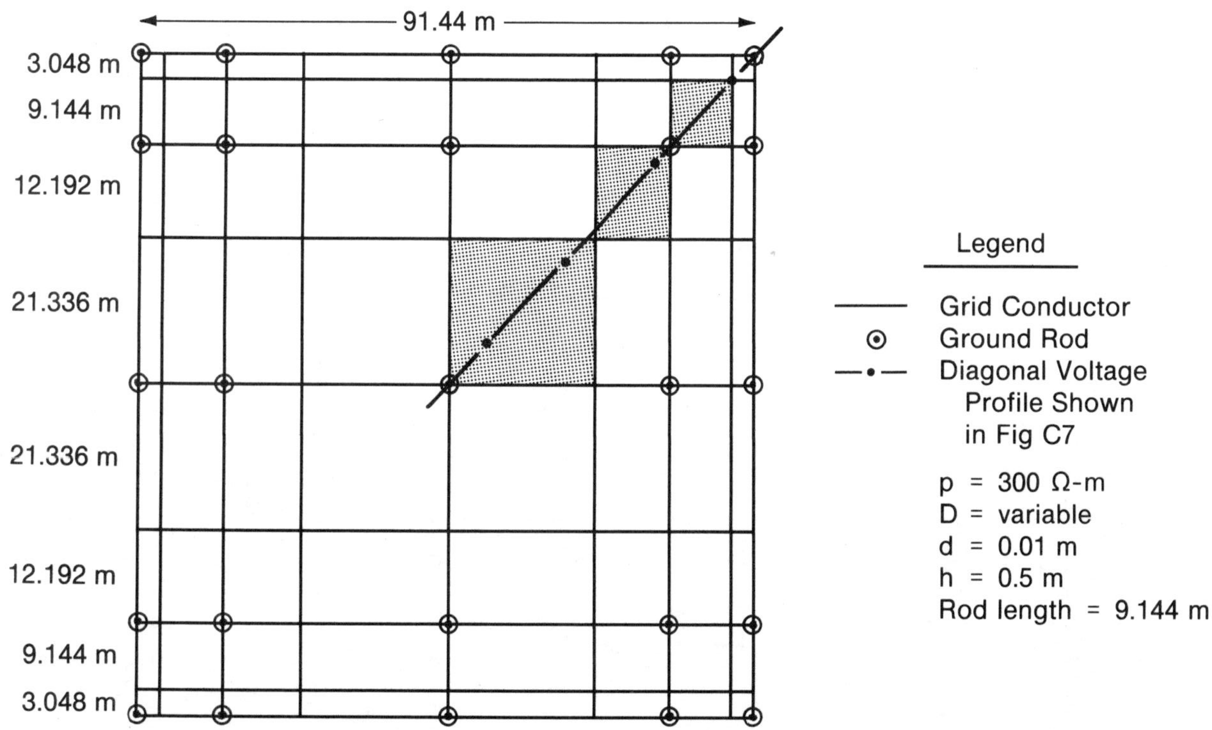

Fig C6
Unequally Spaced Square Grid With Twenty-Five 9.144 m Rods

C5. Unequally Spaced Grid With Ground Rods in Uniform Soil — Exhibit 2

The SGA package was also used to model a square grid with unequally spaced conductors, shown in Fig C6.

The computer output included the grid resistance, a surface voltage profile, the step voltage because of a more dense spacing between the peripheral conductors and the corner mesh voltage.

As shown in Fig C7, the corner mesh voltage is only 9.29% of the ground potential rise, while the maximum touch voltage, occurring above the largest interior mesh, is 17.08% of the ground potential rise.

The maximum touch voltage thus did not occur in the corner mesh. For other choices of conductor spacings, the maximum touch voltage may occur above some other meshes. Therefore, for unequal spacings, the touch voltages must be investigated over the entire grid, and the simplified criterion for checking the

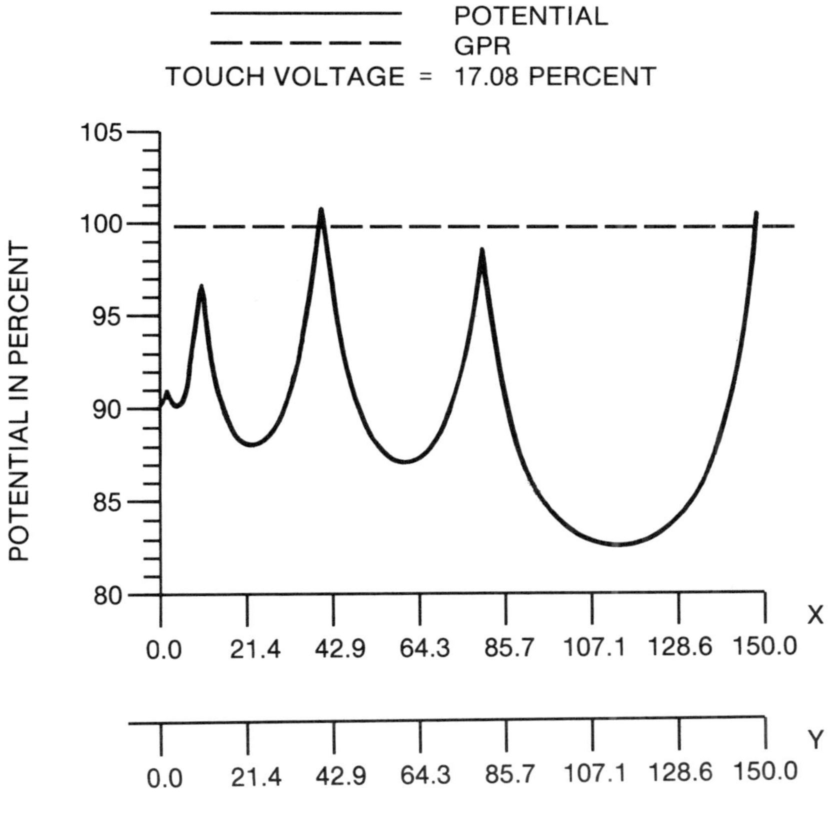

Fig C7
Diagonal Voltage Profile for an Unequally Spaced Grid of Fig C6

corner mesh voltage alone is not sufficient. On the other hand, the resistance R_g is not too dependent on the exact configuration of grid conductors and ground rods. (For instance, were R_g estimated by Eq 40 for a combined length of grid conductors and ground rods $L = 18 \cdot 19.44$ m + $25 \cdot 9.144$ m = 1874.5 m, the calculated value of 1.61 Ω would be less than 14% higher than the value of 1.416 Ω calculated by the computer.)

Appendix D
Special Considerations for Gas-Insulated Substations: Equivalent Circuits and Calculation Notes

(This Appendix is taken from work by Sverak and Dodds contained in "Safe Substation Grounding—Part II," *IEEE Transactions on Power Apparatus and Systems*, vol PAS-101, no 10, sect 8, pp 4006-4023; and additional input from J. M. Nahman.)

D1. Basic Concepts

Before examining the basic concepts of GIS bus equivalents, it is useful to clarify the difference between a circulating current and eddy currents and their role in producing the sheath voltages.

With enclosures both grounded and bonded between phases, the voltage occurring on the outer sheath is of two kinds: a common mode voltage induced longitudinally and a transverse mode voltage caused by eddy currents. The two voltage modes can be assumed superposed. Analyzing the transverse mode first, any nonmagnetic tubular shield surrounding a conductor will contain to some degree an alternating magnetic field produced by currents flowing in the inner conductor that, therefore, will somewhat reduce the effect of that field on other outside conductors. This shielding effect results solely from the presence of eddy currents circulating in a perpendicular plane with respect to the conductor, as if the enclosure consisted of circular segments with no longitudinal current flow between them. And, since the currents produce their own magnetic fields opposing and counterbalancing the field of an inner conductor, a transverse voltage will occur across the enclosure, whether or not it is grounded. At 60 Hz, however, no nonmagnetic enclosure of practical thickness provides enough shielding to be of any importance; the induced voltages are no more than a few volts and the relative shielding effectiveness would usually be less than 10%. Since the typical GIS enclosure of continuous design is effective by 80-90%, the predominant shielding effect results from the longitudinal circulation of currents in a closed path via the enclosures of individual phases and their bonds. With this insight, the following assumptions can reasonably be made:

(1) For continuous enclosures, the effect of eddy currents may be completely neglected as a contributing voltage factor

(2) Coupling from an inner conductor of one phase to any other outside conductor can be viewed as unaffected by the eddy current shielding effect of its own enclosure

The tolerable potential difference between two points of contact can be calculated in terms of the circuit constants and the allowable body current by Thevenin's theorem (Helmholz equivalent source theorem). The body current resulting from a touch should be equal to that which the preexisting voltage would cause to flow through the body resistance, assumed here as equal to 1000 Ω, in series with the equivalent network connected to the points of contact. Actually, the real network always includes at least two paths in parallel: the permanent, direct path through the enclosures and supporting structures to the ground, and a parallel path of the *accidental circuit*, with current passing either from hand-to-hand or hand-to-feet. However, a simplified approach, analogous to that used in Section 6 (Figs D4, D5, D8, D9), is also adequate in this application.

Now, concerning solely the longitudinal effects, consider two elementary circuit models (a) and (b) of Fig D1. Either sketch illustrates a system of two coupled circuits described by the following general equations:

$$V_s = Z_i I_1 + Z_m I_2 \qquad \text{(Eq D1)}$$

$$0 = Z_m I_1 + Z_e I_2 \qquad \text{(Eq D2)}$$

where

Z = self-impedance of the phase conductor with earth return
Z_e = self-impedance of the outer sheath
Z_m = mutual impedance between the phase conductor and sheath
S = spacing between the coupled circuits

Solving Eq D2 for I_2 and substituting the result into Eq D1 will yield expressions for the apparent circuit impedance Z as seen by the source and the relative magnitude, and the direction of I_2 with respect to I_1:

$$I_2 = -\frac{Z_m}{Z_e} I_1 \qquad \text{(Eq D3)}$$

and

$$V_S = I_1 Z \qquad \text{(Eq D4)}$$

Fig D1
Elementary Coupled Circuit Concepts

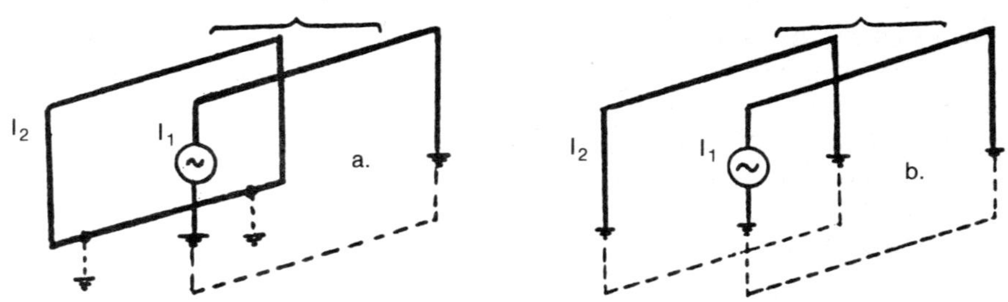

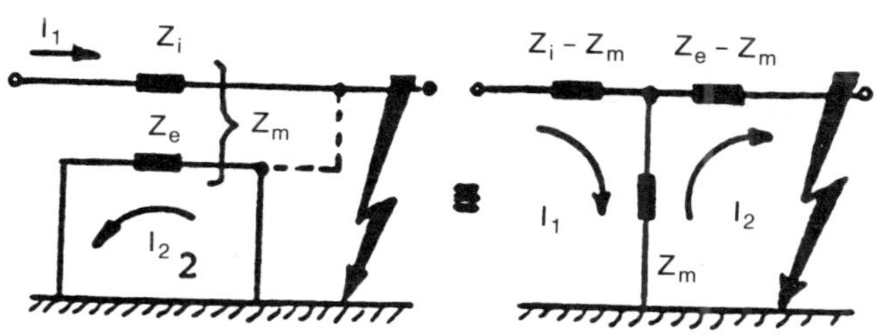

Fig D2
***T*-Equivalent of an Ideal Transformer**

$$Z = Z_i - \frac{(Z_m)^2}{Z_e} \qquad \text{(Eq D5)}$$

In turn, these results can both be interpreted and manipulated in terms of the well-known *T*-equivalent of an ideal transformer per Fig D2, with the following implications:

(1) If the current return path includes both the ground connection and bonded enclosure loops, the circuit impedance is Z, as determined by Eq D5

(2) If only a small or no current flows into the ground connection and most of it returns via the sheath, the impedance is closer to Z' defined below:

$$Z' = Z_i + Z_e - 2\,Z_m \qquad \text{(Eq D6)}$$

This condition is equivalent to eliminating the ground branch of the *T*-equivalent in Fig D2.

It is important to realize that all conventions per Eqs D1 and D2 hold. Referring to Fig D1(a), if, for instance, both conductors are assumed to have the same geometric mean radius GMR, and be placed horizontally in a distance $h = S_0/2$ above ground, then X_i, X_e, and X_m can be expressed with respect to a ground plane, as

$$X_e = X_i = C \ln (S_0/\text{GMR}) \qquad \text{(Eq D7)}$$

and

$$X_m = C \ln (S_0/S) \qquad \text{(Eq D8)}$$

Using Eq 24, it follows that

$$X = C\,[2 \ln (S_0/\text{GMR}) - 2 \ln (S_0/S)] = 2\,C \ln (S/\text{GMR}) \qquad \text{(Eq D9)}$$

which, of course, is the inductance of a closed loop of two long parallel conductors separated by distance S.

D2. Internal Faults

As illustrated in Figs D3, D4, and D5, on a single-bus basis, there are three possible locations of a fault within the bus with respect to the current source:

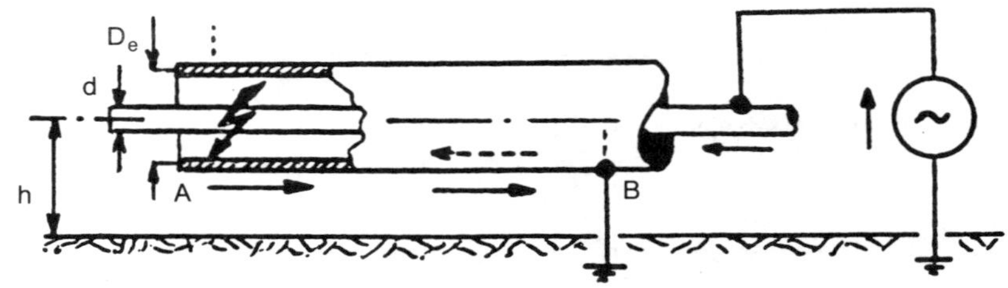

Fig D3
Internal Fault Remote from the Enclosure Grounding Point and from the Fault Current Source Feed Point

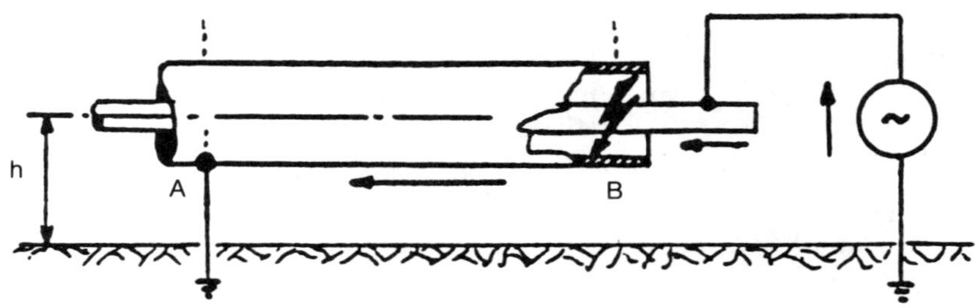

Fig D4
Internal Fault Remote from the Enclosure Grounding Point but Near the Fault Current Feed Point

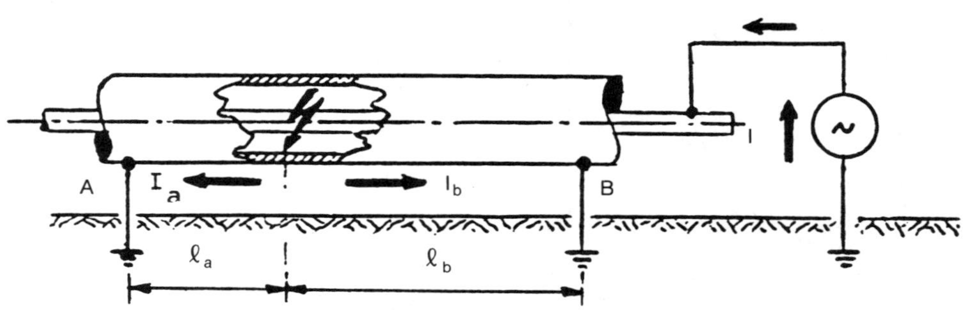

Fig D5
Internal Fault Between Two Grounding Points

(1) *Internal Fault*: Flashover occurs at point A, enclosure grounded at point B; Fig D3.

Based on Eqs D6 - D9, the respective resistive and inductive voltage drops are as follows:

$$X_e = C \ln (S_0/\text{GMR}) \qquad \text{(Eq D10)}$$

$$X_i = C \ln (S_0/r_0) \qquad \text{(Eq D11)}$$

$$2X_m = 2C \ln [S_0/(\text{GMR} - r_0)] = C \ln [(S_0^2/(\text{GMR} - r_0)^2] \qquad \text{(Eq D12)}$$

Consequently,

$$V_{eR} = l\, R_e I \qquad \text{(Eq D13)}$$

and

$$V_{eL} = \text{j}lI\,(4.61\,\omega \cdot 10^{-7}) \log_{10} \frac{(\text{GMR} - r_0)^2}{r_0\,\text{GMR}} \qquad \text{(Eq D14)}$$

where

r_0 = effective radius of inner bus, taken as $r_0 = 0.9\ d/2$, d being the actual diameter in m
GMR = geometric mean radius of enclosure, GMR = $D_e/2$, D_e = inner radius + outer radius in m
I = current in inner bus in A
R_e = enclosure resistance per unit length in Ω/m
l = enclosure length in m
h = height above ground in m
ω = $2\pi f$, f being frequency in Hz

If both the sheath resistance and inductance are considered, the total voltage drop along the enclosure is

$$V_e = (V_{eR}^2 + V_{eL}^2)^{1/2} \qquad \text{(Eq D15)}$$

However, since only a minimum magnetic field exists outside the enclosure and most of the flux is retained inside the sheath acting as a coaxial cable, it follows that

$$X_e + X_i - 2X_m \approx 0 \qquad \text{(Eq D16)}$$

and V_e reduces into

$$V_e = V_{eR} \qquad \text{(Eq D17)}$$

This voltage drop is in phase with the fault current.

(2) *Internal Fault*: Enclosure grounded at point A, flashover occurs at point B; Fig D4.

In contrast to condition (1), here the resistance can be neglected since the effect of enclosure inductance is predominant. The voltage drop is

$$V_e = V_{eg} = \text{j}l\,(4.61\ \omega \cdot 10^{-2}) \cdot \log_{10} (2h/\text{GMR})I \qquad \text{(Eq D18)}$$

(3) *Internal Fault*: Enclosure grounded at both ends. Flashover occurs between the endpoints A and B; Fig D5.

For a fault anywhere between A and B, the fault current I_i will divide according to impedances to ground that it sees. Hence, the longitudinal voltage drop will be a maximum whenever the following conditions are met for the respective left and right side currents I_A, I_B:

$$V = Z_A I_a \qquad \text{(Eq D19)}$$

$$V \approx Z_B I_b \qquad \text{(Eq D20)}$$

$$I_a + I_b = I \qquad \text{(Eq D21)}$$

With z_A, z_B determined on per unit length basis, Z_A, Z_B and the total enclosure length between ground points A and B, are:

$$Z_A = l_a z_A \qquad \text{(Eq D22)}$$

$$Z_B = l_b z_B \qquad \text{(Eq D23)}$$

$$l_{ab} = l_a + l_b \qquad \text{(Eq D24)}$$

Applying expression Eq D18 and formula Eq D16 to determine Z_A, Z_B, it can be shown that for unknown l_a, l_b, which satisfy the implicit constraint that $l_a + l_b = l_{ab}$, the solution is

$$l_a = \frac{1 - z_A/z_B}{1 - z_A/z_B} \; l_{ab} \qquad \text{(Eq D25)}$$

Fig D6
Example of GIS Bus Installation

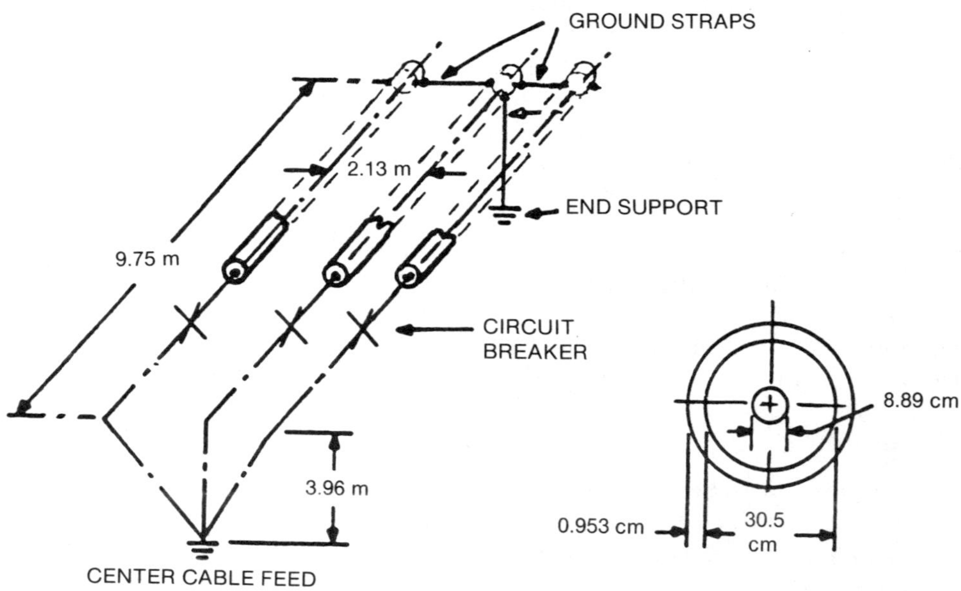

$$l_b = \frac{1 - \sqrt{z_B/z_A}}{1 - \sqrt{z_B/z_A}} \; l_{ab} \qquad \text{(Eq D26)}$$

In a case of $z_A = z_B$, the current would split evenly, with an internal fault occurring in the center:

$l_a = l_b = 0.5\ l_{ab}$

Example 1 — Internal Fault. Referring to Fig D6, consider the illustrated half of a GIS 242 kV bus, consisting of a cable feed, a breaker, and an overhead transmission line termination at the elevated end, which has been described in the literature [B84].

In addition to the basic dimensions shown in Fig D6, the following design data are given: f = 60 Hz, GMR (sheath) = 0.157 m, GMR_g (ground strip and sheath ties) = 0.0227 m, d(bus) = 0.089 m, R_e (enclosure resistance for unit length) 3.16 $\mu\Omega$/m, and current in the inner bus I_i = 50 kA. Spacing between phases is S = 2.13 m.

For infinite ground plane at zero depth (ground grid slab with reinforcing steel bars tied-in), $S_0 = 2h$ = 7.92 m. Using $r_0 = 0.9\ (d/2)$ = 0.04 m, we get on a per meter basis:

$$Z_A = \text{j}4.61\,(377)\,10^{-7} \log_{10}(7.92/0.157) = \text{j}0.296 \cdot 10^{-3}\ \Omega/\text{m}$$

$$\begin{aligned} Z_B &= 10^{-6}\,(3.16 + \text{j}\,173.3 \log_{10}[(0.157 - 0.04)^2/(0.157)\,(0.04)] \\ &= (3.16 + \text{j}\,58.8)\,10^{-6} \approx \text{j}0.059 \cdot 10^{3}\ \Omega/\text{m} \end{aligned}$$

Assuming a fault in the center phase to be the worst case for a given arrangement, the equivalent enclosure length (including the length of grounding strips converted to enclosure length), is[35]

$$\begin{aligned} l_{AB} &= 9.75\ \text{m} + 3.96\ \text{m} + 3.96\ \text{m} \log_{10}(7.92/0.227)/\log_{10}(7.92/0.157) \\ &= 13.71\ \text{m} + 3.96\ \text{m}\ (1 + 1.49) = 23.57\ \text{m} \end{aligned}$$

Determining

$$l_A = 23.57\ \text{m}\ [1 - (296/59)^{1/2}]/(1 - 296/57) = 7.3\ \text{m and}$$

$$l_B = l_{AB} - l_A = 23.57\ \text{m} - 7.3\ \text{m} = 16.3\ \text{m},$$

for the assumptions of

$z_A = x_A; z_B = x_B$:

$$X_A = l_A x_A = 7.3\,(0.296) \cdot 10^{-3} = 2.15 \cdot 10^{-3}\ \Omega$$

$$X_B = l_B x_B = 16.3\,(0.059) = 0.96 \cdot 10^{-3}\ \Omega$$

and

$$I_A = 2.15/(2.15 + 0.96)\ 50\ \text{kA} = 34.5\ \text{kA}$$

$$I_B = 0.96/(2.15 + 0.96)\ 50\ \text{kA} = 15.5\ \text{kA}$$

[35] Conversion formula, used for this purpose, is

$l_{eq} \log_{10}(2h/\text{GMR}) = l_{tie} \log_{10}(2h/\text{GMR}_{tie})$

Hence, the maximum voltage drop is

$$0.00215\ \Omega\ (15.5\ \text{kA}) \approx 0.00096\ \Omega\ (34.5\ \text{kA}) \approx 33.4\ \text{V}$$

D3. External Faults

For the typical arrangement of a continuous design (enclosures of individual phases bonded together at both ends), it is not unreasonable to assume that all sheath currents can return via the adjacent enclosures of other phases and no induced currents will flow into and close via the ground.

With three identical buses in flat configuration, the enclosures can be viewed as forming two loops, each superposed on the other, and each separately coupled with the energized conductor. When the outer phase conductor is energized as shown in Fig D7, the two corresponding loop equations are:

$$(Z_{M1} - Z_{M2})\, I_0 + Z_{\text{loop}}\, I_1 = 0 \qquad \text{(Eq D27)}$$

$$(Z_{M2} - Z_{M3})\, I_0 + Z_{\text{loop}}\, I_2 = 0 \qquad \text{(Eq D28)}$$

NOTE:In Eqs D27 and D28 the mutual coupling between the two loops is not taken into account. Theoretically, designating the mutual impedance of loop1 and loop2 as Z_{ML}, the complete equations for the case in Fig D7 would be

$$(Z_{M1} - Z_{M2})\, I_0 + Z_{\text{loop1}}\, I_1 - Z_{ML}\, I_2 = 0 \qquad \text{(Eq D27a)}$$

$$(Z_{M2} - Z_{M3})\, I_0 - Z_{ML}\, I_1 + Z_{\text{loop2}}\, I_2 = 0 \qquad \text{(Eq D28a)}$$

However, the experiments with a scaled-down physical model, described in [B42], indicate that the effect of Z_{ML} is likely to be very minor and can be neglected.

If I_0 (that is, the current flowing in the inner bus toward the point of external fault) is given, loop currents I_1 and I_2 are defined by their ratio to I_0:

Fig D7
Equivalent Model of Ungrounded Enclosure Circuits

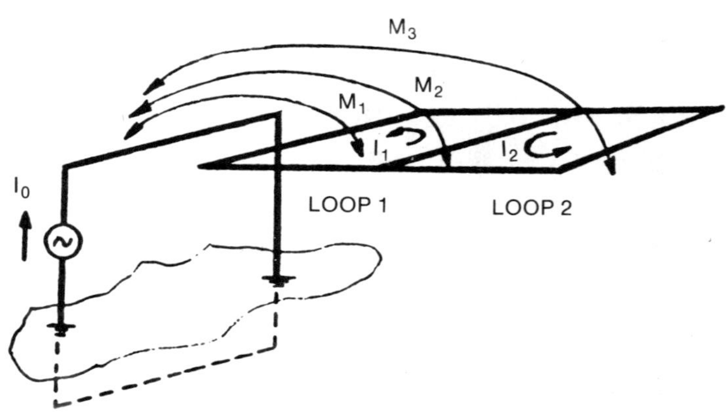

$$(-I_1/I_0) = Z_{M1} - Z_{M2})/Z_{\text{loop}} = (l'_m/l'_0)\ \frac{\log_{10} \lfloor S/(\text{GMR} - r_0)\rfloor}{2 \log_{10} (S/\text{GMR})} \qquad \text{(Eq D29)}$$

$$(-I_2/I_0) = Z_{M2} - Z_{M3})/Z_{\text{loop}} = (l'_m/l'_0)\ \frac{\log_{10} (2S/S)}{2 \log_{10} (S/\text{GMR})} \qquad \text{(Eq D30)}$$

where, as before, S denotes spacing between two adjacent buses, and

$$l'_0 = l_{\text{bus}} + ½\, l_{\text{ties}} \qquad \text{(Eq D31)}$$

Alternately, if the center phase conductor is assumed as feeding the fault, an approximate value of the enclosure current can be determined from Eq D29, with l'_0 replaced by l''_0 or, as is usually the case, assuming the enclosure current to be 10-15% higher than that calculated for the outer phase.

$$l''_0 = l_{\text{bus}} \qquad \text{(Eq D32)}$$

$$l''_0 \approx l_m \qquad \text{(Eq D33)}$$

Once the induced currents are determined independently for the single-line-to-ground fault of each phase, superposition might be used to determine the net value of current flowing in the enclosures during a phase-to-phase or three-phase fault. The fault currents are 180 degrees out of phase for the phase-to-phase and 120 degrees for the three-phase fault.

Example 2 — External Fault. Given the same bus arrangement and design data as those used in Example 1, estimate the probable enclosure currents for a close external fault, with 50 kA supplied either from (a) outer phase, and (b) center phase.

The related coupled enclosure and loop lengths are:

l'_m = 9.75 m 3.96 m = 13.71 m

l'_0 = 13.71 m + 0.5 (2.13 m) 149 = 15.296 m [for condition (a)]

l''_0 = 13.71 m [for condition (b)]

For a loop formed by any two adjacent bus enclosures and their ties, the loop self-impedance is

$Z'_{\text{loop}} = 1.74 \cdot 10^{-4}\ 2 \log_{10} (2.13/0.157)\ 15.296 = 6.028\ \text{m} \cdot 10^{-3}\ \Omega$
[for condition (a)]

$Z''_{\text{loop}} = 0.394 \cdot 10^{-3}\ 13.71 = 5.402\ \text{m} \cdot 10^{-3}\ \Omega$ [for condition (b)]

Using Eqs D27 and D28, the equivalent source voltages for condition (a) are:

$(Z_{M1} - Z_{M2})\, I_0 = 13.71\ \{1.74 \cdot 10^{-4} \log_{10} [2.13/(0.157 - 0.04)]\}\ (50)\ 10^3 = 150.31$ V

$(Z_{M2} - Z_{M3})\, I_0 = 13.71\ [1.74 \cdot 10^{-4} \log_{10} (2)]\ (50)\ 10^3 = 25.906$ V

Applying Eqs D29 and D30, the loop currents are:

$I_1 = 150.31/(6.028 \cdot 10^{-3}) = 24.935$ kA

$I_2 = 35.906/(6.028 \cdot 10^{-3}) = 5.956$ kA

$I_1 + I_2 = 30.891$ kA

In a rough estimate, for the external fault occurring at a distant point of the center phase, condition (b) is

$I_{1\ \text{center phase}} = 150.31/(5.402 \cdot 10^{-3}) = 27.82$ kA

Finally, for condition (a), the current in the enclosure of a faulted phase can be put equal to $I_1 = 24.9$ kA, $I_1 - I_2 = 18.9$ kA anticipated in the center enclosure, and $I_2 = 5.9$ kA in the other outer enclosure. The maximum sheath voltage along the outer phase enclosure of the faulted bus can be estimated as approximately 150.3/2 V, or ≈ 75 V.

For the case of a fault current in the center phase, condition (b), the sheath voltage can be estimated to reach, at most,

$$\frac{27.8}{24.9}\ 75\ \text{V} \approx 84\ \text{V}$$

D4. Two General Limiting Cases for a Grounded and Ungrounded Bus

To generalize the conditions that develop during a line-to-ground fault external to the GIS bus consisting of a number of enclosure runs bonded at both ends, consider the following cases:

First, as has already been illustrated in Fig D1, let us assume again that both the energized bus circuit and any coupled enclosure circuit have a common path of ground returns, and let

Z_{ii} = self-impedance of ith conductor with a ground return
Z_{ij} = mutual impedance between ith and jth conductors with common ground return

Using index "0" to denote the energized bus, the voltage drop along each conductor and the ratio between circuit currents I_0 and I_N in such a system are defined by the following equations:

Fig D8
Simplifying Concept of an Ungrounded Enclosure Loop

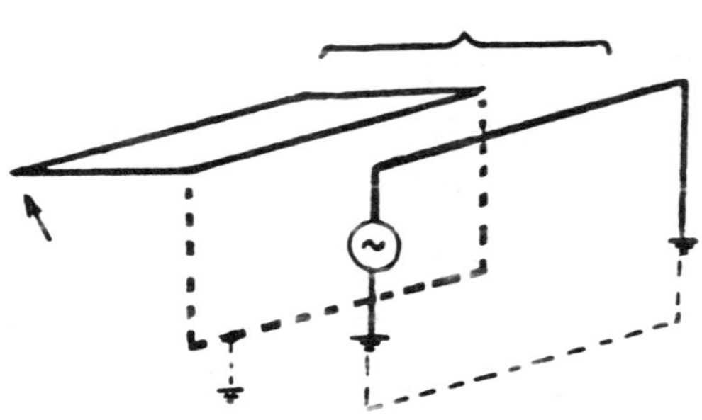

$$E_s = Z_{00} I_0 + Z_{0N} I_N \quad \text{(Eq D34)}$$

$$0 = Z_{N0} I_0 + Z_{NN} I_N \quad \text{(Eq D35)}$$

Noting $Z_{N0} = Z_{0N} = Z_M$,

$$\frac{I_N}{I_0} = \frac{Z_M}{Z_{NN}} \quad \text{(Eq D36)}$$

Alternately, if the enclosure circuit is assumed ungrounded and forms a closed loop, as depicted in Fig D8, the above equations change to:

$$E_s = Z_{00} I_0' + (Z_{01} - Z_{0\dot{a}}) I_N' \quad \text{(Eq D37)}$$

$$0 = (Z_{10} - Z_{20}) I_0' + (Z_{12} - Z_{21}) I_N' \quad \text{(Eq D38)}$$

Similarly, as before, let

$$\frac{I_N'}{I_0'} = \frac{Z_M'}{Z_{NN}'} \quad \text{(Eq D39)}$$

where Z'_{NN} is the self-impedance of the sheath loop consisting of two conductors,

$Z'_{NN} = Z_{12} - Z_{21}$; $I_N = I_1 = I_2 = I_{12}$, and $Z'_N = Z_{10} - Z_{20}$

Expanding the above concepts for a system with multiple buses, assuming one phase conductor energized and other phase conductors floating or carrying no current, but with all enclosures bonded together and grounded, the above procedures can be unified, leading to the following general form:

$$\begin{bmatrix} E_s \\ \hline 0 \\ \vdots \\ 0 \end{bmatrix} = \begin{bmatrix} Z_{00} & Z_{01} & Z_{02} & \cdots & Z_{0N} \\ \hline Z_{10} & Z_{11} & Z_{12} & \cdots & Z_{IN} \\ \vdots & \vdots & \vdots & & \vdots \\ Z_{N0} & Z_{N1} & Z_{N2} & \cdots & Z_{NN} \end{bmatrix} \begin{bmatrix} I_0^* \\ \hline I_1^* \\ \vdots \\ I_N^* \end{bmatrix} \quad \text{(Eq D40)}$$

In such a notation, the *primitive* currents I^* can be related to the actual currents I by some convenient *transformation* matrix C, containing appropriate arrays of 0, 1, -1 values:

$$[I^*] = [C]\,[I], \text{ and } [I] = [C]^T\,[I^*] \quad \text{(Eq D41)}$$

If the current of energized bus I_0 is given (maximum fault assumed), the above matrix system ZI can be partitioned along the shown lines, and a reduced system solved for the enclosure currents $I_1, I_2, \cdots, I_n$.

For instance, if a three-phase system is assumed and a faulted bus is as shown in Fig D9, only three equations need to be calculated to determine the sheath currents.

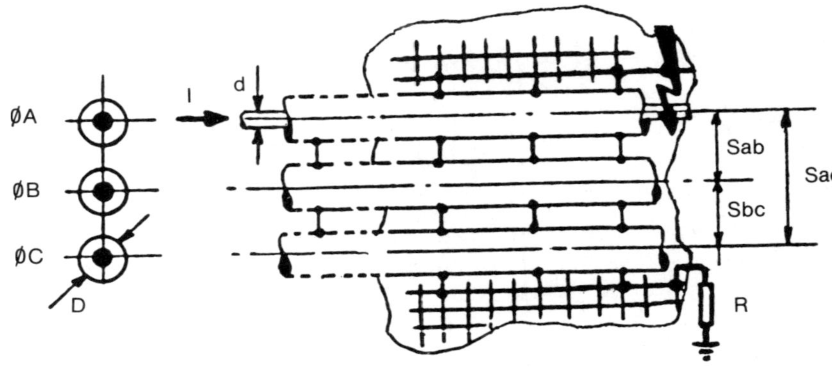

Fig D9
Flat Arrangement of GIS Enclosures With Multiple Grounding and Bonding

However, even a 3 · 3 matrix containing complex elements becomes a 6 · 6 array in actual calculations. Thus, to simplify further, assume for the moment only the inductive terms as significant, which allows us to introduce the following typical identities, characterizing a flat spacing of three identical enclosures:

$$X_{10} I_0 + X_{ss} I_1 + X_{1s} I_2 + X_{2s} I_3 = 0 \quad \text{(Eq D42)}$$

$$X_{1s} I_0 + X_{1s} I_1 + X_{ss} I_2 + X_{1s} I_3 = 0 \quad \text{(Eq D43)}$$

$$X_{2s} I_0 + X_{2s} I_1 + X_{1s} I_2 + X_{ss} I_3 = 0 \quad \text{(Eq D44)}$$

where for $i, j = 1,2,3$, $Z_{ss} = Z_{ii}$ and $Z_{ij} = Z_{ji}$

$Z_{1s} = Z_{ij}$ for $i - j = 1$

$Z_{2s} = Z_{ij}$ for $i - j = 2$

Subsequent addition of Eqs D42-D44 and division by three, results in an equation of an average single circuit, rendering the introduction of geometric mean concepts.

$$I_0 (X_{10} + X_{1s} + X_{2s})/3 + (I_1 + I_3)(X_{ss} + X_{1s} + X_{2s})/3 + I_2 (X_{ss} + 2X_{1s})/3 = 0$$

where (Eq D45)

$$(X_{10} + X_{1s} + X_{2s})/3 = X_{0\,\text{avg}}\,(\text{GMD}_0) \quad \text{(Eq D46)}$$

$$(X_{ss} + X_{1s} + X_{2s})/3 = X_{2\,\text{avg}}\,(\text{GMD}_2) \quad \text{(Eq D47)}$$

$$(X_{ss} + 2\,X_{1s})/3 = X_{1\,\text{avg}}\,(\text{GMD}_1) \quad \text{(Eq D48)}$$

Recognizing $I_1 + I_2 + I_3 = I_N$, the total sheath current I_N can be determined from Eq D49, where the right side expression now includes an added resistive term $\boldsymbol{R}$, representing the resistance of a return path to the current source. With the average reactance for $(X_{ss} + X_{1s} + X_{2s})/3$ and $(X_{ss} + 2\,X_{1s})/3$ represented by means of a term GMMD, it follows that the I_N/I_0 ratio can be expressed as

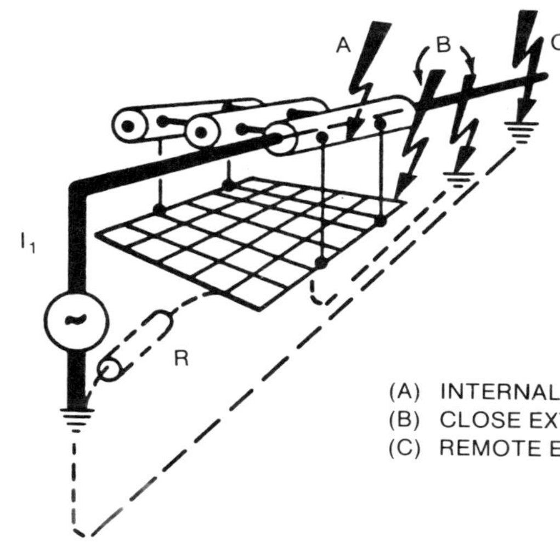

Fig D10
Ground Faults Within and Outside GIS

$$\frac{-I_N}{I_0} = \frac{j\, l_m\, u \log_{10}(S_0/\mathrm{GMD}_0)}{j\, l_0\, u \log_{10}(S_0/\mathrm{GMMD}) + R} \tag{Eq D49}$$

where

$$\mathrm{GMD}_0 = [(\mathrm{GMR} - r_0)\,(S)\,(2S)]^{1/3} \tag{Eq D50}$$

$$\mathrm{GMMD} = [(\mathrm{GMD}_0)\,(\mathrm{GMD}_1)\,(\mathrm{GMD}_2)]^{1/3} = [\mathrm{GMR}^3\,(S)^4\,(2S)^2]^{1/9} \tag{Eq D51}$$

$$= \sqrt[3]{1.587\ \mathrm{GMR}\ S^2}$$

S_0 = distance to such a plane or point of the current return path over or to which all flux can be assumed integrated. For the assumption of an infinite metallic plate at the earth's surface, S_0 is equal to twice the height of the GIS above the ground; $S_0 = 2h$

GMR-r_0 = distance from the inner wall of its enclosure

$u\,(f)$ = $2.3965\, f \cdot 10^{-3}$, where f is frequency in Hz

GMR = geometric mean radius of the outer sheath (enclosure)

D5. Recommendations

Generally, if the source neutral is remote and not connected to the grid, or as in the usual case, a combination of enclosure bonding and grounding results in much lower impedances of the current path via enclosure horizontal ties in comparison to that of the grounding strips, the direct use of Eqs D29 and D30 is adequate. Alternatively, if the source neutral is tied to the grid and the impedance of vertical grounding strips is comparable to (or less than) the combined impedance of bonding links, then Eq D49 will yield higher enclosure current. For instance, if $R = 2\ \Omega$ is assumed for Example 2, by using Eq D49 we get:

$$\text{abst}\ (I_n) = 50\ \text{kA} \cdot (j2.1207\ \Omega)/(2 + j2.342\ \Omega)^{1/2} = 34.4\ \text{kA}$$

which is higher than the highest loop current of 27.8 kA calculated previously for the bus fault of the center phase. Hence, if the simplified method of loop currents is used in estimating the current flow in enclosures, and the fault current return path includes both a metal connection and the earth, (that is, nonzero R in Fig 14, for instance $R = R_g$, where R_g is the ground resistance of a grid), the sum of currents induced in the sheath should be checked by Eq D49 and, if necessary, the results of simplified loop calculations proportionally adjusted.

Appendix E
Parametric Analysis of Grounding Systems

(This Appendix is taken from F. Dawalibi and D. Mukhedkar, "Parametric Analysis of Grounding Grids," *IEEE Transactions on Power Apparatus and Systems*, vol PAS-98, no 5, pp 1659-1668, Sept/Oct 1979, and "Influence of Ground Rods on Grounding Grids," *IEEE Transactions on Power Apparatus and Systems*, vol PAS-98, no 6, pp 2089-2098, Nov/Dec 1979.)

In order to efficiently design a safe grounding system it is necessary to have knowledge of how various parameters affect the performance of the grounding system. Some of these parameters include: grid conductor spacing and arrangement, number of ground rods, location and length, and soil resistivity parameters (that is, homogeneous or multilayered with various surface layer thickness and values of K, the reflection factor coefficient).

This appendix gives a brief discussion of how the above parameters affect the behavior of grounding systems for uniform soil resistivity and for two-layer soil resistivity. There are many other parameters that may affect the performance of the grounding system, but it is not within the scope of this appendix to discuss these parameters, as they are described elsewhere in the literature [B12], [B34], [B36], [B97]. These references include recent EPRI projects performed by Ohio State University and the Georgia Institute of Technology, where several parameters were varied over wide ranges using both scale and computer models.

E1. Uniform Soil

E1.1 Current Density — Grid Only. For a grounding system consisting only of grid conductors, the current along any one of the conductors is discharged into the earth in a fairly uniform manner. However, a larger portion of the current is discharged into the soil from the outer grid conductors rather than from the conductors at or near the center of the grid (refer to Figs E1 and E2). An effective way of making the current density more uniform between the inside and periphery conductors is to employ a nonuniform conductor spacing, with the conductor spacing larger at the center of the grid and smaller toward the perimeter. However, analysis of grids with this type of spacing cannot be accomplished using the simplified methods of this guide, but must be done using techniques similar to those described in the references.

E1.2 Resistance — Grid Only. For a given area to be grounded, the effect on resistance of increasing the number of meshes in a grid-only system becomes minimal. That is, as the number of meshes increases from one, the resistance of the

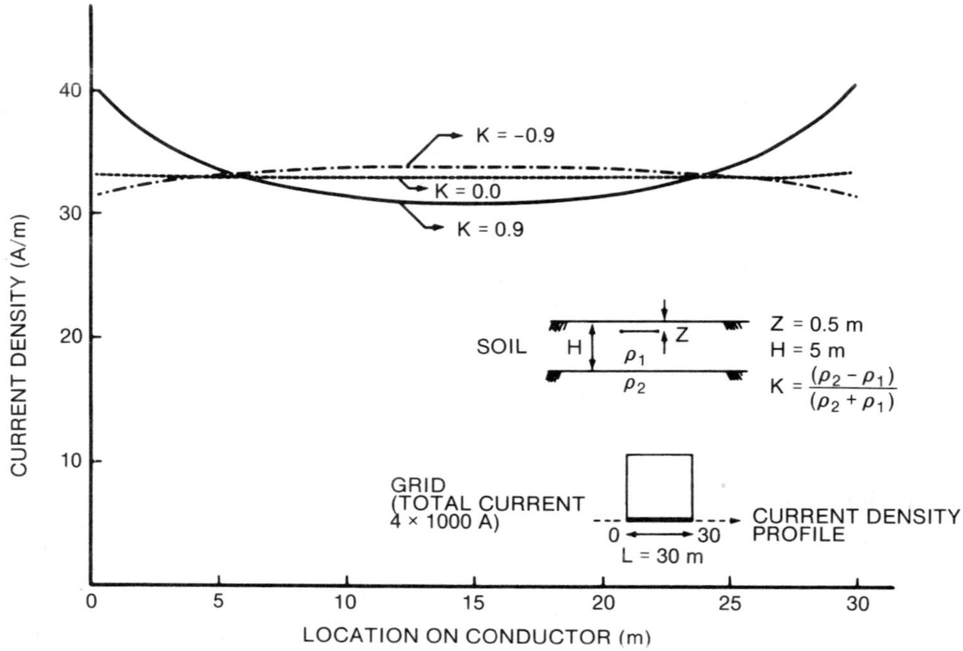

Fig E1
One Mesh Grid Current Density

grid decreases. However, this decrease quickly becomes negligible for large numbers of meshes (or small parallel conductor spacing). See Figs E3 and E4.

As shown in Fig E5, the resistance also shows a gradual decrease with burial depth, until it approaches one-half its resistance value at the surface as the depth increases to infinity. But for typical variations of burial depth found within the industry (that is, approximately 0.5 - 1.5 m), this change in resistance with depth is negligible for uniform soil.

E1.3 Step and Touch Voltages — Grid Only. Since most of the current in a uniformly spaced grid is discharged into the earth from the outer conductors, the worst touch and step voltages occur in the outer meshes, especially in the corner meshes. Increasing the number of meshes (decreasing the conductor spacing) tends to reduce the touch and step voltages until a saturation limit is reached. Beyond this number of meshes, reducing the conductor spacing has minimal effect on reducing the voltages (refer to Figs E6 - E9). This saturation limit is the vertical component of voltage caused by the depth of burial of the grid, and is changed only with a change in depth of the grid.

The grid burial depth also influences the step and touch voltages significantly as shown in Figs E10 and E11. For moderate increases in depth the touch voltage decreases, due mainly to the reduced grid resistance and corresponding reduction in the grid potential rise. However, for very large increases in depth, the touch

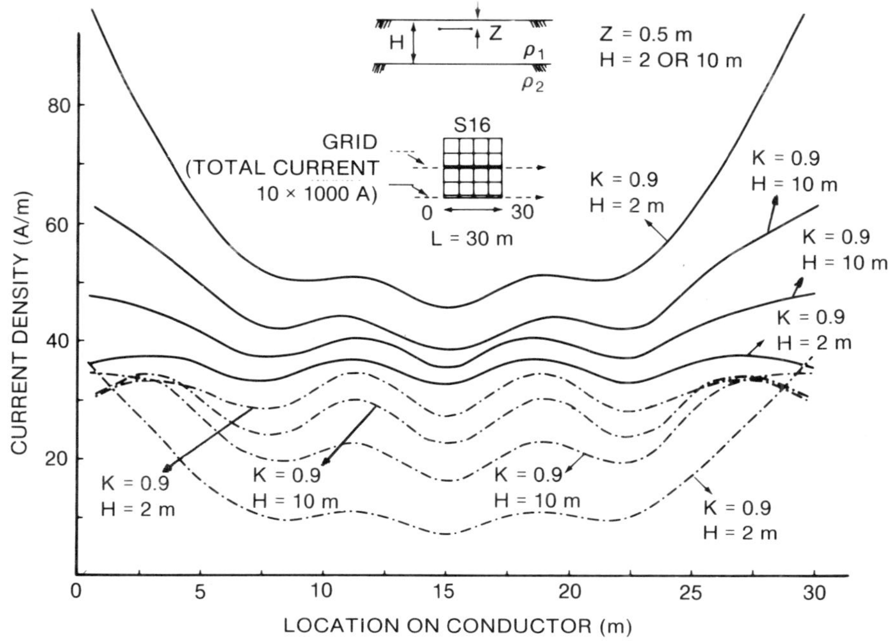

Fig E2
Sixteen Mesh Grid Current Density

voltage may actually increase. The reduction in grid potential rise reduces to a limit of approximately half its value at the surface as the depth of the grid approaches infinity, while the earth surface potential approaches zero at infinite depths. Therefore, depending on the initial depth, an increase in grid burial depth may either increase or decrease the touch voltage, while the step voltage is always reduced for increased depths.

E1.4 Ground Rods Only. For systems consisting only of ground rods the current has been found to discharge into the earth at a fairly uniform rate along the length of the rod with a gradual increase with depth and with slightly higher increases in current density near the ends (refer to Fig E12). As in the case of the grid conductors, the current density is greater in the rods near the periphery of the grounding system than for those in the center (refer to Figs E13 and E14). Thus, the step and touch voltages are higher near the outer ground rods.

Increasing the length of the rods is effective in reducing the resistance of the system and, therefore, reducing the step and touch voltages. Increasing the number of rods also reduces the resistance until the grounded area is saturated, and is even more effective in reducing the step and touch voltages as shown in Table E1. This is true because in addition to the lower resistance and lower ground potential rise, the spacing between the rods is reduced, which tends to make the earth surface potential more uniform. The comments above on the effects of grid burial depth also apply to the effects of the top-of-the-rod depths.

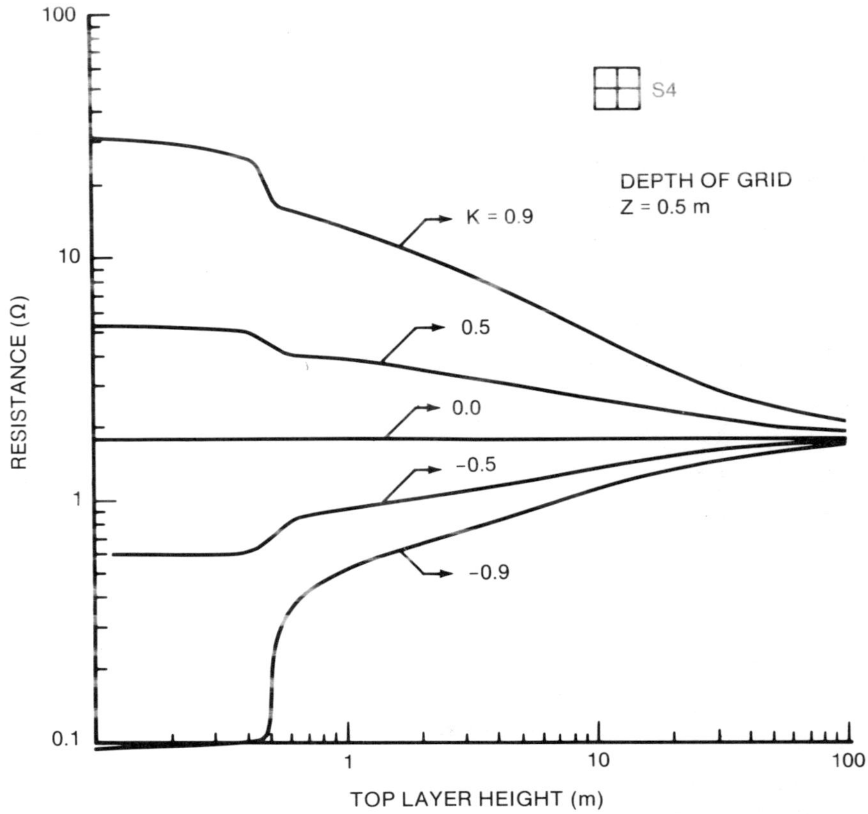

Fig E3
Four Mesh Grid Resistance

E1.5 Grid and Ground Rod Combinations. When a combination of grid conductors and ground rods are used in a grounding system, the number and length of ground rods may have a great influence on the performance of the grounding system. For a given length of grid conductor or ground rod, the ground rod discharges much more current into the earth than does the grid conductor, as shown in Figs E15-E18. This current in the ground rod is also discharged mainly in the lower portion of the rod. Therefore, the touch and step voltages are reduced significantly compared to that of grid alone.

E1.6 Conclusions. In general, a uniformly spaced grounding system consisting of a grid and ground rods is superior to a uniformly spaced grounding system consisting only of a grid with the same total conductor length. The variable spacing technique discussed earlier might be used to design a grounding system consisting of a grid only, with lower step and touch voltages than a uniformly spaced grid and ground rod design of equal length. However, this variable spacing technique might also be used to design a better grounding system using nonuniformly spaced grid conductors and ground rods. It shall be emphasized that this type of design shall be analyzed using the detailed analysis techniques in the references.

E2. Two-Layer Soil

The performance of a grounding system in multilayered earth can differ greatly from the same system in uniform soil. In addition to other parameters, the performance is affected by the resistivity and thicknesses of the soil layers and the burial depth of the grounding system. The following discussion will consider only two-layer earth models, due to the complexity and numerous combinations possible for additional layers. For an explanation of two-layer earth analysis of grounding systems, refer to 11.5 of this guide.

For brevity of the discussion, the following variables are defined:

ρ_1 = resistivity of upper layer of soil

ρ_2 = resistivity of lower layer of soil

K = reflection factor coefficient $\dfrac{\rho_2 - \rho_1}{\rho_2 + \rho_1}$

H = height of upper layer of soil

E2.1 Current Density — Grid Only. For grounding systems consisting only of grid conductors, the current density is highly dependent on both K and H, as shown in Figs E1 and E2. For negative values of K ($\rho_1 > \rho_2$), the current density is fairly uniform over the entire grid with slightly higher densities in the conductor between intersection points on the grid, and is slightly higher for outer conductors than for conductors near the center of the grid. As the height of the top layer increases, this higher current density in the outer conductors becomes more dominant. This can be explained as follows. For small values of H, most of the current discharged from the grid goes downward into the low resistivity soil, while for large values of H most of the current remains in the high resistivity layer of soil, assuming that the grid is in this upper layer. As H increases, the model approaches that of uniform soil with a resistivity equal to that of the upper layer. Therefore, as in the case of the uniform soil model discussed previously, the outer grid conductors discharge a larger portion of the current into the earth than do the center conductors.

For positive values of K ($\rho_1 < \rho_2$), the current has a much higher tendency to remain in the low resistivity soil, even for moderately small values of H. As H increases the current density rapidly approaches that of a uniform soil, with higher current densities in the periphery conductors.

E2.2 Resistance — Grid Only. The resistance of a grid-only system may vary greatly as a function of K and H and, thus, may be higher or lower than the same grid in a uniform soil, as shown in Figs E3 and E4. In general, the resistance of a grid is lowest if it is in the most conductive layer of soil. As H increases the resistance of the grid approaches that of a grid in uniform soil of the same resistivity as the upper layer. Assuming that the grid is located in the upper soil layer with resistivity equal to ρ_1, the following can be generalized:

(1) For negative values of K ($\rho_1 > \rho_2$), the resistance of the grid will be higher than that of an identical grid in uniform soil with resistivity ρ_2

(2) For positive values of K ($\rho_1 < \rho_2$), the resistance grid will be lower than that of an identical grid in uniform soil resistivity ρ_2

E2.3 Step and Touch Voltages — Grid Only. The step, touch voltages, and mesh voltages may also vary significantly with K, H, and grid depth. They may be very much higher or lower than a corresponding uniform soil model. See Figs E6–E9.

For grids buried near the surface of the earth, increasing the number of meshes is an effective means of reducing the mesh voltages. However, as the grid depth increases, the effectiveness of this method of reducing the mesh voltages decreases until at some characteristic grid depth, the mesh voltages begin to increase. The reasons for this phenomenon are identical to those described previously for uniform soil. For a very large number of meshes (that is, small spacing between parallel conductors), the touch voltages are relatively unaffected by H and K.

Fig E4
Sixteen Mesh Grid Resistance

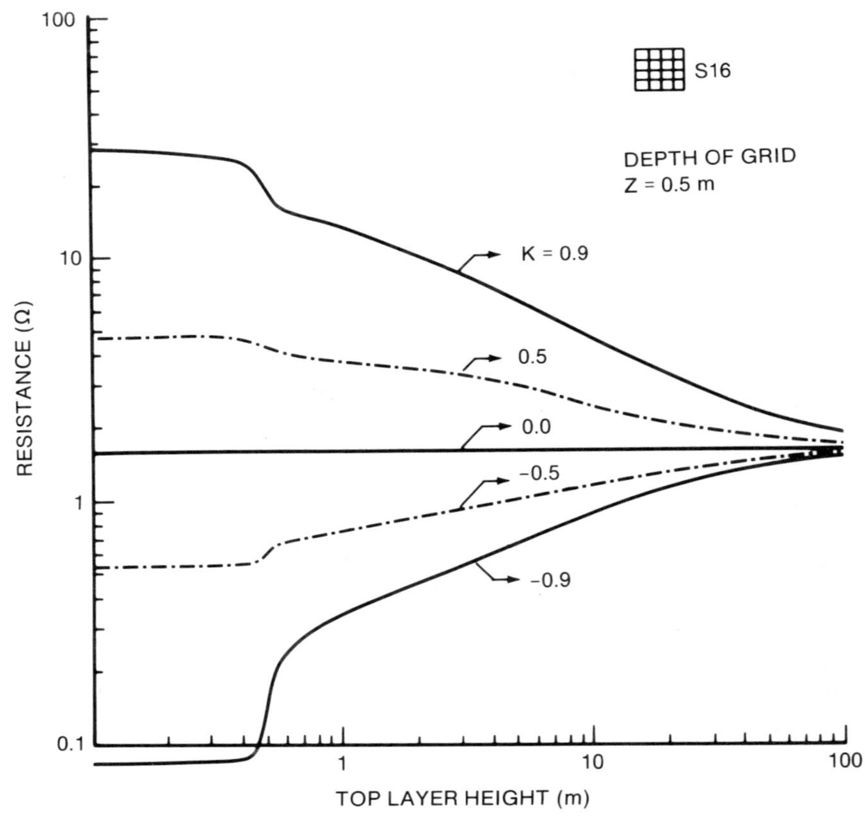

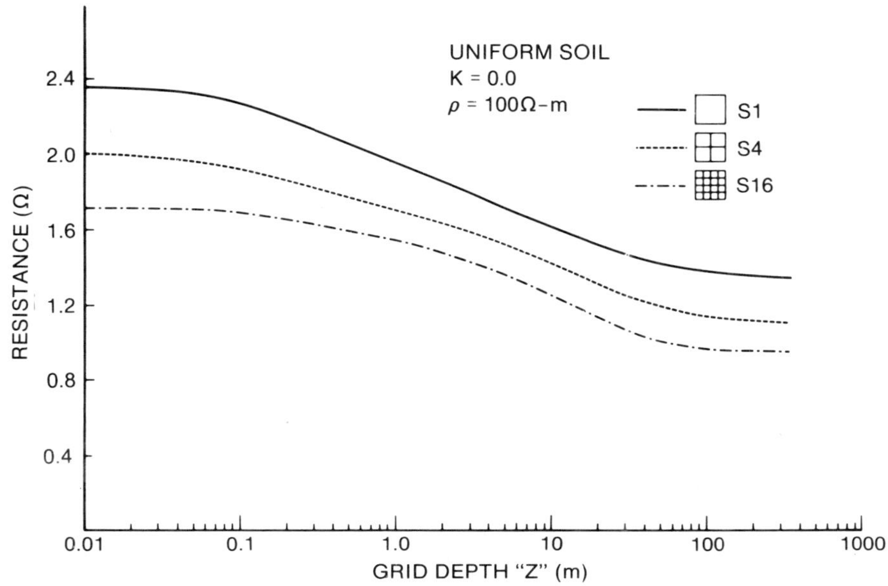

Fig E5
Grid Resistance Versus Grid Depth

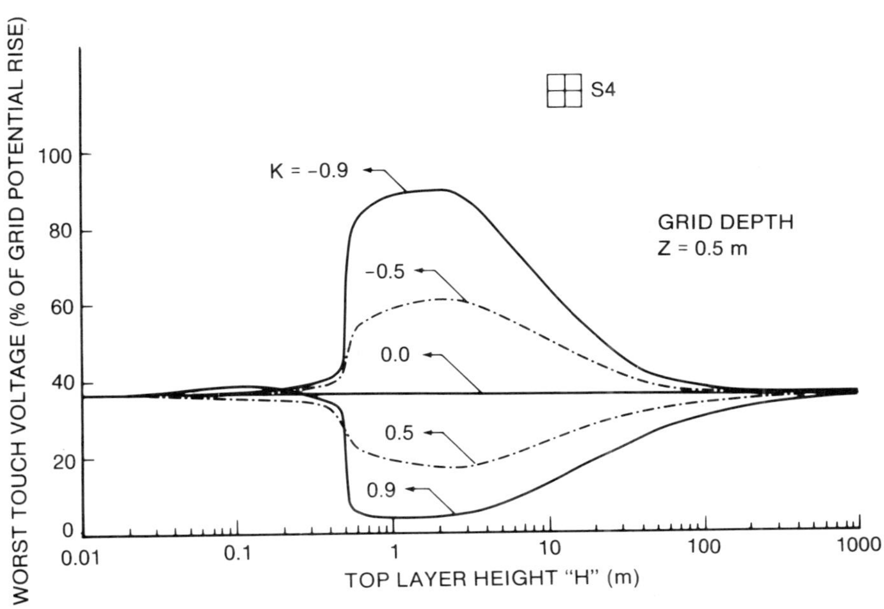

Fig E6
Four Mesh Grid Touch Voltage

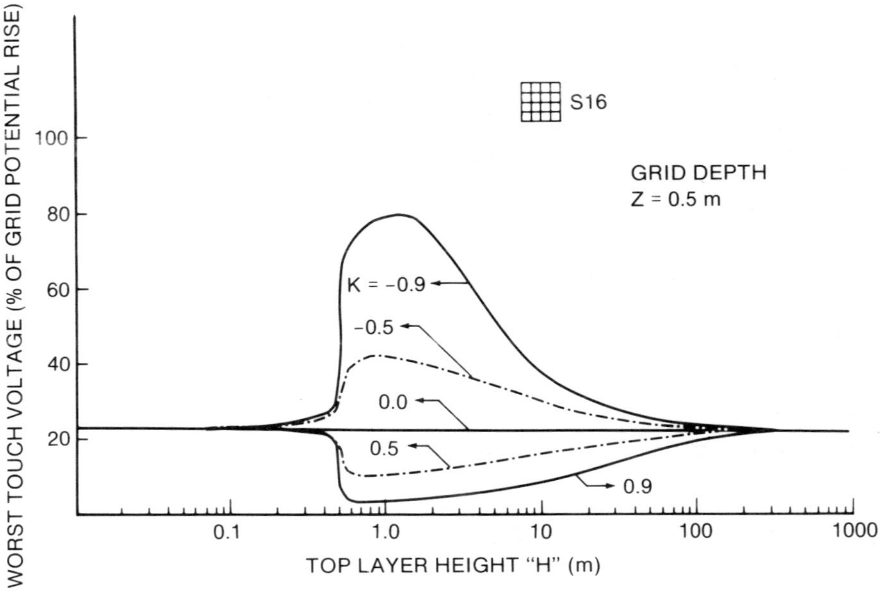

Fig E7
Sixteen Mesh Grid Touch Voltage

For negative values of K ($\rho_1 > \rho_2$), the highest touch potential occurs when H is slightly greater than the grid depth. For positive values of K ($\rho_1 < \rho_2$), the highest touch potentials occur when H is less than the grid depth, or when H is much greater than the grid depth.

One way of reducing the touch voltage without increasing the total amount of conductor is to omit the cross-connecting conductors (except at the ends) and reduce the spacing between the remaining parallel conductors. It must be noted, however, that while the touch voltage is reduced, the step voltage is increased when this design is used.

E2.4 Ground Rods Only. The behavior of a grounding system consisting only of ground rods may vary greatly from that in uniform soil. The major differences are due to the fact that the current density in each rod can be much higher in the portion of the rod located in the lower resistivity layer, depending on the value of K. As the absolute value of K increases, so does the percentage of the current discharged from the portion of the rod located in the lower resistivity layer of soil, as shown in Fig E12.

Assuming that the rod extends through the top layer into the bottom layer of soil, the current density in the portion of the rod in either layer is essentially uniform with a slight increase near the boundary of that layer. There is an abrupt change in current density, however, at the surface layer depth H. For rods that are mainly in the low resistivity layer, there is an appreciably higher current density in the outer rods as compared to rods near the center of the design, but

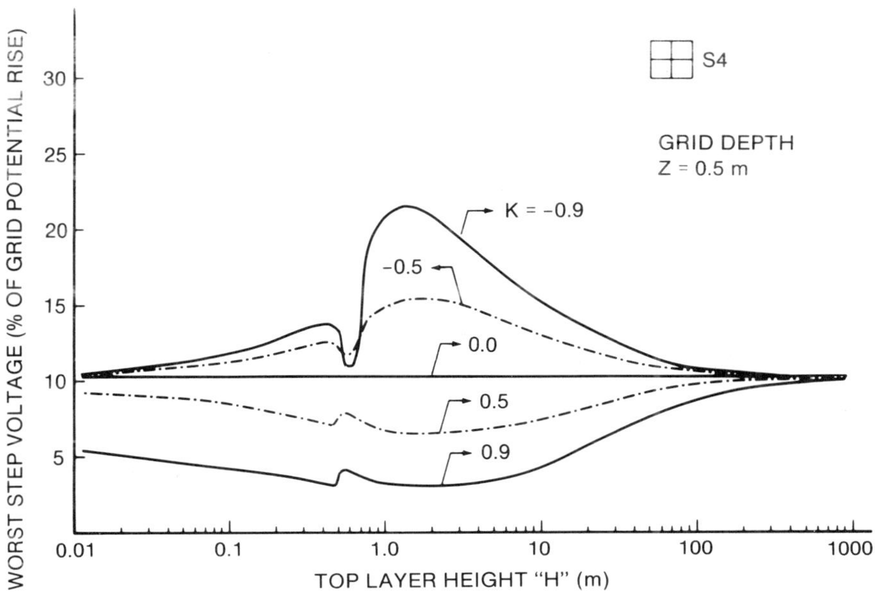

Fig E8
Four Mesh Grid Step Voltage

for rods mainly in the high resistivity layer the difference in the current density of the outside and inside rods is much less. See Fig E14.

As in the case of the grid, positive values of K ($\rho_1 < \rho_2$) generally give a higher resistance and negative values of K ($\rho_1 > \rho_2$) generally give a lower resistance for a system of ground rods as compared to the identical grounding system in uniform soil with a resistivity of ρ_1. However, as the surface layer height increases, the resistance of the rods for all values of K approaches that of the uniform soil model. See Table E1.

E2.5 Grid and Ground Rod Combinations. Depending on the values of K and H, adding ground rods to a system of grid conductors can have a tremendous effect on the performance of the grounding system. For negative values of K ($\rho_1 > \rho_2$) and for values of H limited so that the rods extend into the more conductive soil, the majority of the current is discharged through the rods into the lower layer of soil. Even for large values of H where none of the rod extends into the more conductive soil, the current density is higher in the ground rods than in the grid conductors, as shown in Figs E17 and E18.

If K is positive ($\rho_1 < \rho_2$), the current density for the portion of the ground rods in the upper layer is still higher than that of the grid conductors. For positive values of K, the effects of the ground rods become largely dependent on H, or on the length of the rods in the more conductive layer. Depending on the magnitude of K and H, the lengths of the rods are effectively shortened so that they may not contribute significantly to the control of step and touch voltages. However, for

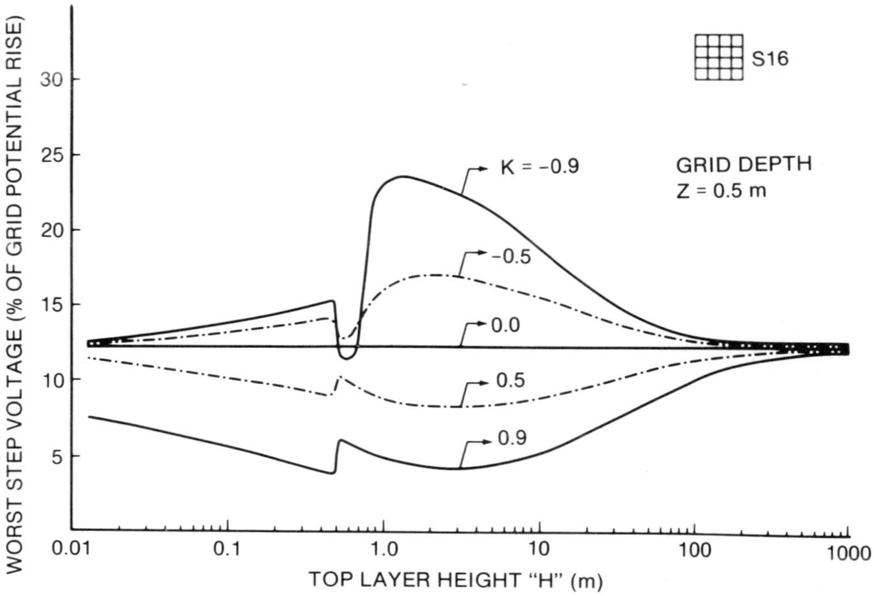

Fig E9
Sixteen Mesh Grid Step Voltage

moderate positive K values and large H values, the ground rods can be used to effectively improve the step and touch voltages.

If K is negative ($\rho_1 > \rho_2$), the step and touch voltages are reduced significantly with the addition of ground rods to a system of grid conductors. For small to medium values of H, relatively all of the current is discharged into the lower soil layer, thereby reducing the step and touch potentials. As H increases, the performance of the grounding system approaches that of an identical system in uniform soil of resistivity ρ_1.

E3. Summary

The two-layer parameters H and K discussed above can have considerable influence on the performance of the grounding system. A system designed using the uniform soil techniques can give results for step and touch potentials and station resistance ranging from highly pessimistic to highly optimistic, depending on the specific values of various parameters. Table E2 summarizes the effects of a two-layer soil environment on touch voltage of adding a ground rod to a grid, and on the touch voltage for a grid-rod combination.

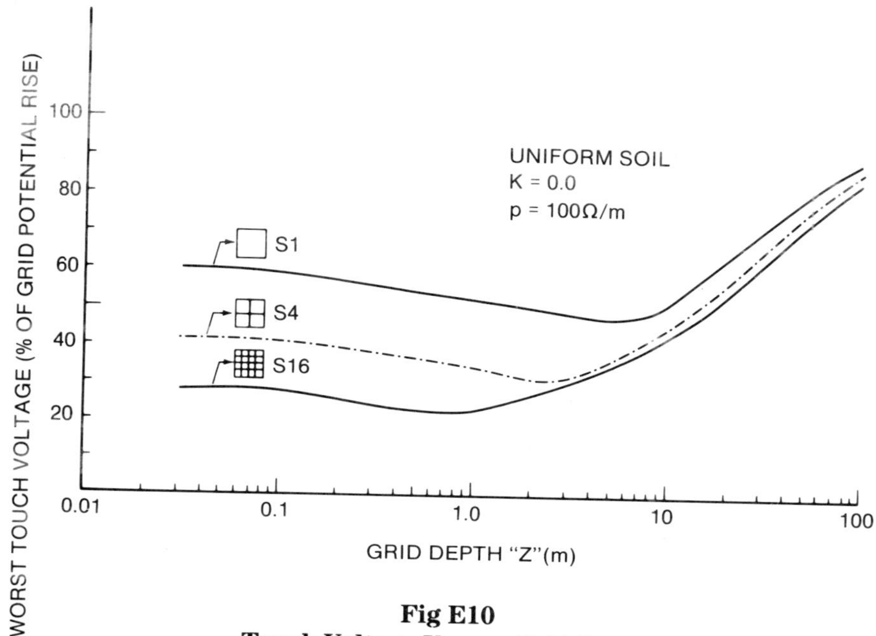

Fig E10
Touch Voltage Versus Grid Depth

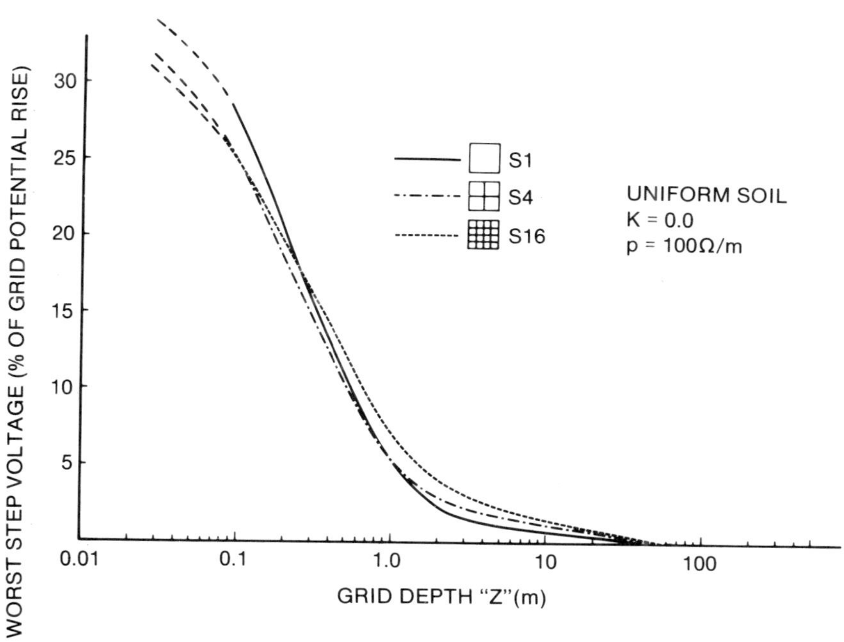

Fig E11
Step Voltage Versus Grid Depth

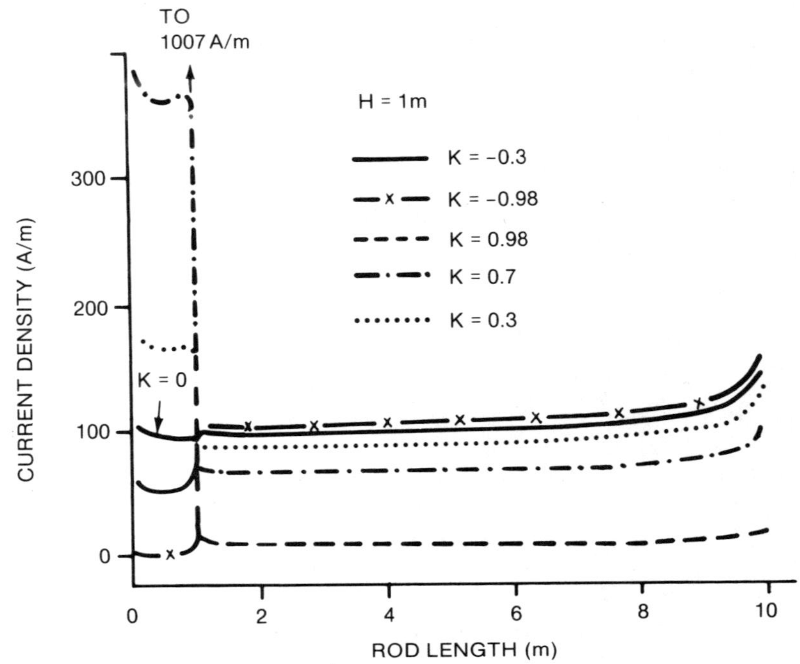

Fig E12
Single Rod Current Density

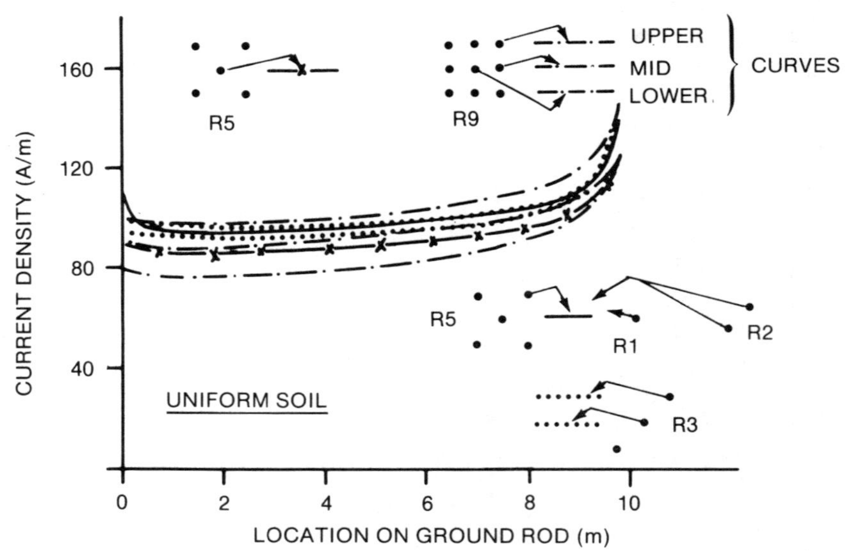

Fig E13
Multiple Driven Rod Current Density

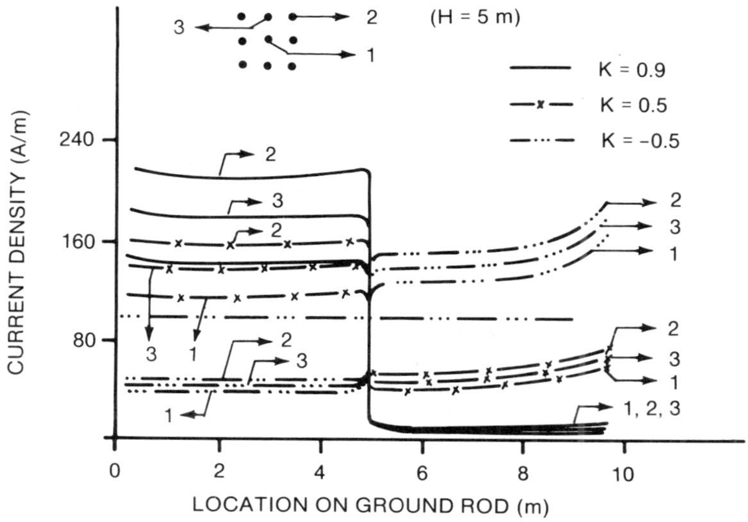

Fig E14
Current Densities in Multiple Drive Rods

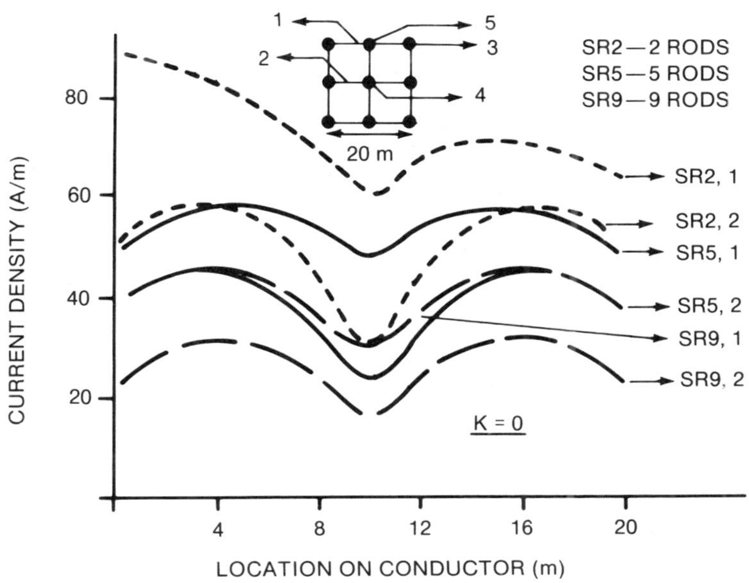

Fig E15
Grid Current Densities—Rods and Grid

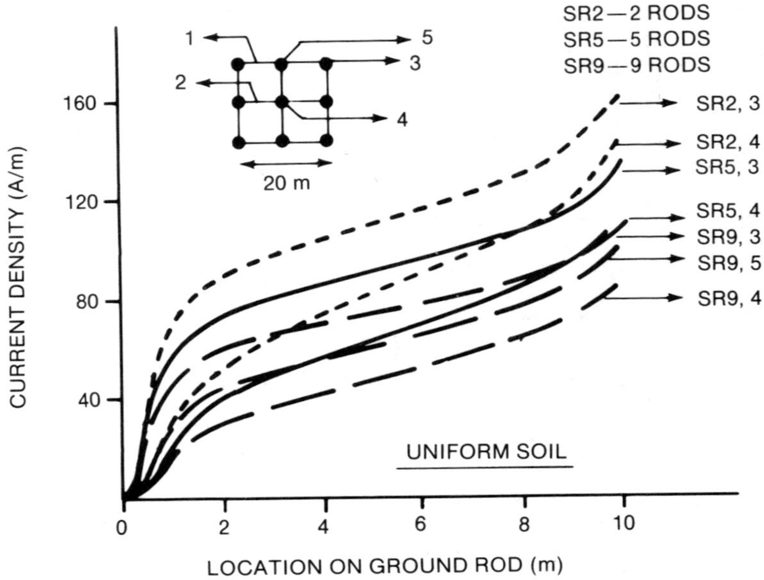

Fig E16
Rod Current Densities—Rods and Grid

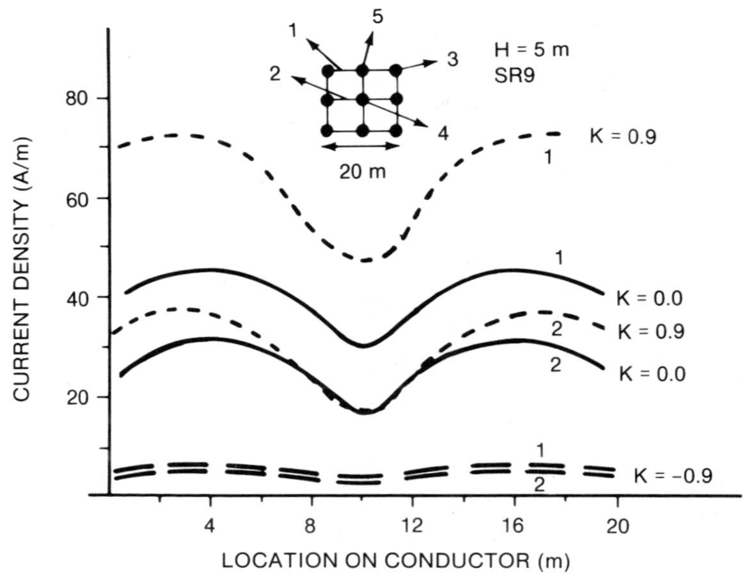

Fig E17
Rod and Grid Current Density—9 Rods and Grid in Two-Layer Soil

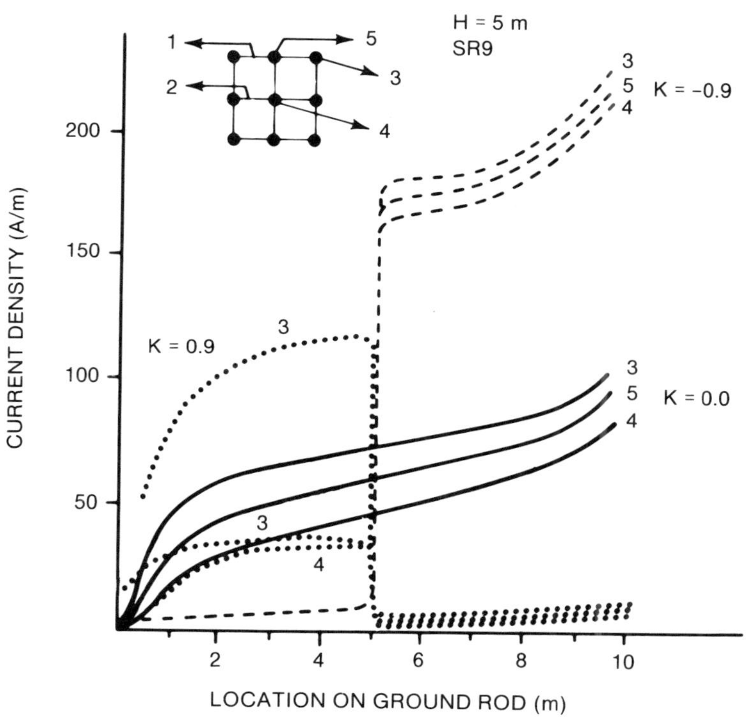

Fig E18
Rod and Grid Current Density—9 Rods and Grid in Two-Layer Soil

Table E1
Touch Voltages for Multiple Driven Rods

	R1	R2	R3	R5	R9
ELECTRODE TYPE	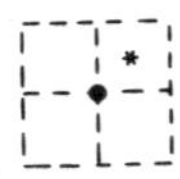	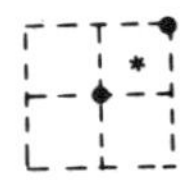	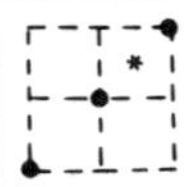	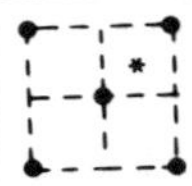	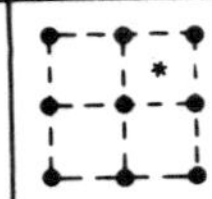
RESISTANCE (Ω)	11.85	6.43	4.52	3.01	2.16
TOUCH* VOLTAGE (%)	84.7	72.0	68.2	59.1	40.8

(A) UNIFORM SOIL

REFLECTION FACTOR K	-0.9	-0.5	UNIFORM SOIL (0.0)	0.5	0.9
RESISTANCE (Ω)	0.169	0.926	2.16	4.21	8.69
TOUCH* VOLTAGE (%)	51.1	47.4	40.8	31.8	19.3

(B) R9 IN TWO-LAYER SOIL (H = 5 m)

Table E2
Touch Voltages for Grid and Ground Rod Combinations in Two-Layer Soil

	S4	SR1	SR2	SR5	SR9
ELECTRODE TYPE	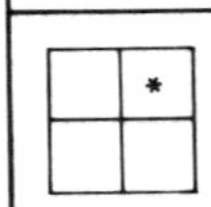	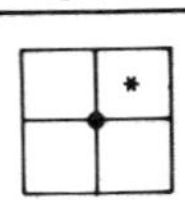	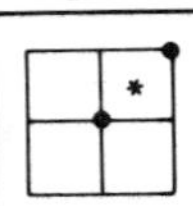	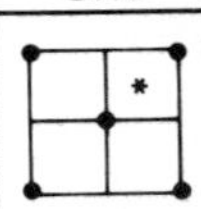	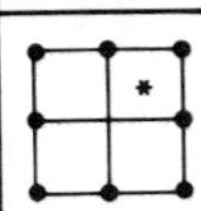
RESISTANCE (Ω)	2.58	—	2.28	2.00	1.81
TOUCH* VOLTAGE (%)	35.0	—	31.0	25.0	21.0

(A) UNIFORM SOIL

REFLECTION FACTOR K	-0.9	-0.5	UNIFORM SOIL (0.0)	0.5	0.9
RESISTANCE (Ω)	0.164	—	1.81	3.50	7.78
TOUCH* VOLTAGE (%)	35.0	—	21.0	13.4	6.6

(B) SR9 IN TWO-LAYER SOIL (H = 5 m)

Appendix F
Alphabetical Index of Definitions

auxiliary ground electrode. See Section 7.2, p 50

continuous enclosure. See Section 8.2, p 55

dc offset. See Section 13.1, p 92
dc offset factor. See Section 13.1, p 92
decrement factor. See Section 1.3, p 19

effective asymmetrical fault current. See Section 1.3, p 20
enclosure currents. See Section 8.2, p 55

fault current division factor. See Section 13.1, p 92

gas-insulated substation. See Section 8.1, p 55
ground. See Section 1.3, p 19
ground current. See Section 1.3, p 19
ground electrode. See Section 7.1, p 49
ground mat. See Section 7.1, p 49
ground potential rise (GPR). See Section 6.1, p 43
ground return circuit. See Section 1.3, p 19
grounded. See Section 1.3, p 19
grounding grid. See Section 7.1, p 49
grounding system. See Section 7.1, p 49

initial symmetrical ground fault current. See Section 1.3, p 19

main ground bus (GIS). See Section 8.3, p 56
maximum grid current. See Section 13.1, p 91
mesh voltage. See Section 6.1, p 43

noncontinuous enclosure. See Section 8.2, p 55

primary ground electrode. See Section 7.7, p 50

step voltage. See Section 6.1, p 43
subtransient reactance. See Section 13.1, p 92
symmetrical grid current. See Section 13.1, p 91
synchronous reactance. See Section 13.1, p 92

transferred voltage. See Section 6.1, p 43
transient reactance. See Section 13.1, p 92

***X/R* ratio.** See Section 13.1, p 92

Appendix G
Auxiliary Information

G1. Bibliography Not Cited in Text

ARCONE, S. A. *Computer Program to Determine the Resistance of Long Wires and Rods to Non-Homogeneous Ground.* US Army Cold Reg. Res. and Eng. Lab., CRREL Rep no 77-2, Jan 1977.

AREKELYAN, A. M., GABRIELYAN, R. O., and ARUTYUNYAN D. V. Quick Calculation of Parameters of Earthing Rods in Inhomogeneous Media. *Transactions of Electrical Technology USSR*, no 3, 1976, pp 58-61.

AREKELYAN, A. M., ARUTYUNYAN, D. V., and GABRIELYAN, R. O. Calculating the Resistance and the Potential Distribution of a System of Annular Electrodes in a Non-Uniform Two-Layer Medium. *Elektrichestvo (USSR)*, vol 52-7, no 10, 1971.

ASTEN, A. E. W. and TAYLOR, H. G. *Protection to Animals from Voltage Gradients Around the Earth Electrodes.* CIGRE Report 210, 1937.

BAER, G. and FISCHER, G. Die Erdungsanlage fuer das Kavernenkraftwerk Saeckingen und die 220/380 kV Umspannanlage Kuehmoos (Joint Grounding Installation for Saeckingen Pumped Storage Station and for Keuhmoos 220/380 kV Transformer Station). *Elektrizitaetswirtschaft*, vol 68, no 12, June 9, 1969, pp 394-401.

BAJOREK, J. Wplyw Uwarstwienia Gruntow Na Wlasnosci Statyczne Dlugich Uziomow (Effect of Ground Stratification on the Static Properties of Long Grounding Electrodes). *Przegl Elektrotech*, vol 49, Aug 1973, pp 360-363.

BELLASCHI, P. L. Impulse and 60-cycle Characteristics of Driven Grounds. *AIEE Transactions*, vol 60, Mar 1941, p 123.

BELLASCHI, P. L., ARMINGTON, R. E., and SNOWDEN A. E. Impulse and 60-cycle Characteristics of Driven Grounds — II, *AIEE Transactions*, vol 61, 1942, pp 349-363.

BENSTED, D. J., DAWALIBI, F., and WU, A. Y. The Application of Computer-Aided Grounding Design Techniques to a Pulp and Paper Mill Grounding System. *IEEE Transactions on Industry Applications*, vol IA-17, Jan/Feb 1981.

BERGER, K. *Le Comportment des Prises de Terre Sous Courants de Choc de Grande Intensite (Behavior of Grounds under High-Potential Impulse Currents).* CIGRE Report 215, 1946.

BODIER, M. G Prospection Systematique des Gradients de Potential au Voisinage et a L'interieur d'un Poste de Transformation (Systematic Investigation of Potential Gradients In and Around a Transformation Station). *Bulletin Societé* Francaise des Electriciens, July 1951.

BODLE, D. W. *Earth Potential Distributions Associated with Power Grounding Structures*, IEEE Conference Paper 62-205.

BOGAJEWSKI, W. Zmiany Rezystwnosci Betonow Warunkach Gruntowych (Variations of Resistivity of Concretes in a Soil Environment). *Arch. Inz. Ladowej*, vol 24, 1978, pp 237-256.

BRANDOLINI, A. and UBALDINI, M. B. Verso una Generalizzazione del Progetto Automatico di un Sistema di Messa a Terra (Toward a Generalization of Computer Grounding System). *Elettrotecnia*, Univ. of Pavia, Italy, vol 57, Feb 1970, pp 67-83.

BRANDOLINI, A. Studio Sui Modelli di Dispersione in Regime Variabile e Prove Comparative Tradispersori di Forma Diversa (Study of Electric Grounding Models for Impulse Currents and Comparative Tests with Differently Shaped Electrodes). *Elettrotecnia*, vol 62, no 10, Oct 1975, pp 885-896.

BRIDGES, J. E., FORD, G. L., SHERMAN, I. A. and VAINBERG, M., Eds. *Electrical Shock Safety Criteria*, Proceedings of the First International Symposium on Electrical Shock Safety Criteria, 1985.

BRUNE, O. and EATON, J. R. Experimental Studies in the Propogation of Lightning Surges on Transmission Lines. *AIEE Transactions*, vol 50, Sept 1931, p 1132.

BULLARD, W. R. Grounding Principles and Practice IV — System Grounding. *Electrical Engineering*, vol 64, Apr 1945, pp 145-151.

CALDECOTT, R. and KASTEN, D. G. Scale Model Studies of Station Grounding Grids. *IEEE Transactions on Power Apparatus and Systems*, vol PAS-102, Mar 1983, pp 558-566.

Computation of Zero-Sequence Impedances of Power Lines and Cables. Engineering Report no 37, Joint Subcommittee on Development and Research of EEI and Bell Telephone System, July 22, 1936 (with Reports 26-38 as Engineering Reports, volume IV of the above joint committee, Jan 1937, pp 301-331.

CONGER, A. C., SEELY, R. K., and CLAGGET, W. H. Grounding Effectiveness at Grand Coulee 230 kV Switchyards Verified by Staged Fault Tests. *AIEE Transactions*, vol 70, part II, 1951, pp 1547-1552.

CURDTS, R. B. Some of the Fundamental Aspects of Ground Resistance Measurement. *AIEE Transactions*, vol 77, part I, 1958, pp 760-767.

DAWALIBI, F. and MUKHEDKAR, D. Ground Electrode Resistance Measurements in Non-Uniform Soils. *IEEE Transactions on Power Apparatus and Systems*, vol PAS-93, Jan/Feb 1974, pp 109-115.

DAWALIBI, F., BENSTED, D., and MUKHEDKAR, D. Soil Effects on Ground Fault Currents. *IEEE Transactions on Power Apparatus and Systems*, vol PAS-100, July 1981, pp 3442-3450.

DAWALIBI, F., MUKHEDKAR, D., and BENSTED, D. Measured and Computed Current Densities in Buried Ground Conductors. *IEEE Transactions on Power Apparatus and Systems*, vol PAS-100, Aug 1981, pp 4083-4092.

DEVGAN, S. S. and WHITEHEAD, E. R. Analytical Models for Distributed Grounding Systems. *IEEE Transactions on Power Apparatus and Systems*, vol PAS-92, Sept/Oct 1973, pp 1763-1781.

DOHLER, A. The Effect of Overhead Line Protective Sections on Dimensioning the Earthing Systems of Transformer Stations. *Energietechnik*, vol 32, no 7, July 1982, pp 267-269.

DUKE, C. A. and SMITH, L. E. The Technique and Instrumentation of Low-Impedance Ground Measurement. *AIEE Transactions*, vol 77, part I, 1958, pp 767-770.

DUNKLE, W. F. A Method to Calculate Electrode Resistance. *Electrical World*, May 2, 1955, p 62.

DUNKLE, W. F. Steel Ground Grid Works Well. *Electrical World*, May 2, 1955, pp 63-65.

DWIGHT, H. B. Calculation of Resistance to Ground. *Electrical Engineering*, vol 55, Dec 1936, p 1319.

DWIGHT, H. B. *Electrical Elements of Power Transmission Lines*. New York: MacMillan, 1954, pp 161-171.

EATON, J. R. Grounding Electrical Circuits Effectively, Part II — Calculations and Installation, *G.E. Review*, vol 44, 1941, pp 397-404.

ELEK, A. Proper Grounding Reduces Hazards. *Electrical World*, Feb 16, 1959, pp 78-82.

ENDRENYI, J. Fault-Current Analysis for Station Grounding Design. *Ontario Hydro Research Quarterly*, second quarter, 1967, pp 16-21.

ERBACHER, W. Probleme der Erdung von Hochstspannungsstationen unter besonderer Berucksechtigung der Messung (Grounding Problems of High-Voltage Stations with Special Consideration to Measurement). *Osterreichische Zeitschrift for Elektrizatswirtschaft*, vol 8, no 1, Jan 1955, pp 1-12.

EYRE, B. E, WILKINSON, F., CHENERY, D. A., and BERZ, I. Earthing Practices on the LEB Distribution System. IEE Conference on International Medium Voltage Earthing Practices, London, England, Mar 1972, pp 217-224.

FAGAN, E. J. and LEE, R. H. The Use of Concrete Enclosed Reinforcing Rods as Grounding Electrodes. *IEEE Transactions on Industry and General Applications*, vol IGA-6, no 4, July/Aug 1970.

FALLAHI, A. A. Substation Grounding in Rocky Terrain. IEEE Catalogue number 81CH1664-2, 1981.

FAWCETT, GRIMMET, SHOLTER, and TAYLOR. Practical Aspects of Earthing. *IEE Journal*, vol 87, Oct 1940, p 357.

FEYDT, M. Zur Verteilung Gemessener Hoechstwerte der Beruehrungs und Schrittspannungen in Imspannwerken (On the Distribution of Measured Maximum Values of Contact Voltages and Surface Voltage Gradients in Transformer Stations). *Elektrie*, vol 36, no 7, 1982, pp 350-354.

FUJIMOTO, N., DICK, E. P., BOGGS, S. A., and FORD, G. L. Transient Ground Potential Rise in Gas Insulated Substations — Experimental Studies. *IEEE Transactions on Power Apparatus and Systems*, vol PAS-101, Oct 1982, pp 3603-3609.

GIAO, F. N. and SARMA, M. P. On the Potential and Field Distribution Around a Ground Electrode for HVDC Transmission. *IEEE Transactions on Power Apparatus and Systems*, vol PAS-90, Nov/Dec 1971, pp 2793-2801.

GILCHRIST, F. G. *Substation Grounding.* 1982 Rural Electric Power Conference, IEEE, Apr 1982.

GILLIES, D. A. and RAMBERG, H. C. Methods for Reducing Induced Voltages in Secondary Circuits. *IEEE Transactions on Power Apparatus and Systems*, vol PAS-86, 1967, pp 907-916.

GROSS, E. F. B. and HOLLITCH, R. F. Grounding Grids for High-Voltage Stations III, Resistance of Rectangular Grids. *AIEE Transactions*, vol 75, part III, 1953, pp 926-935.

GROSS, E. F. B. and WISE, R. B. Grounding Grids for High-Voltage Stations — II. *AIEE Transactions*, Vol 74, part III, 1955, pp 801-809.

GUPTA, B. R. and THAPAR, B. Impulse Impedance of Grounding Systems. *IEEE Transactions on Power Apparatus and Systems*, vol PAS-98, abstract, Jan/Feb 1979, p 8.

GUPTA, B. R. and THAPAR, B. Impulse Impedance of Grounding Grids. *IEEE Transactions on Power Apparatus and Systems*, vol PAS-99, Nov/Dec 1980, pp 2357-2362.

HAMMAM, M. S. and BAISHIKI, R. S. A Range of Body Impedance Values for Low Voltage, Low Source Impedance Systems of 60 Hz. *IEEE Transactions on Power Apparatus and Systems*, vol PAS-102, May 1983, pp 1097-1105.

HARRINGTON, E. J. and MARTIN, T. M. C. Placement of Protective Grounds for Safety of Linemen. *AIEE Transactions*, vol 73, part III-B, 1954, pp 950-955.

HORN, R. S. Are Your Substations Safe? *Power Engineering*, Oct 1953, p 86.

HORN, R. S. Ground Your Power Station Safely. *Power Engineering*, Jan 1956, p 85.

HUEY, R. M. and RAJARATNAM, K. Identification of Electrical Parameters in Large Earth Grids. *IEEE Transactions on Power Apparatus and Systems*, vol PAS-93, Jan/Feb 1974, pp 187-195.

HUMPHRIES, J. D. Earth Electrodes System for Large Electric Stations. *Proceedings of the IEE*, vol 104, part A, Oct 1957, pp 383-399.

JENSEN, C. Grounding Principles and Practice: II — Establishing Grounds. *Electrical Engineering*, vol 64, Feb 1945, pp 68-74. (Also included in AIEE Publication S-2.)

JENSEN, C. H. *Deep-Driven Grounds*. Presented to EEI, T and D Committee, May 1951.

JONES, G. C. and JAGER, C. R. *Potential Gradients Within and Around Substation Ground Fields*. Presented at AIEE Southern District Meeting, Nov 3-5, 1948.

JOY E. B., PAIK, N., BREWER, F. E., WILSON, R. E., WEBB, R. P., and MELIOPOULOS, A. P. Graphical Data for Ground Grid Analysis. *IEEE Transactions on Power Apparatus and Systems*, vol PAS-102, Sept 1983, pp 3038-3048.

KINYON, A. L. Correlation of Measured and Calculated Substation Ground Grid Resistance. *AIEE Transactions*, vol 78, Part I, 1959, pp 698-701.

KOMARAGIRI, K. G. and MUKHEDKAR, D. Selected Non-IEEE Bibliography on Grounding. *IEEE Transactions on Power Apparatus and Systems*, vol PAS-100, June 1981, pp 2993-3001.

KOSTENKO, M. V. and CHALILOV, F. C. Investigation of the Effectivity of Additional Elongated Earthing Strips for the Lightning Protection of the Transformer Stations with High Specific Earth Resistance. *Elektrie*, vol 35, no 8, 1981, pp 402-405.

KOUTEYNIKOFF, P. Numerical Computation of the Grounding Resistance of Substations and Towers. *IEEE Transactions on Power Apparatus and Systems*, vol PAS-99, May/June 1980, pp 957-965.

KOVATS, F. J. Estimating Substation Earth Grid Voltage Rises Due to Fault Currents. *Inst. Eng. Australia Eng. Trans.*, vol EE-13, no 1, 1977, pp 43-48.

LANTZ, M. J. Factors Determining the Potential Rise in Station Ground Mats. *Electrical World*, Dec 14, 1953.

LARSEN, S. L. and NORDELL, D. E. Measurement of Substation Ground Resistance and Its Use in Determining Protection for Metallic Communication Circuits. *IEEE Transactions on Power Apparatus and Systems*, vol PAS-94, Sept/Oct 1975, pp 1666-1673.

LAURENT, P. G., RUDENBERG, R., SCHWARZ, S. J., FRENCH, R. X., KAUFMAN, R. H., HAYES, J. S., and STEVENS, R. F. Discussions of Working Group Report. *AIEE Transactions*, vol 77, part III, 1958, pp 685-692.

LEE, R. H. Electric Safety in Industrial Plants. *IEEE Spectrum*, June 1971, pp 51-55.

LISINKER, L. S. and TSELEBROVSKII, Y. V. Allowing for Substation Earthing Non-Equipotentiality in Calculating Contact Voltage. *Elektrichestvo*, no 3, Mar 1978, pp 85-88.

LITRIDES, S. J. All-Steel Network Grounds Substation. *Electrical World*, May 2, 1955.

LOUCKS, W. W. and LEMIRE, W. A. R. Transmission and Distribution Grounding in the Hydro-Electric Power Commission of Ontario. *AIEE Transactions*, vol 70, part II, 1951, p 1493.

MacKENZIE, W. F. Impedance and Induced Voltage Measured on Iron Conductors. *AIEE Transactions*, vol 73, part I, 1954, p 557.

MANILOV, V. E. and TORPOV, A. K. Measurement of the Resistance of the Ground Connections of a 220-110 kV Substation with Actual Short-Circuit Currents. CIGRE Report no 332, 1937.

McCROCKLIN, Jr., A. J. and WENDLANDT C. W. Determination of Resistance to Ground of Grounding Grids. *AIEE Transactions*, vol 71, part III, 1952, p 1062.

MELIOPOULOS, A. P., WEBB, R. P., and JOY, E. B. Analysis of Grounding Systems. *IEEE Transactions*, vol PAS-100, Mar 1981, pp 1039-1048.

MELIOPOULOS, A. P. and MOHARAM, M. G. Transient Analysis of Grounding Systems. *IEEE Transactions on Power Apparatus and Systems*, vol PAS-102, Feb 1983, pp 389-399.

MELIOPOULOS, A. P., WEBB, R. P., JOY, E. B., and PATEL, S. Computation of Maximum Earth Current in Substation Switchyards. *IEEE Transactions on Power Apparatus and Systems*, vol PAS-102, Sept 1983, pp 3131-3139.

MRIG, L. C. Method of Calculating Plant Switchyard Ground Potential. Control of Power Systems Conference and Exposition, Mar 1-2, 1978, IEEE, 1978, pp 96-98.

MUKHEDKAR, D. and GERVAIS, Y. Earthing Techniques — Part I. *Engineering Journal*, vol 53, no 10, Oct 1970, pp 51-55. Part II, vol 54, no 10, Oct 1971, pp 19-20.

MUKHEDKAR, D., GERVAIS, Y., and DAWALIBI, F. Modeling of Potential Distribution Around a Grounding Electrode. *IEEE Transactions on Power Apparatus and Systems*, vol PAS-92, no 5, 1973, pp 1455-1459.

NAHMAN, J. and SKULETICH, S. Irregularity Correction Factors for Mesh and Step Voltages of Grounding Grids. *IEEE Transactions on Power Apparatus and Systems*, vol PAS-99, Jan/Feb 1980, pp 174-180.

NAHMAN, J. and SKULETICH, S. Resistance to Ground and Mesh Voltages of Ground Grids. *Proceedings of the IEE*, vol 126, Jan 1979, pp 57-61.

NIEMANN, J. Umstellung von Hochstspan-nungs-erdungsanlangen auf den Betrieb mit Starr gerrdetem Sternpunkt (Changeover from High-Tension Grounding Installation to Operation with a Grounded Star Point). *Electrotechnische Zeit*, vol 73, no 10, May 15, 1952, pp 333-337.

NORINDER and PETROPOULOS. *Characteristics of Pointed Electrodes*. CIGRE paper no 310, 1948 Session.

OLSON, A. B. and DELYANOV, A. G. Calculation of the Grounding Grids in Multi-Layer Soil. Tula Polytech Inst., Soviet Union, *Elektrichestvo*, no 5, May 1971, pp 23-26.

PALMER, L. S. *Examples of Geoelectrical Surveys*. IEE paper 2791M, Dec 1958, pp 231-244.

PERRY, E. W. and LUONGO, A. M. Compression Connectors for Grounding Grid Systems. *Plant Engineering*, vol 29, Dec 1975, pp 105-107.

PETERS, C. S. Ground Connections for Electric Systems. US Bureau of Standards, Technical Paper no 108, June 1918.

PETERSON, L. C. Problems with Large Ground Fault Currents in Utility Substations In or Near Industrial Sites. *IEEE Transactions on Industry Applications*, vol 1A-9, Nov/Dec 1973, pp 680-686.

POHLHAUSEN, K. Grundlagen der Bemessung von Slarkstromerdern (Basic Concepts for the Design of Grounding Systems). *V.D.E. Fachberichte*, 1927, pp 39-41.

POWELL, K. B. Charts Determine Substation Grounds. *Electrical World*, vol 189, no 2, Jan 1978, pp 56-58.

PREIS, K. Berechung von Erdausbreitungswiderstaenden mit hilfeder Finiten Elemente Methode (Application of the Finite Element Method to the Calculation of Ground Resistance). *Arch Elektrotech*, vol 65, no 6, 1982, pp 349-354.

PRESTIGIACOMO, E. Earthing Installations in Electric Power Stations. *Elettrificazione*, no 9, Sept 1976, pp 395-401.

PROCHORENKO, S. V. Probability Consideration to the Evaluating of the Earthing Systems of High-Voltage Transformer Plants. *Elektrie*, vol 36, no 7, 1982, pp 346-348.

PURDY, A. B. Accurate Equations for the Characteristics of Rod Electrodes in a Homogeneous Medium. *IEEE Transactions on Power Apparatus and Systems*, vol PAS-98, abstract, July/Aug 1979, p 1141.

Recommended Grounding Practices for Single Polarity Direct-Current Structures. Conversion Subcommittee of AIEE Substations Committee, Paper no 57-719, vol 76, part III, 1957.

ROBERTSON, N. C. and VALENTINO, A. R. *Step Potentials Near a Buried-Wire Grounding Electrode*, IEEE preprint, paper A79402-9, 1979.

ROGERS, E. J., HALL, R. H., and WHITE, J. F. Fault Induced Voltages on Metallic Fencing Located in the Vicinity of a High Voltage Substation. *IEEE Transactions on Power Apparatus and Systems*. vol PAS-101, Mar 1982, pp 746-750.

RUDENBERG, R. Uber den raumlichen Verlauf von Erdschlusstromen (Distribution of Short-Circuit Currents in Ground). *Elektrotechnischen Zeitschrift*, vol 31, 1921; also in *Bulletin Schweiz. Elek. Ver.*, vol 12, 1921, pp 276-283.

RUDENBERG, R. *Transient Performance of Electric Power Systems*. New York: McGraw-Hill, 1950, chapters 24 and 25, pp 310-339.

RUDENBERG, R. Sternpunktserdung bei Hochspannungs Leitungen einige grundsatzliche Betrachtungen (Basic Considerations Concerning Systems). *Elektrotechnischen Zeitschrift*, vols 11 and 12, 1926.

RYABKOVA, E. Y. The Surge Characteristics of Substation Grounding Devices. *Elektrichestvo*, no 3, Mar 1973, pp 67-70.

RYDER, R. W. Earthing Problems. *IEE Journal*, vol 95, part II, Apr 1948, p 175.

SAKATA, H. and WATANABE, F. Ground Electrode and Soil Potential Distribution. *Bull. Nagoya Inst. Technol.*, vol 25, 1973, pp 237-244.

SANDHOLM, F. and VALJUS, J. The Two-Layer Soil Model As an Aid to Earth Electrode Design. *Saehkoe Electrical and Electronics*, vol 56, no 2, Feb 1983, pp 16-21.

SCHWARZ, S. J. Let's Look to Station Grounding. *Electrical World*, May 2, 1955.

SMITH, H. M. Utility Grounding Practices. National Cable Television Association Convention, Anaheim, CA, June 17-20, 1973, pp 37-42.

STEVENS, R. F. Optimum Diameter Spacing and Burial Depth of Ground Grind Conductors. *IEEE Transactions on Power Apparatus and Systems*, vol PAS-80, part III, June 1961, pp 313-316.

SZATKOWSKA-KARPA, B. and JABLONSKI, W. Calculation of the Electrical Data for Substation Earthing Installations. *Energietechnik*, vol 32, no 7, July 1982, pp 262-264.

SZOMMER, P. On the Earthing of Reinforced Concrete Foundations. *Villamossag*, vol 25, no 12, Dec 1977, pp 372-375.

TAYLOR, H. H. *Earth Current Voltage Gradients Around Substations and Power Stations*. CIGRE Report no 322, 1935.

THAPAR, B. and NAGAR, R. P. Irregularity Correction Factor for Grounding Grids. *Journal Inst. Eng.*, Elec. Eng. Div., vol 58, Aug 1977, pp 36-41.

THAPAR, B. and PURI, K. K. Mesh Potentials in High-Voltage Grounding Grids. *IEEE Transactions on Power Apparatus and Systems*, vol PAS-86, 1967, pp 249-254.

TIBENSKY, M. S. and PERFECKY, L. J. Methods for RMS Symmetrical Station Ground Potential Rise Calculations for Protection of Telecommunications Circuits Entering Power Stations. *IEEE Transactions on Power Apparatus and Systems*, vol PAS-100, Dec 1981, pp 4785-4794.

TOWNE, H. M. Impulse Characteristics of Driven Grounds. *G.E. Review*, vol 31, Nov 1928, p 605.

TOWNE, H. M. Lightning Arrestor Grounds — Parts I, II, and III. *G.E. Review*, vol 35, Mar/Apr/May 1932, pp 173, 215, 280.

TSIRELL, Y. A. Earthing Equipment for Cable Networks. *Elek, Stantsii*, vol 43, no 1, 1972, pp 68-70.

VERMA, R. and MUKHEDKAR, D. Impulse Impedance of Buried Ground Wire. *IEEE Transactions on Power Apparatus and Systems*. vol PAS-99, Sept/Oct 1980.

Voltage Gradients Through Ground Under Fault Conditions. AIEE Working Group 56.1, paper no 58-98, *AIEE Transactions*, vol 77, part III, 1958, pp 669-685.

VORONINA, A. A., MEN'SHOV, B. G., and MIRONENKO, E. S. Analiticheskii Raschet Soprotivleniya Rastekaniyu Zazemlitelei Podstantsii V. Duukhsloinom Grunte S Gorizontal Noi Granitsei Razdela (Analytical Calculation of Flow Resistances of Substation Grounding Connections in Two Layer Soil with a Horizontal Interface). *Moscow Inst. of Telecommunications, Electronics and Automatics, Izv Vyssh Uchebn Zaved, Energ*, Feb 1973, pp 19-25.

WASILUK, W. Badanie Wplywu Dlugotrwale Przeplywajacego Pradu Przemiennego Na Rezystancje Ukladu Uziom-grunt (Examiniation of the Effect of Long-Lasting Alternating Current Flow on the Resistance of the System Ground Electrode-ground). *Arch Elektrotech*, vol 30, no 2, 1981, pp 349-359.

WHITEHEAD, L. E. Ground Grid Nomogram Gives Quick Estimates of Mesh Potential. *Power Engineering*, June 1971, pp 48-49.

WILLHEIM, R. and WATERS, M. Neutral Grounding in High-Voltage Transmission. New York: Van Nostrand, 1956.

ZABORSZKY, J. and RITTENHOUSE, J. W. Design Charts for Determining Optimum Ground Rod Dimensions. *AIEE Transactions*, vol 72, part III, 1953, p 810.

ZAMORANI, G. and SUBIOLI, A. Impianto di Terra con Puntazze e Conduttori Isolati di Interconnessione (Grounding Systems with Grounding Rods and Isolated Cable Interconnections). *Elettrotecnia*, June 1974, pp 507-510.

ZASTROW, O. W. Electrical Grounding Design for Safety and Corrosion Control. *West Virginia University Experiment Station Bulletin*, no 113, 1974, p 367.

ZINK, J. H. Comments on the IEEE Guide 80 for Safety in Substation Grounding. IEEE preprint, paper A79549-7, 1979.

ZUKERMAN, L. G. Simplified Analysis of Rectangular Grounding Grids. *IEEE Transactions on Power Apparatus and Systems.* vol PAS-98, Sept/Oct 1979, pp 1777-1785.

ZUPA, F. P. and LAIDIG, J. F. A Practical Ground Potential Rise Prediction Technique for Power Stations. *IEEE Transactions on Power Apparatus and Systems.* vol PAS-99, Jan/Feb 1980, pp 207-216.

G2. Abstracts of References Not Readily Available

OLLENDORF, F., *Erdstrome (Ground Currents)*, Springer-Verlag, 1928. Published in German. No English translation available. Out of print but available at a few US libraries.

One of the most comprehensive basic texts on problems of grounding. Excellent illustrations of current flow and potential fields in ground for various types of electrodes. Method of images well illustrated.

Steady Flow in the Earth. Description of earth currents. Earth conductivity for some common earth materials. Power and radio ground currents and grounding of each. Fundamental equations for steady-state fields in the earth. Derivation of the potential field equation in rectangular and curvilinear coordinates. The behavior of steady flow at the boundary of two mediums. Derives potential field equation for a point current source in an infinite homogenous earth for points far removed from source. Defines "distributed ground resistance." Introduces method of images for obtaining the potential field of a current source located below the earth's surface. Physiological effects of magnitudes of the field along the earth's surface. Defines step potential, contact potential, potential funnel, danger current, and danger zone. Discusses body resistances and "distributed resistance" of human foot. Use of an electrostatic model to represent the conduction field around ground electrode.

Simple Ground Electrode in Homogeneous Earth. Hemispherical ground electrode. Circular plate electrode. Formula for the distributed ground resistance of the human foot. Pipe (rod) electrode. Long horizontal bar electrode. Comparison with rods. Annular electrode. Comparison with rods. For each of the above electrode types, the book derives equations for potential field (potential funnel) and distributed ground resistance, step potential, maximum step potential, ground electrode current that will produce hazardous potential field, danger zone radius, and contact potential (electrode potential). Typical calculations made.

Multiple Ground Electrodes in Homogeneous Earth. Fundamental properties. Effect of finite spacing. "Figure of merit" of multiple ground electrode system. Single ground electrodes with large spacing. Application of theory of plane potential. Derivation of potential equation for a line source of current. Application of Laplace's equation to a current source in a medium. Substitution of a single-rod electrode for multiple-rod electrodes arranged in a circle with small spacing. Equivalent radius of substitute electrode, resistance, and figure of merit. Equivalent radius of a plane plate. Shown to be one-fourth its length. Applied to row of ground electrodes closely spaced to derive formulas for distributed resistance and figure of merit. Center electrodes in a row shown to contribute very little to reducing resistance. Similarly, central portion of a flat circular plate contributes little. Curves included.

Buried Electrodes. Compares characteristics of buried electrodes and surface electrodes from the standpoint of danger current, contact, and step potentials. Electrodes buried at great depths. Equations for potential distribution and step potential on earth's surface using method of images. Formulas for maximum body current, danger current, and danger distance. Permissible danger current increases as the fourth power of the electrode depth below the surface. Formula for danger zone for large currents. Deeply buried electrodes not of much practical use for grounding, especially for the larger current values, have relatively small advantage over electrodes at or near the surface. Reduction of resistance with small depths of burial. Approximate equation for electrode potential and ground resistance of a buried spherical electrode. Resistance rapidly approaches a limiting value as depth increases. Equations for circular plate and rod electrodes.

Effect of Ohmic Voltage Drop in the Grounding System. Effect of finite resistance of metals employed in grounding systems. (As grounding system is made larger and larger, ohmic resistance imposes a lower limit to the value of distributed ground resistance). Distribution of ground-fault currents along high-voltage overhead lines. Equations for distributed ground resistance and tower ground current for a line-to-tower short circuit at one end of a transmission line with overhead ground wire. Numerical example shows only a fraction of the fault current flows through the faulted tower. Calculation of danger zone. Above repeated for a fault in the middle of a long line. Though the formulas were derived for an infinitely long line, they may be applied to a line of twenty towers or more with an error of less than one percent. Use of overhead transmission line ground wires for reducing station ground resistance. Ground resistance of long conductors. Long conductors (sheaths) in contact with the earth in extensive high-voltage cable systems and electric railways. Uncertainties regarding contact resistance between sheaths or rails and earth. Approximate equation for ground resistance of infinitely long line in infinite homogeneous earth. Radius of the equivalent sphere and the length of the equivalent cylinder. Method of images used to obtain ground resistance for conductor at or near the earth's surface. Typical values calculated. Potential field around long conductors. Approximate solution at the fault point for potential distribution, step potential, maximum body current, danger current, and danger zone for a conductor at or near the surface. Danger zone extends further than in the case of ground electrodes and also depends upon conductor material. Formula and curve for distributed resistance with conductor buried at moderate depths. Equation for longitudinal attenuation of current due to leakage into ground. For normal values, the conductor current is reduced to ⅓ in a distance of about 1 km. Typical calculations and practical discussion included.

Inhomogeneities in the Earth. Effects of current overload of the ground electrodes. Contact resistance. Surface ground electrodes in stratified earth. Resistance of deep ground electrodes in stratified earth. Conduction through damp building walls. The principle of three-dimensional images. Effects of surface water reservoirs. Current conduction in long conductors.

Time Varying Fields. Return circuit of capacitive ground fault currents. Plane eddy currents in the earth. Boundary conditions for plane eddy currents. Spatial distribution of eddy currents in the earth.

Quasistationary Earth Currents. Return currents for cables in sea water. Return-circuit field in homogeneous earth for a single conductor carrying alternating current (approximate solution). Return currents in homogeneous earth for a single conductor carrying alternating current (rigorous solution). Field in stratified earth with high conductivity in the top layer. Division of fault ground current between ground wire and earth. Return conduction of alternating current fault ground currents through long conductors buried in the earth. Induction from alternating-current ground currents in low-voltage communication circuits. Conduction in homogeneous earth. Conduction in earth with a high-conductivity top layer. Determination of soil conductivity by use of two long parallel wires. Ground bar for radio antennas. Operation with counterpoises and ground conductors. Resonant frequency of a single conductor with earth return.

Effect of Ground Current on Phenomena of Steady Ground Faults in High-Capacity Networks. Equivalent diagram of line-to-ground fault. The equivalent conductor for a circuit fault to ground. The capacitive fault ground current. Ground-fault current neutralizers. The ground short-circuit current.

Transient Electromagnetic Phenomena in the Earth. Basic equations for surge phenomena. Transient field in earth for a short conductor with suddenly applied current. Transient currents in earth for long conductors with suddenly applied current. Overvoltages when switching alternating current.

Fundamentals for the Calculation of Temperature Rise of Ground Electrodes. Principles of heat conduction including boundary conditions. Steady-state temperature field for a spherical ground electrode at great depth. Same for circular plate and for bar electrode. The steady-state critical loading of hemispherical electrode on the earth's surface with air flow. The steady-state critical loading of hemispherical ground electrode with motionless air. Dynamic time constant of a ground electrode.

SUNDE, E.D., *Earth Conduction Effects in Transmission Systems,* Van Nostrand, New York, 1949

This is a comprehensive study of fundamental mathetical basis of earth conduction effects. Chapters 2 and 3, which bear directly on the subject of the present guide, are summarized very briefly below. Other chapters are listed by title only; however, in many cases these contain valuable information on closely related problems.

Basic Electromagnetic Concepts and Equations; Earth Resistivity Testing and Analysis. Describes and develops general equations for the four-electrode method normally used; develops equation for the simplified case with electrodes at surface of uniform earth, also for electrodes in infinite medium. Develops equations for two-layer stratification and three-layer stratification. Includes curves plotted for the functions appearing in these equations, to simplify use of the latter. Discusses "arbitrary stratification" into n layers of different resistivities; also case of exponential variation of resistivity with depth. Shows how actual conditions can be approximated by a two-layer, or if desired, by a more complicated theoretical stratification, through comparison of computed data for the latter with actual measurements at various electrode spacings.

Techniques and instrumentation for earth resistivity measurement discussed in detail. Use of double commutating dc instrument to eliminate effect of stray currents and give direct reading in ohms. Often desirable to use sufficient range of probe spacings so that fairly accurate estimate for still greater spacings may be made by extrapolation.

Resistivity discussed in relation to geology. Typical resistivity values given for material of different geological periods and formations.

Resistance of Grounding Arrangements. Discusses resistances of various types of electrodes and combinations thereof, including spheres, round plates, vertical and horizontal rods or cables, radial wires, and combinations of wires and ground rods. Discusses equivalent radius of flat conductors. Discusses effect of variation of earth resistivity with depth. Includes curve showing effect of variation of resistivity with depth on resistance of horizontal ground 150 m long, and of 0.4 cm diameter buried at surface. Discusses effect of salt treatment.

Discusses earth potential near grounds with following conclusion:

For a vertical ground rod or pipe of length L the earth potential at the surface at a distance of y from the rod equals

$$U = \frac{1}{2\pi L} \log \frac{2L}{y} \quad \text{when } y < L$$

$$= \frac{1}{2\pi y} \quad \text{when } y > L$$

For a horizontal wire length L buried at a depth d, the potential at a point on the surface at the horizontal separation y from the wire and at a longitudinal distance x from a line through the midpoint of the wire is given by

$$U(x, y) = \frac{I\rho}{2\pi L}$$

$$= \log \frac{\left[(x + L/2)^2 + y^2 + d^2\right]^{1/2} + x + L/2}{\left[(x - L/2)^2 + y^2 + d^2\right]^{1/2} + x - L/2}$$

$$U(0,y) = \frac{I\rho}{\pi L} \log \frac{\left[L/2)^2 + y^2 + d^2\right]^{1/2} + L/2}{(y^2 + d^2)^{1/2}}$$

$$U(0,0) = \frac{I\rho}{\pi L} \log \frac{[L/2)^2 + d^2]^{1/2} + L/2}{d}$$

$$\cong \frac{I\rho}{\pi L} \log \frac{L}{d} \quad \text{when } d < L$$

$$U(L/2,0) \cong \frac{I\rho}{2\pi L} \log \frac{2L}{d} \quad \text{when } d < L$$

The maximum earth potential equals $U(0,0)$.

For a number of rods of wires, the resultant potential is obtained by adding the potential due to each as figured in above, using appropriate values for the current in each rod.

Equations are developed on the effects of heating of the soil.

Theory and techniques of ground resistance measurements are discussed in detail.

Mutual Impedance of Insulated Earth Return Conductors. Subject discussed with special regard to inductive interference. Effects of high- and low-frequency and transient currents considered for various conductor locations. Equations and curves developed.

Propogation Characteristics of Earth Return Conductors.

Direct-Current Earth Conduction and Corrosion Protection.

Power System Earth Conduction and Inductive Interference.

Lightning Protection of Cable and Transmission Lines. The bibliography included in this book contains eighteen references on the subject of "earth resistivity testing and analysis" that should be helpful to anyone wishing a more thorough study of this subject. Additional references on "resistance of grounding arrangements" largely duplicate those listed in the present guide. Extensive reference lists are also provided in connection with other chapters.

BODIER, G., *La Securite des Personnes et la Question des Mises a la Terre dans les Postes de Distribution (Personal Safety and Distribution Substation Grounding)*, France, Bulletin de la Societé Francaise des Electriciens, 6th series, volume VII, no 74, pp 545-562, Oct 1947, Discussion: 6th series, vol VIII, no 81, pp 255-278, May 1948.

When low-resistance grounds are impractical, local gradients must be modified.

Danger of fatality exists above some current value that lies between 20 and 100 m. Computation of body current in terms of resistances of body and ground beneath feet. Latter computed by assuming foot equivalent to circular plate of 7 cm radius.

Body resistance varies between 600 to 800 Ω (bare feet on damp soil) and 5000 to 10 000 Ω (dry shoes on dry ground); normal value order of 2000 Ω for fairly good contact. From two hands to two feet: with damp skin, 1800 Ω; between the two feet (dry skin), 2300 Ω; with nailed shoes and damp feet, 6500 Ω. Assumed for calculations, 2000 Ω, though with skin punctures resistance could be reduced to 500 Ω.

Tables show, for thirteen types of soil, wet and dry resistivity and computed permissible values of step and touch potential.

Touch voltages can be suppressed around a metal structure by a metallic platform connected to it. However, if short circuit is great enough and resistance of part of structure shunted by body is high enough, there may be danger. This can be corrected if necessary.

Few studies made of the problem of safety from impulse currents. Duration very short (order of 100 μs); physiological effects much different, and often limited to minor disturbances. Calculations for body current passing through the body differ because of large capacitative effects. Probably sufficient to take account only of commercial frequencies, if order of magnitude of impulse current is not too much greater

Effect of earth layers of different resistivity. If the resistance of the ground system as a whole, R, is known, the mean resistivity ρ_m can be determined from the equation $\rho_m = 2\,\pi\,r\,R$, where r is radius of equivalent hemisphere electrode. The value of ρ_m can also be calculated if depth and resistivity of layers are known.

While it is sometimes assumed that the top-layer gradients will be be appreciably modified from the values computed from the mean resistivity, experience shows that these can vary considerably with heterogeneous soil. Equation developed for step potential around hemispherical electrode in terms of surface and mean resistivity. Family of curves given, showing the ratio of total ground current to shock current, plotted against ratio of surface to average resistivity. (Curve for each of six different distances from the electrode and for body resistances of 500 Ω and 2000 Ω). These show safety notably improved when surface resistance is above the average (for example, ground covered with gravel, asphalt, or *dry* concrete). Worse for conditions where surface resistivity is less than average (for example, fertilized country fields).

Curves cannot be used quantitatively for other electrode types or systems, or where limited horizontal spread of a layer might affect situation.

Ways to reduce danger: (1) increase area of ground system, (2), provide metallic mats with very shallow burial at special points such as switch handles, (3) insulate ground lead for some distance into ground, (4) provide barriers (preferably insulating material) around ground lead to block access to area where gradient is high.

Suggests use of "simulated person" to determine danger potentials by actual test (see [27] by same author).

Relative merits of separate and common grounds (common ground safer). Discusses ground resistance limitations for systems ungrounded, solidly grounded, and grounded through an inductance.

Ground cannot be called "good" or "bad" unless one knows the maximum ground current, soil resistivity at surface and in depth, the arrangements of electrodes, depth of burial, and type of system grounding.

Safety best achieved in outdoor station by sufficiently close mesh buried grid; in indoor station by ground bus around inside of building connected to many electrodes spaced around outside. Insulating barriers and coarse gravel surfaces are effective aids.

The very extensive discussion is concerned mainly with values of body resistance and tolerable shock currents and with the advisability or otherwise of connecting low-tension neutrals to the general station ground under various circumstances.

Appendix H
Fundamental Considerations on Ground Currents

Reinhold Rüdenberg
(*Electrical Engineering*, January 1945)

In this article are considered the main scientific problems related to the conduction of electric currents to the earth and through the ground. While the overall performance of an electric system in which one or more points are grounded depends on the behavior of the wire network and the ground, this article is confined to the underground flow of the electric currents, which is not so easily measurable and therefore not so well understood as the behavior of the currents in the overhead network.

The earth is a body of three dimensions and therefore the beautiful simplicity of the linear wires by which electric currents usually are directed is lost. In the earth the currents spread out in the entire space, and it is necessary to follow their paths in order to analyze their performance in the underground.

The earth has been used as a conductor for electric currents since the beginning of engineering. However, after a brief period of preference for sending return currents through the ground, great difficulties and hazards were found from this in all branches of electrical engineering. Nowadays the earth is used mainly for fixing the neutral point of the electric system, and in many instances the inclusion of the earth as part of the circuit cannot be avoided.

Resistance of the Soil

Geological Strata. The electric conductivity of the materials constituting the earth's surface is very low compared with the high conductivity of the metals. Two main constituents of the earth, silicon oxide and aluminum oxide, actually are excellent insulators, and the conductivity of the ground is due in large measure to salts and moisture embedded between these insulators. On the other hand, even a semiconductor may carry a considerable amount of current if only the cross section is large enough, and in this respect the earth, by its great depth, presents no limitations.

Because of the high resistivity, all currents flowing through the ground suffer a considerable voltage drop. Therefore we must break with the popular concept that the potential of the ground is always zero. A substantial electric-field strength, or potential gradient, can develop and this may affect extended regions of the earth's surface. We have to distinguish between two different zones which will be considered in detail, namely, the spaces in proximity to the ground electrodes and the long path between such electrodes.

The ground under the surface of the earth is by no means homogeneous, and this makes a rigorous analysis of the distribution of currents very difficult if not impossible. Figure 1 represents a typical geological cross-section of the underground indicating the great heterogeneity of the conducting body through which electric currents may pass. Random changes caused by weather and seasons, by rain, frost, and by other temperature variations, with their influence on the conductivity of the soil, increase the difficulty of analysis. It would not be wise, therefore, to develop here the laborious considerations for a very high degree of accuracy. However, a quantitative analysis of the electric phenomena in the ground is necessary in order to make numerical calculations and to draw definite conclusions if the electric system from time to time shall not experience technical trouble or cause serious accidents.

Variation of Earth Resistivity. The resistivity of the soil depends on many parameters. It depends on the type of soil and therefore varies with distance as well as with depth, as indicated by Figure 1. Further, the resistivity is much smaller below the subsoil water level than above it. In frozen soil, as in the surface layer in winter, it is particularly high. It is important, therefore, for good ground electrodes to avoid the frost limit and to reach the subsoil water plane. Figure 2 shows the variation of resistivity ρ of a certain soil as measured (a) versus moisture content, (b) versus temperature change, and (c) versus added salt in percent of weight. It is seen by the logarithmic scales used that these three parameters affect the earth resistivity by several powers of ten, thus very large variations may occur.

Attempts have been made to correlate the resistivity with the geological age of the strata. Although older material in general has higher resistivity, many exceptions were found. In the actual strata under the surface, loam, clay, and limestone usually have lower resistivity; sandy and rocky materials have higher resistivity. While for power networks and for radio waves the resistivity near the surface is more important, for telephone interference by action at a distance the resistivity of the deeper layers has greater significance.

Spherical Ground Electrodes

Current Paths and Grounding Resistance. There will be considered first the simpler electrode elements and later some of the more complex combinations used in practice. For quantitative calculations the meter-kilogram-second system will be used which leads directly to the practical electric units.

The simplest possible electrode is shown in Figure 3, namely, a sphere in the ground which is symmetrical in all directions. It may be entirely embedded in the ground, or only the lower hemisphere embedded in the half space of ground under the surface plane of the earth, a case which will be considered. If a current I flows through this electrode, spreading out radially in the ground, the current density at distance x from the center of the hemisphere is

$$i = \frac{I}{2\pi x^2} \tag{1}$$

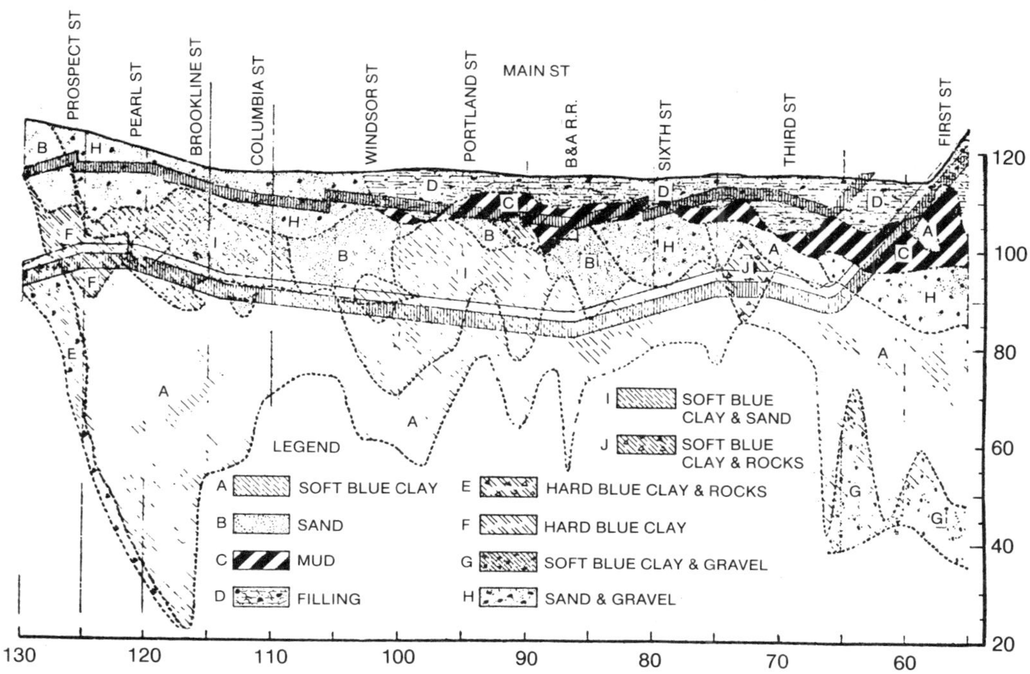

Fig 1
Subsoil Strata under Main Street, Cambridge, Massachusetts

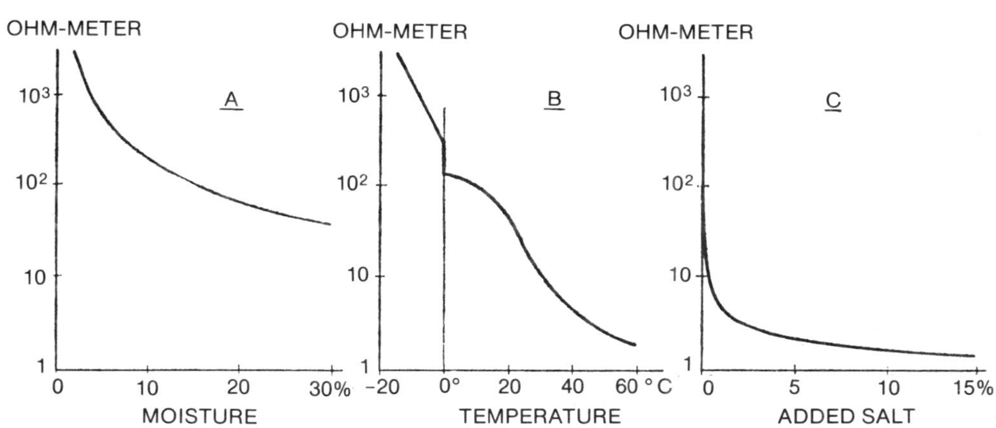

Fig 2
Resistivity of the Earth Depending on Moisture, Temperature and Salt Contents

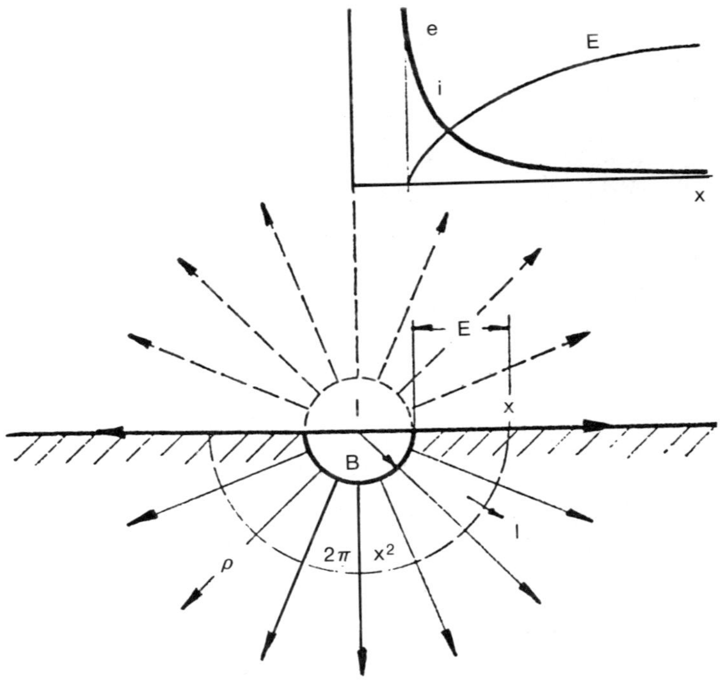

Fig 3
Radial Flow of Current from Spherical Electrode to Ground

According to Ohm's law such a current produces in the resistivity ρ of the soil an electric-field strength

$$e = \rho i = \frac{\rho I}{2\pi x^2} \tag{2}$$

The voltage, as line integral of the field strength from the surface of the conducting sphere of radius B to the distance x, is therefore

$$E = \int_B^x e dx = \frac{\rho I}{2\pi} \int_B^z \frac{dx}{\mathrm{x}^2} = \frac{\rho I}{2\pi} \left[\frac{1}{B} - \frac{1}{x} \right] \tag{3}$$

Current density, field strength, and voltage, in their dependence on distance, are represented graphically near the top of Figure 3.

The total voltage between the spherical electrode and a far distant point with $x = \infty$ is, according to equation 3,

$$E = \frac{\rho I}{2\pi B} \tag{4}$$

and therefore the resistance experienced by the streamlines of current diverging from the hemisphere is

$$R = \frac{E}{I} = \frac{\rho}{2\pi B} \tag{5}$$

Any additional resistance due to incomplete contact of the electrode and the ground occurs only rarely in practice.

If we consider, for a numerical example, a hemispherical electrode of radius B = 1 meter embedded at the surface of the earth of resistivity $\rho = 10^2$ ohm-meter, the ground resistance of this electrode will be

$$R = \frac{10^2 \text{ ohm-m}}{2\pi \cdot 1\,\text{m}} = 16 \text{ ohms}$$

As the curves in Figure 3 show, this resistance is distributed over the entire half space; however, the major part of it is concentrated in proximity to the electrode.

If two such electrodes are used through which the current enters and leaves the ground, the streamlines of the current develop as superposition of two hemispherical current distributions. This is shown in Figure 4a under the assumption that the resistivity of the soil is uniform in space. If a voltage of 110 volts is impressed between the electrodes, one having a diameter of 10 meters and the other of 1 meter, and separated by a distance of 1,000 meters, the streamlines and equipotential lines develop as shown in Figure 4a, and concentrate mainly around the smaller electrode.

If, on the other hand, a more highly conducting layer should exist at the surface of the earth, for example after a heavy rain, then the streamlines do not develop in depth but remain near the surface. Their shape within this layer and the distribution of potential are shown in Figure 4b for electrodes of the same dimensions and distance apart as before. Here the voltage does not concentrate so heavily near the electrodes but spreads out over larger areas of the surface.

Field Strength on Surface. If a man or an animal is walking through such a surface field of electric potention, for example in the neighborhood of a faulty transmission tower, as shown in Figure 5, the body diverts some current from the earth and may suffer damage from the potential gradient. The tower footing carrying a current I to the ground can be represented for the present purposes by an equivalent hemisphere of radius B, as shown by the dotted line in Figure 5. The field strength on the surface of the ground is then given by equation 2 and increases toward the tower to a maximum value at distance B

$$e_B = \frac{\rho I}{2\pi B^2} \tag{6}$$

With the figures of the former example and with a tower current to ground I = 100 amperes, the maximum field strength at the surface is

$$e = \frac{10^2 \text{ ohm-m} \cdot 100 \text{ amp}}{2\pi \cdot (1\,\text{m})^2} = 1{,}600 \text{ volts per meter}$$

a remarkably high value.

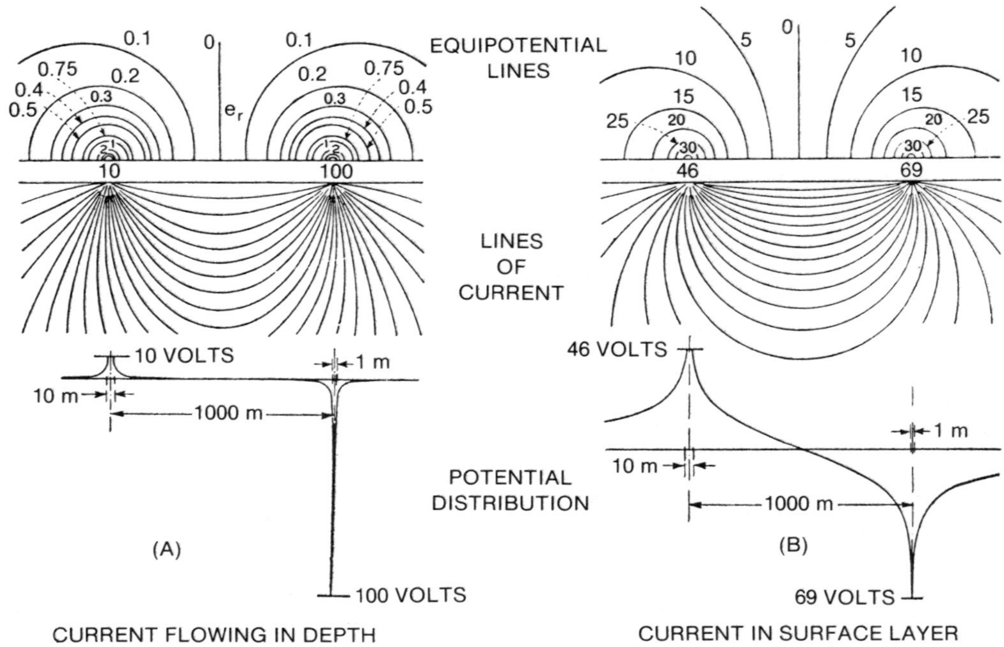

Fig 4
Flow of Current Between Two Ground Electrodes

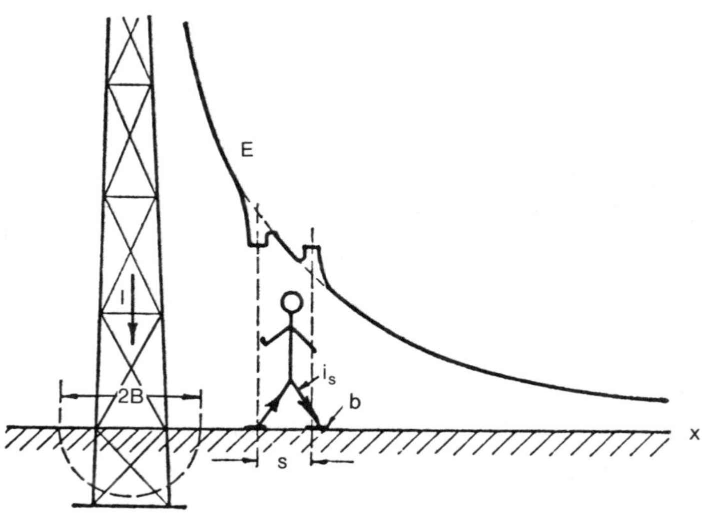

Fig 5
Potential at the Surface of the Earth Near a Current-Carrying Electrode

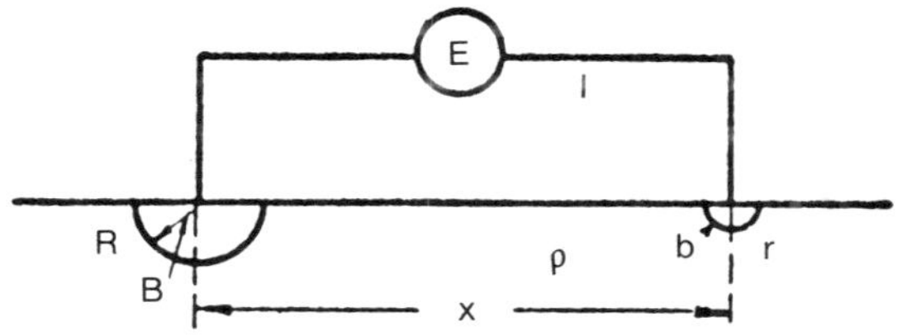

Fig 6
Measurement of the Resistivity of the Soil

This field strength produces a current in any living creature stepping over the surface. The maximum possible current develops if the internal body resistance is small compared with the foot resistance on the ground, which is

$$r = \frac{\rho}{2\pi b} \tag{7}$$

if b signifies the equivalent radius of the creature's foot. For a man's foot this has been measured in an electrolytic trough as b = 7 centimeters on the average. Since the voltage drop by the body current through the resistance $2r$ due to both feet is given by the voltage taken over the step length s

$$2ri_s = es \tag{8}$$

and therefore the maximum body current becomes

$$i_s = \frac{es}{2r} = \frac{\pi bes}{\rho} = \frac{sb}{2x^2}\ I \tag{9}$$

if expressed in terms of the tower current I.

With the numerical values already used, there is obtained for a man taking a step one meter long at a distance of three meters from the center of the tower, a body current

$$i_s = \frac{1\,\text{m} \cdot 0.07\,\text{m} \cdot 100\ \text{amp}}{2 \cdot (3\,\text{m})^2} = 0.39\ \text{ampere}$$

and this maximum current is certainly deadly.

Measurements by Probes. Measurements of ground resistance require some consideration. Care should be taken that probes or auxiliary electrodes, as in Figure 6, are sufficiently far removed from the spreading-out zone of the main current, if the probes are to measure the entire resistance experienced by the current. The voltage E will be independent of the distance x between the main and auxiliary electrodes only if x is much greater than both b and B.

By application of equation 3 to both systems of current distribution of Figure 6 around the hemispheres of radii B and b, superposition gives the voltage between the two electrodes. Thus the total resistance is

$$\frac{E}{I} = \frac{\rho}{2\pi}\left[\frac{1}{B} + \frac{1}{b} - \frac{2}{x}\right] \tag{10}$$

which is smaller than the sum of the resistances R and r proper of the two electrodes. By equation 10, ρ can be determined from the measured quotient of voltage and current if the dimensions are known and if ρ is constant over the ground. If this latter cannot be assumed, the resistivity around the smaller electrode nevertheless can be measured if its radius b is made so small that it is much less than both B and x, so that not only the last term but also the first term in the bracket of equation 10 becomes negligible. This gives the resistivity

$$\rho = 2\pi b\,\frac{E}{I} \tag{11}$$

as determined only by the quotient of voltage and current and the linear dimensions b of the probe. In this way numerous measurements have been performed showing that there are very great differences in the resistivity of the natural soil.

If no specific measurements for a definite spot of ground are made, the figures of Table I may be used as average resistivity over the underground.

Two Parallel Ground Electrodes. If two electrodes are arranged mutually within their spaces of high field strength, as in Figure 7, they experience mutual influence. For each of the spheres in Figure 7 the voltage from the sphere of radius b through distance y is, by equation 3,

$$E = \frac{\rho I}{4\pi}\left[\frac{1}{b} - \frac{1}{y}\right] = Vb - V \tag{12}$$

Since the voltage is determined by the difference in the bracket, which depends only on distances, the voltage can be expressed by the difference of two potentials, as on the right-hand side of equation 12. The potential V of a spherical source of current in space is therefore

$$V = \frac{\rho I}{4\pi y} \tag{13}$$

and such potentials always superpose if more than one source is present.

In the center plane between the two spheres of Figure 7 the potential is therefore

$$V_x = 2\,\frac{\rho I}{4\pi y} \tag{14}$$

if both spheres carry equal currents I to the ground. On the other hand, at the surface of either sphere the self-potential is determined by the radius b, the mutual potential by the average distance $2s$ between the centers of the spheres. The total potential is therefore

$$V = E = \frac{\rho}{4\pi}\left(\frac{I}{b} + \frac{I}{2z}\right) = \frac{\rho I}{4\pi b}\left(1 + \frac{b}{2z}\right) \tag{15}$$

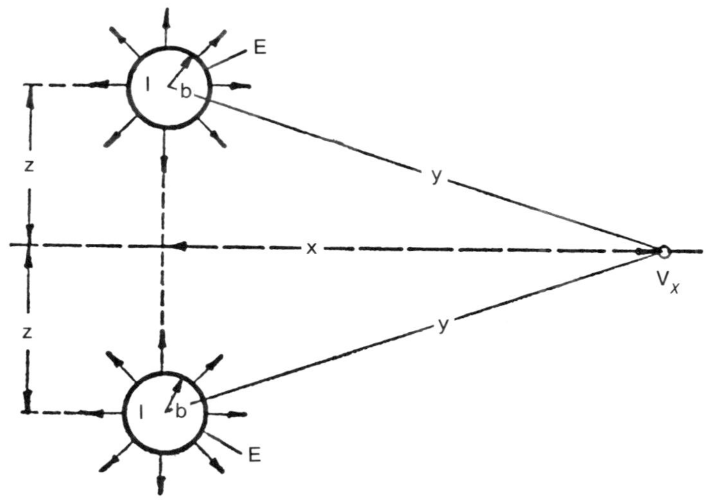

Fig 7
Two Spherical Ground Electrodes in Proximity

Hence the resistance of two spheres in parallel in the ground, carrying together the current $2I$, is

$$R = \frac{E}{2I} = \frac{\rho}{4\pi b} \cdot \frac{1}{2}\left(1 + \frac{b}{2z}\right) \tag{16}$$

The first quotient, by comparison with equation 5, gives the ground resistance of one sphere. The last factor for far-distant electrodes with $z = \infty$ becomes ½; but for close electrodes, for example, with $z = b$ becomes ½ (1 + ½) = ¾. Thus distant spheres are mutually independent in their resistances and their parallel resistance can be computed according to ordinary rules. Close electrodes, however, experience an increase of their ground resistance by mutual interference.

Depth Electrode. Figure 8 shows a spherical electrode buried in depth. Considering the arrangement of Figure 8 as half that of Figure 7, we obtain the resistance of such a depth electrode as twice that of equation 16, namely,

$$R = \frac{E}{I} = \frac{\rho}{4\pi b}\left(1 + \frac{b}{2z}\right) \tag{17}$$

Near the surface, for example at $z = b/2$, we have $\left(1 + \frac{b}{2b/2}\right) = 2$, and the resistance becomes the same as with equation 5. Under the surface the resistance decreases and in greater depth the correction term $b/2s$ vanishes, thus cutting the resistance in half.

With a depth electrode the space of high current concentration is not accessible to creatures, thus the danger of stepping near the electrode decreases. At any

Table I
Average Resistivity of the Ground

Type of Ground	Resistivity ρ, ohm-meters
Wet organic soil	10
Moist soil	10^2
Dry soil	10^3
Bed rock	10^4

point of the surface, as seen by Figure 8,

$$\frac{x}{y} = \sin \alpha = \frac{dy}{dx} \tag{18}$$

where x is the distance from the axis of the depth electrode and α the corresponding central angle. Thus the field strength at the surface is

$$e = \frac{dVz}{dx} = \frac{dVz}{dy} \sin \alpha \tag{19}$$

Substituting the potential Vz from equation 14, we obtain the field strength at the surface

$$e = \frac{\rho}{2\pi} \frac{d\,(1/y)}{dy} \sin \alpha = \frac{\rho I}{2\pi} \frac{\sin \alpha}{y^2} \tag{20}$$

Now, for α near 90 degrees, $\sin \alpha / y^2 = 1/x^2$; while for small values of α, $\sin \alpha / y^2 = x/z^3$. Therefore the field strength at large distances decreases exactly like that around a hemisphere, while vertically over the depth electrode the field strength becomes zero and rises linearly for small values of x. At an angle determined by $\tan \alpha = 1/\sqrt{2}$, the field strength is a maximum, namely,

$$e = \frac{\rho I}{3\sqrt{3}\pi z^2} \tag{21}$$

The ratio of this value to the maximum value of equation 6 for a surface electrode is

$$\frac{2}{3\sqrt{3}} \left(\frac{b}{z}\right)^2 = 0.39 \left(\frac{b}{z}\right)^2 \tag{22}$$

which shows that a depth electrode, for sufficiently large depth z, minimizes the danger for stepping creatures.

Rod and Wire Electrodes

Driven Rod at Earth Surface. Spherical or hemispherical electrodes are not convenient for actual use. In practice rod or wire electrodes, having a relatively small cross section compared with the length, are preferred. We may subdivide with good approximation such a rod electrode driven into the ground, as in

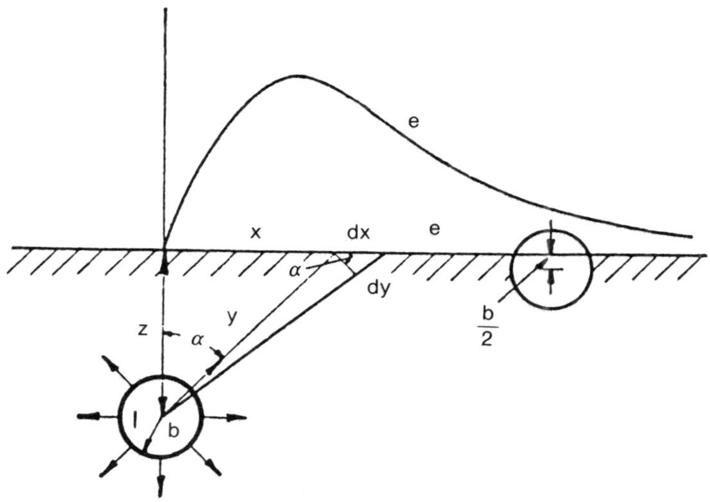

Fig 8
Field Strength on the Surface Over a Depth Electrode

Figure 9, into a large number n of nearly spherical elements, which over the length l of the rod in the ground have a mutual distance

$$dl = \frac{l}{n} \tag{23}$$

each feeding a current I/n into the ground. If y is the distance from any element to a point at the surface of the earth and α is the angle from y to the axis of the rod, the small diagram in Figure 9 shows that

$$\sin \alpha = \frac{y d\alpha}{dl} \tag{24}$$

The potential dV of every element is given by equation 13 with current I/n. By substitution of the distance y from equation 24, and use of equation 23, the incremental potential at the surface of the earth becomes

$$dV = \frac{\rho I/n}{4\pi y} = \frac{\rho I}{4\pi n} \; \frac{n}{l} \; \frac{d\alpha}{\sin \alpha} \tag{25}$$

If the limiting value of angle α is denoted by β, as shown in Figure 9, the potential in the center plane of a rod of length $2l$, containing $2n$ spheres and thus including the fictitious image above the ground, is given by the integral of dV from $+\beta$ to $-\beta$,

$$V = \frac{\rho I}{4\pi l} \int_{+B}^{-B} \frac{d\alpha}{\sin \alpha} = \frac{\rho I}{2\pi l} \log_e \left(\cot \frac{\beta}{2} \right) \tag{26}$$

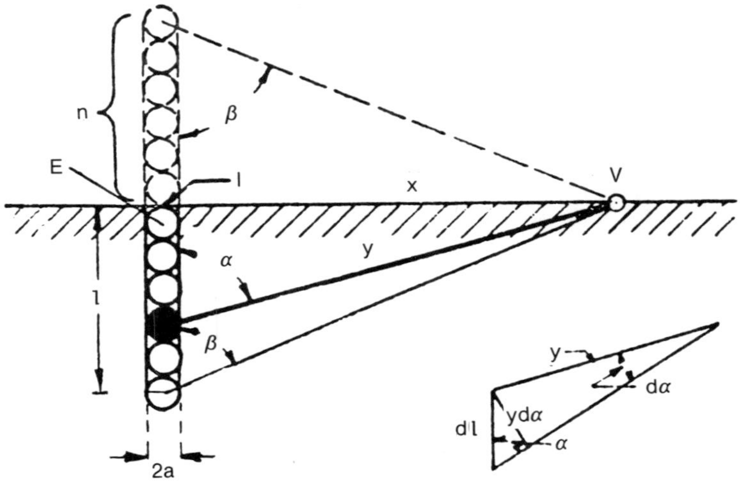

Fig 9
Development of the Potential Around a Rod Electrode

This result would be rigorous if the density of the current emerging from the rod into the earth were uniform over the length, which actually is only a fair approximation.

The electric potential in the symmetry plane of the driven rod thus is dependent on only four parameters, namely: resistivity ρ of the ground, current I flowing into the rod, its length l within the ground, three data which are always given for a definite ground electrode, and the angle of vision β between the axis of the rod and the distance from the bottom of the rod to the point under consideration at the surface. For any point in this center plane β is the only variable parameter.

For large distance x of the point considered from the axis of the rod, the logarithm in equation 26 simplifies to

$$\log_e \left(\frac{1 + \cos \beta}{\sin \beta} \right) \approx \cos \beta \approx \frac{l}{x} \tag{27}$$

and the potential is

$$V_\infty = \frac{\rho I}{2 \pi x} \tag{28}$$

which is identical with that for a hemisphere.

On the other hand, for the surface of the rod, where the potential V is identical with the voltage E of the electrode, Figure 10 shows that for small radius a as compared to length l

$$\cot \frac{\beta}{2} \approx \frac{2l}{a} \tag{29}$$

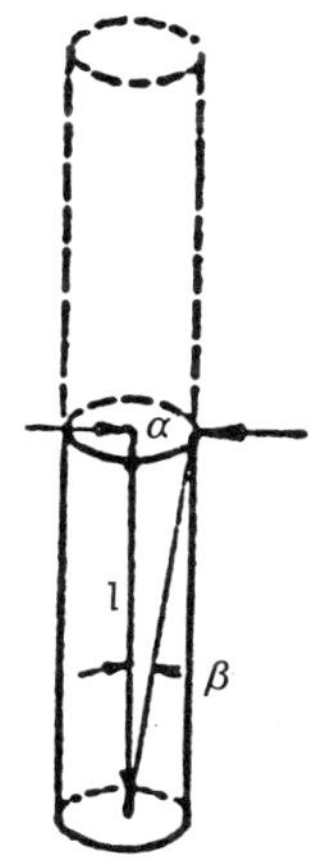

Fig 10
Dimensions of a Driven Rod

Hence the resistance as quotient of voltage and current is

$$R = \frac{E}{I} = \frac{\rho}{2\pi l} \log_e \left(\frac{2l}{a}\right) \tag{30}$$

While the shape of the rod, determining the ration l/a, is of minor significance since it forms only the argument of a logarithm, the length l of the rod is of major importance, for the ground resistance is nearly inversely proportional to the length.

A rod of radius a = 2.5 centimeters and of length l = 6 meters, dimensions often used in practice, driven into moist soil will present a resistance

$$R = \frac{10^2 \text{ ohm-m}}{2\pi \cdot 6\text{m}} \log_e \left(\frac{2 \cdot 6}{2.5 \cdot 10^{-2}}\right) = 2.65 \cdot \log_e 480 = 16 \text{ ohms}$$

This value equals that for a hemisphere two meters in diameter.

Field Strength Around Rod Electrode. On the surface of the ground there develops, according to Figure 11, a field strength

$$e = \frac{dV}{dx} = \frac{dV}{d\beta}\frac{d\beta}{dx} \tag{31}$$

where the angle β is introduced as the variable. It is correlated to x by

$$\left.\begin{aligned} \tan\beta &= \frac{x}{l} \\ \frac{dx}{dB} &= \frac{l}{\cos^2\beta} \end{aligned}\right\} \tag{32}$$

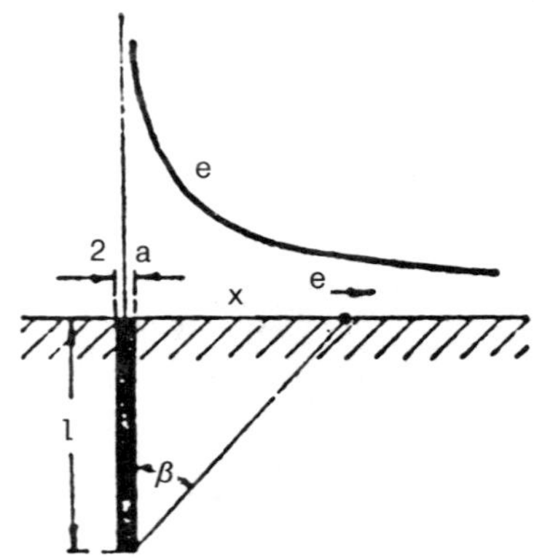

Fig 11
Field Strength on the Surface Around a Driven Rod

where the last expression is merely the derivative of the first. Differentiation of equation 26 with respect to β and substitution of equation 32 into equation 31 gives the field strength

$$e = \frac{\rho I}{4\pi l} \frac{2}{\sin\beta} \frac{\cos^2\beta}{l} = \frac{\rho I}{2\pi l} \frac{\cos\beta}{x} \tag{33}$$

which is decreasing with both increased x and β.

For large distance x, $\cos\beta = l/x$, approximately, and the field strength

$$e_x = \frac{\rho I}{2\pi x^2} \tag{34}$$

becomes identical with that for a hemisphere. For small x, however, $\cos\beta \approx 1$, and the field strength

$$e_o = \frac{\rho I}{2\pi l x} \tag{35}$$

becomes much larger than for a hemisphere. This is due to the high concentration of the current around the small circumference of a rod.

Along the length of the rod the density of the current from rod to earth is nearly constant over the major part of the length, but actually increases at the bottom to about double the value at the center plane, due to the point effect of the end of the rod.

Simple Forms of Ground Electrodes. In a similar way the potential and the ground resistance of other forms of electrodes can be derived by superposition of spherical or cylindrical elements. Table II gives a survey of some simple forms of ground electrodes. We see that a flat strip follows a relation quite like that of a circular rod and a strip thus is equivalent to a rod or a wire of a diameter equal to half the width of the strip.

Table II
Resistance of Simple Forms of Ground Electrodes

	Sphere: $R = \frac{\rho}{4\pi B}$	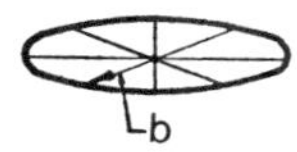	Disk: $R = \frac{\rho}{8b}$
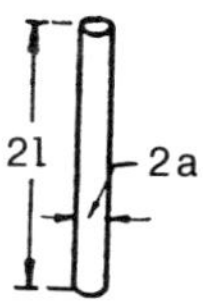	Rod: $R = \frac{\rho}{4\pi l} \log_\varepsilon \left(\frac{2l}{a}\right)$	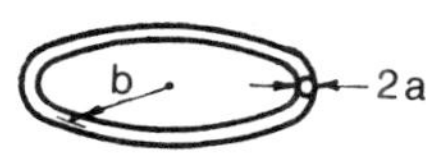	Ring: $R = \frac{\rho}{4\pi^2 b} \log_\varepsilon \left(\frac{8b}{a}\right)$
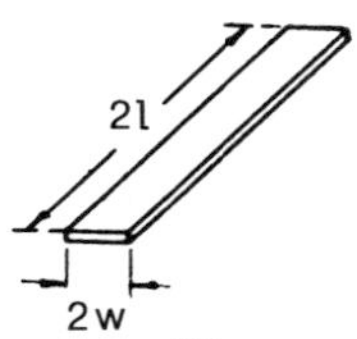	Strip: $R = \frac{\rho}{4\pi l} \log_\varepsilon \left(\frac{4l}{w}\right)$	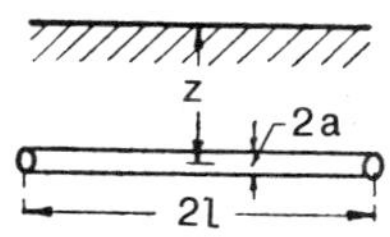	Deep wire: $R_2 = R\left[1 + \frac{\log_\varepsilon \left(\frac{l}{z}\right)}{\log_\varepsilon \left(\frac{2l}{a}\right)}\right]$
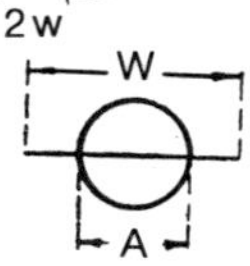	Equivalent rod and strip: $A = W/2$	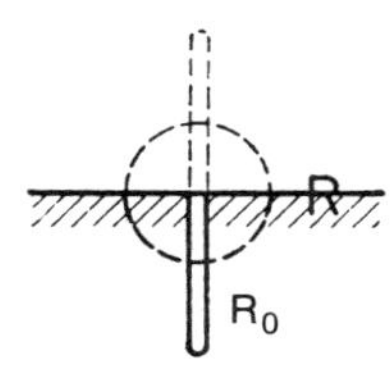	Surface electrode: $R_0 = 2R$

The resistance of a ring of wire, having a periphery $2\pi b$, is only slightly greater than that of a straight wire of length $2l = 2\pi b$, the increase being due to the absence of the ends of the rod with their higher current density.

The resistance of a ground electrode near the surface of the earth is always twice the resistance of that electrode in greater depth, because the distribution of the current through the upper semispace is cut off.

As all the equations show, the resistance is determined primarily by the largest dimension of the electrode and only to a minor extent by the smaller dimensions, like cross section or thickness. Thus the surface of the electrode is unimportant and only the linear extension matters.

The simplest method for determining the resistance of a complicated shape of ground electrode, as for example the tower footing of a high-voltage line, or of a radio antenna, often consists in measuring the resistance of a scale model of the structure in an electrolytic trough. Then the resistance of the actual device is lower in the ratio of the linear dimensions and higher in the ratio of the resistivities of the surrounding medium. The radius of the equivalent hemisphere is always a convenient means of comparison.

Four-Point-Star Electrode. For some structures frequently used which are composed of several rods or wires, it is convenient to derive analytical formulas. In a four-point-star electrode, as in Figure 12, the self-potential of every straight

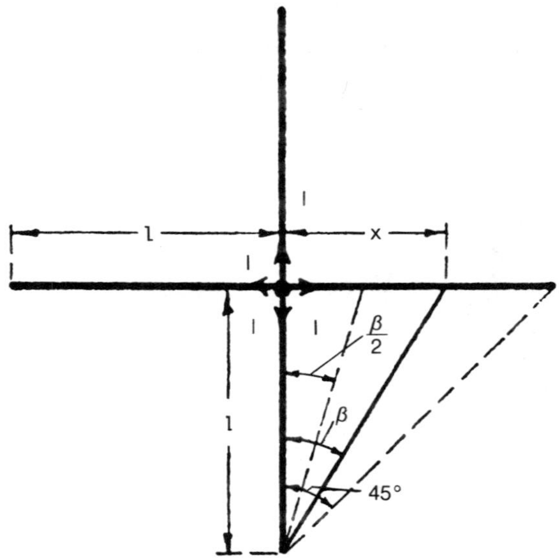

Fig 12
Four-point-star Electrode in the Ground

wire of length $2l$ is, according to equations 26 and 29,

$$V_o = \frac{\rho I}{2\pi l} \log_e \left(\frac{2l}{a} \right) \tag{36}$$

This is to be increased by the mutual potential between the transverse wires of the cross. Equation 26 gives the potential of a wire length $2l$ in its center plane. For angles $\beta \leqq 45$ degrees, as in Figure 12,

$$\cot \frac{\beta}{2} \leqq \frac{2l}{x} \tag{37}$$

The mean value of the logarithm of $2l/x$ over the length l of the wire is

$$\frac{1}{l} \int_o^l \log_e \left(\frac{2l}{x} \right) dx = -2 \int_\infty^2 \log_e \xi \cdot d \frac{1}{\xi} = -2 \int_\infty^2 \frac{\log_e \xi \cdot d\xi}{\xi^2}$$

$$= 2 \left[\frac{\log_e \xi}{\xi} + \frac{1}{\xi} \right]_\infty^2 = \log_e 2 + 1 \tag{38}$$

Thus the mean mutual potential between the crossed wires is

$$V_1 \gtreqless \frac{\rho I}{2\pi l} (\log_e 2 + 1) \tag{39}$$

The voltage E of every conductor is given by the sum of the potentials or

$$E = V_o + V_1 \approx \frac{\rho I}{2\pi l}\left[\log_e\left(\frac{2l}{a}\right) + \log_e 2 + 1\right] \tag{40}$$

and therefore the resistance to ground of the entire four-point-star electrode is

$$R = \frac{E}{4I} \approx \frac{\rho}{8\pi l}\left[\log_e\left(\frac{4l}{a}\right) + 1\right] \tag{41}$$

If such a star of wires is embedded near the surface of the earth at a depth small compared with the length dimension of the electrode, the ground resistance is twice that of equation 41 as indicated in the last example of Table II.

The value of the resistance of equation 41, by the addition of unity to the logarithm in the bracket, is slightly greater than for a straight wire of length $4l$. This increase is due to the mutual influence of the current distributions in the ground caused by neighboring parts of the entire electrode. With more star wires than four, such influence increases, preventing the resistance from being reduced in proportion to the increase of the number of wires. For many wires, finally forming a circular disk, the resistance approaches

$$R = \frac{\rho}{8l} \tag{42}$$

For average values of the ratio radius to length of the wire, a comparison of equation 42 with equation 41 shows that by increasing the number of radial wires to infinity the resistance reduces to only

$$\frac{R^o}{R+} = \frac{\pi}{\log_e\left(\frac{4l}{a}\right) + 1} \approx \frac{\pi}{9.3} \approx \frac{1}{3} \tag{43}$$

of that of the simple four-wire star. The gain with increasing number of wires is not very great.

Multiple Rod Electrodes. Another combined electrode frequently used, namely, multiple driven rods connected in parallel, is shown in Figure 13. The potential at the earth's surface of every rod again is given by the general equation 26. The voltage at each of the electrodes is given by the sum of all the potentials produced by the rod considered and all the other rods. For rod number one, for example, the voltage is

$$E_1 = \frac{\rho}{2\pi l}\left[I_1 \log_e\left(\cot\frac{\beta_1}{2}\right) + I_2 \log_e\left(\cot\frac{\beta_2}{2}\right) + \ldots\right] \tag{44}$$

The currents of the rods are not yet determined but for the sake of simplicity the resistivity of the soil and the length of all the rods have been taken as uniform. As many equations of this type can be developed as there are rods in the electrode, the angles β signifying always the angles of vision from the bottom of each rod toward the top of the rod considered as seen by Figure 13. Therefore, for any

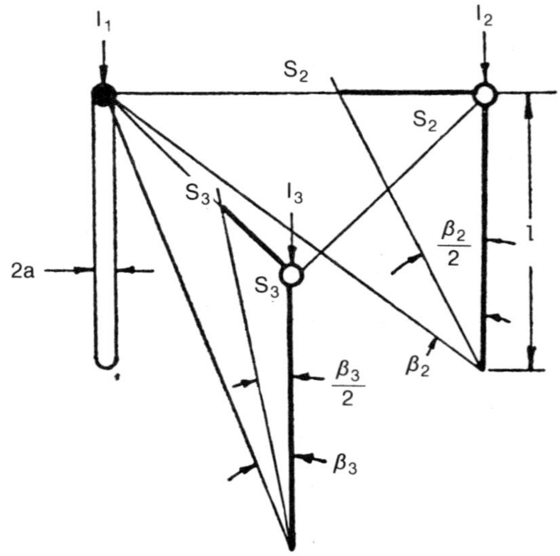

Fig 13
Driven Rods in Parallel

arrangement of driven rods a sufficient number of equations is obtained to determine the distribution of the currents among the individual rods of the electrode.

The first cotangent of equation 44 refers to the rod considered and can be evaluated by equation 29. The further cotangents, which refer to the other rods, can be expressed, as shown in Figure 13, by the ratio of the length l of the rod to a distance s_{n2} cut off by the angle $\beta/2$ on the distance S_n between the rods. Thus

$$\left.\begin{aligned} \cot\frac{\beta_1}{2} &= \frac{l}{a/2} \\ \cot\frac{\beta_n}{2} &= \frac{l}{s_n} \end{aligned}\right\} \tag{45}$$

If the rods are connected by zero resistance and are located symmetrically with respect to one another, the voltages and currents are

$$E_1 = E_2 = E_3 \ldots = E;\ I_1 = I_2 = I_3 \ldots = \frac{I}{n} \tag{46}$$

Therefore, with use of equation 45, all the equations 44 take the form

$$E = \frac{\rho}{2\pi l}\,\frac{I}{n}\left[\log_e\left(\frac{l}{a/2}\cdot\frac{l}{s_2}\cdot\frac{l}{s_2}\cdots\right)\right] \tag{47}$$

For each case equation 47 can be evaluated easily, but three significant examples may be considered in detail.

Table III
Geometric Mean Distances for Parallel-Rod Electrodes

n =	Arrangement	Geometric Mean Distance
2		$A = \sqrt{aS}$
3	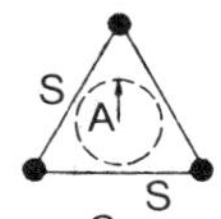	$A = \sqrt[3]{aS^2}$
4	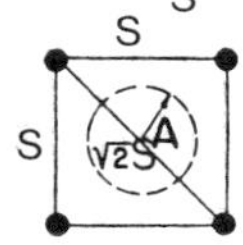	$A = \sqrt[4]{\sqrt{2}aS^3}$

For large ratio S/l, with the rods distant from one another, evidently $\beta_n/2 = 45$ degrees and $s_n = 1$. Thus all the quotients $l/s = 1$, and the combined resistance is

$$R = \frac{E}{I} = \frac{1}{n} \cdot \frac{\rho}{2\pi l} \log_e \left(\frac{2l}{a}\right) \tag{48}$$

In this case the ohmic value is reduced in inverse proportion to the number of parallel rods.

For small ratio S/l, with the rods close together, the angles β are small, and therefore always $s_n = S_n/2$, see Figure 13. Hence, if the number of rods n is placed under the logarithm, the resistance becomes

$$R = \frac{\rho}{2\pi l} \log_e \left(\frac{2l}{\sqrt[n]{aS_2S_3S_4 \ldots}}\right) = \frac{\rho}{2\pi l} \log_e \left(\frac{2l}{A}\right) \tag{49}$$

In the right-hand term we have expressed the nth root of the product of the distances between all of the electrodes and the first electrode including the radius of the first electrode, by the geometric mean distance

$$A = \sqrt[n]{aS_2S_3S_4 \ldots} \tag{50}$$

A comparison of equation 49 with equation 30 shows that the sum of several or of many driven rod electrodes in close proximity acts as if there were only one rod of radius A. In Table III A is evaluated for three simple examples. Since A is under the logarithm in equation 49, the total ground resistance of such closely spaced electrodes becomes diminished only slightly with increased number of rods.

For medium ratio $S/l = 1$, where the rods are driven at distances equal to their length, evidently $\beta_n = 45$ degrees, and thus cot $\beta_n/2 = 2.4$. For example, if $n = 3$ rods, the characteristic part of equation 47 becomes

$$\frac{1}{3} \log_e \left(\frac{l}{a/2} \cdot 2.4 \cdot 2.4\right)$$

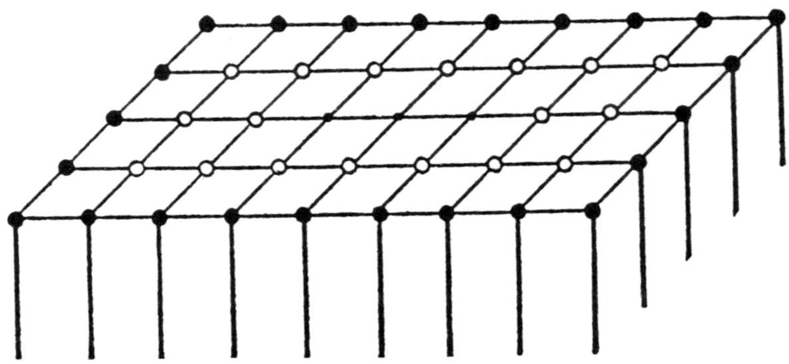

Fig 14
Extended Bed of Rod Electrodes

and for values of l = 6 meters, a = 2.5 centimeters, $\rho = 10^2$ ohm-meters as in the former examples, the resistance is

$$R = \frac{2.65}{3} \log_e (480 \cdot 2.4^2) = 7.0 \text{ ohms}$$

If the rods were far apart, $S = \infty$, the resistance would be $^{16}/_3$ = 5.3 ohms. Thus three rods with separation equal to their length experience a mutual influence which increases their resistance by 32 percent.

In order to obtain low ground resistance in bad soil of high resistivity, it is often necessary to arrange for quite a number of driven rods placed in lines or over an extended area. In such ground bends, as in Figure 14, the inner rods carry lower currents than the outer rods due to the mutual influence of the surrounding rods. In every case the application of the proper number of equations 44 gives the correct solution for the current distribution and the total ohmic value of the ground resistance. Even unequal lengths of the rods and different resistivity around the individual rods easily can be taken into account.

Depth Measurement. Frequently the resistivity ρ varies not only from rod to rod but also with the depth for each rod. In order to obtain a survey of such variation, it is useful to measure the ground resistance during the driving of the individual rods. The ground conductance of a rod, as reciprocal of the resistance, equation 30, is

$$G = \frac{2\pi l}{\rho \cdot \log_e \left(\frac{2l}{a}\right)} \tag{51}$$

As indicated in Figure 15, the increment of conductance with lengthening of the rod by an element dx is therefore

$$dG = \frac{2\pi \cdot dl}{\rho \cdot \log_e \left(\frac{2l}{a}\right)} \tag{52}$$

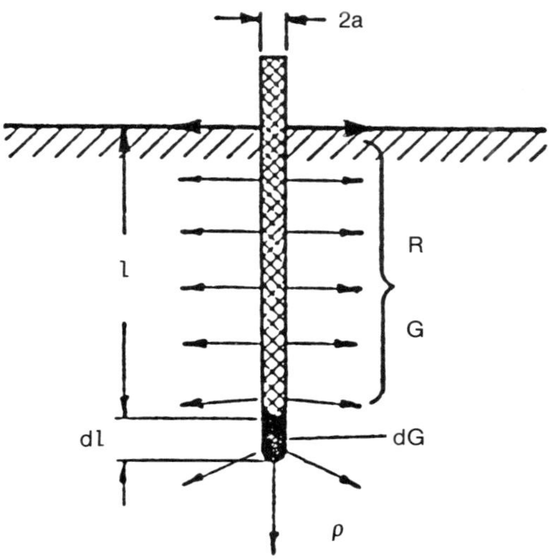

Fig 15
Increment of Ground Conductance with Length of a Rod

The logarithm is not differentiated since it was derived under the condition of constant resistivity and its variation is very small under any circumstances. During the depth measurement the actual resistivity at the lower end of the rod over an increment of length dx is, by equation 52,

$$\rho = \frac{2\pi}{\log_e\left(\frac{2l}{a}\right)} \cdot \frac{dl}{dG} \tag{53}$$

Thus ρ is inversely proportional to the derivative of the conductance which is shown as dashed curve in Figure 16b.

With respect to the resistance R,

$$\left.\begin{aligned} G &= \frac{1}{R} \\ dG &= -\frac{dR}{R^2} \end{aligned}\right\} \tag{54}$$

and thus the resistivity at the lower end, dependent on the derivative of the resistance curve in Figure 16b, is

$$\rho = \frac{2\pi}{\log_e\left(\frac{2l}{a}\right)} \cdot \frac{R^2}{-\frac{dR}{dl}} \tag{55}$$

Figure 16a indicates that a small derivative is measured through a stratum of high resistivity, a large derivative through a stratum of low resistivity.

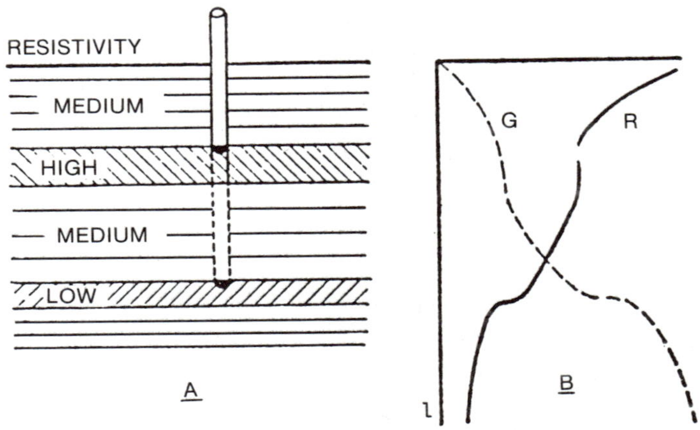

Fig 16
Measurement of Rod Distance Against Depth

If a rod of radius a = 2.5 centimeters is driven step by step into the depth and if at length l = 6 meters, the resistance is measured as R = 16 ohms and the derivative as dR/dl = -0.15 ohm per meter, the resistivity at that depth will be

$$\rho = \frac{2\pi}{\log_e(480)} \frac{16^2}{0.15} = 174 \text{ ohm-meters}$$

By such continuous measurements of resistance during the driving of a bed of rods, it is possible to obtain information about the resistance strata of the earth in which the ground electrodes have to function.

Heating of the Ground

If a ground electrode is to be loaded continuously, or even over a short time only, the temperature rise θ of the soil must be considered in order to avoid an overloading which might evaporate the moisture. The current density about a spherical electrode with radius B embedded in the ground varies over the distance x is

$$i = \frac{I}{4\pi x^2} \tag{56}$$

as is indicated in Figure 17. Thereby an amount of heat ρi^2 is produced in every element of volume, which may be considerable because of the high value of the resistivity ρ. This heat is partly stored in the volume elements of the ground which have an average specific heat $\gamma \approx 1.75 \cdot 10^6\ ws/m^3C$ (watt-seconds/meter^3 · degrees centigrade); partly conducted from higher to lower temperature within the ground, the average heat conductivity being $\lambda \approx 1.2\ w/mC$ (watts/meter · degrees centigrade). Although these two thermal constants of the soil are of major significance, their actual values seldom have been measured accurately.

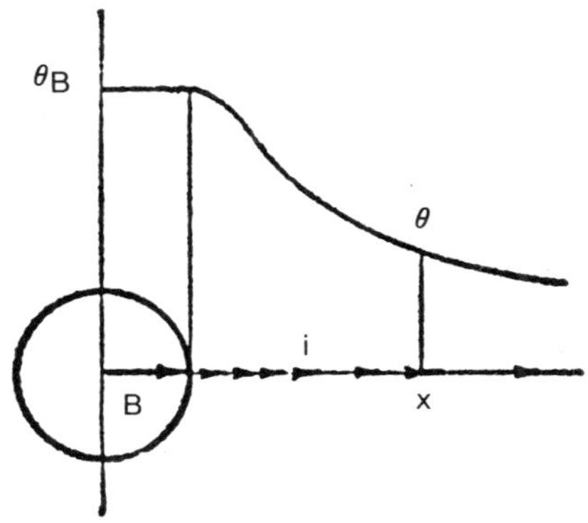

Fig 17
Temperature Distribution Around a Spherical Ground Electrode

The differential equation of the radial heat conduction about a sphere is

$$\gamma \frac{d\theta}{dt} - \frac{\lambda}{x} \frac{d^2(x\theta)}{dx^2} = \rho i^2 \tag{57}$$

It is difficult to obtain a general solution of this equation since i varies with x. However, we can discuss two interesting phases of the heating, giving particular integrals, namely the steady state and the short-time state.

Steady-State Heating. For continuous ground currents the time derivative in equation 57 vanishes, and the differential equation becomes

$$\lambda \frac{d^2(x\theta)}{dx^2} + \frac{\rho}{x^3}\left(\frac{I}{4\pi}\right)^2 = 0 \tag{58}$$

With two simple integrations the solution for the temperature distribution over distance x becomes

$$\theta = \frac{\rho}{\lambda}\left(\frac{I^2}{4\pi}\right) \cdot \frac{1}{x}\left(\frac{1}{B} - \frac{1}{2x}\right) \tag{59}$$

which is plotted against x in Figure 17. The maximum earth temperature at the electrode with $x = B$ is

$$\theta_B = \frac{1}{2} \frac{\rho}{\lambda}\left(\frac{I}{4\pi B}\right)^2 \tag{60}$$

and depends only on the two constants ρ and λ of the ground and on the linear current density I/B. The admissible current to ground from the spherical electrode is therefore

$$I = 4\pi B \sqrt{2 \frac{\lambda}{\rho} \theta} = \frac{1}{R} \sqrt{2\rho\lambda\theta} \tag{61}$$

where the right hand term is obtained by substituting the resistance $R = \rho/4\pi B$. The current-carrying capacity of the electrode is thus determined, in addition to the ground constants, by resistance and temperature rise θ only.

Since the heat flow and the current distribution around electrodes of any shape follow the same mathematical law, namely Laplace's differential equation, the righthand expression of equation 61 is valid not only for spherical electrodes but for any shape of electrode whether it be a rod, a ring, a disk, or a more complicated form. Therefore, if the resistance is known, which differs with the shape, the continuous current-carrying capacity can be determined easily.

The voltage at the ground electrode, measured from the metal to a far distant point is, from the last term of equation 61

$$E = IR = \sqrt{2\rho\lambda\theta} \tag{62}$$

Besides the ground constants E depends only on the temperature rise in the steady state. Conversely, the voltage applied to a ground electrode determines its steady-state temperature. These conclusions are true for any shape of electrode.

As an example, for $\theta = 60$ degrees centigrade in moist soil,

$$E = \sqrt{2 \cdot 10^2 \cdot 1.2 \cdot 60} = 120 \text{ volts}$$

This voltage must not be surpassed if that temperature rise is not to be exceeded.

The admissible current density at the surface of the ground electrode, where the highest temperature exists, determined for a spherical electrode from equation 61 is

$$i = \frac{I}{4\pi B^2} = \frac{1}{B}\sqrt{2\frac{\lambda}{\rho}\theta} \tag{63}$$

For the former example

$$i = \frac{1}{1\text{m}}\sqrt{2\frac{1.2}{10^2}60} = 1.2 \text{ amperes per square meter}$$

Thus the current density permitted for continuous loading of the electrode is fairly small. The soil surrounding the electrode must not heat up to 100 degrees centigrade, since the moisture would then evaporate completely, and the current would be interrupted by the enormously increased resistance.

Short-Time Heating. For short-time loading of the ground electrode the second member of equation 57, the heat-conduction term, plays only a minor role and thus may be neglected. The remaining differential equation then becomes

$$\frac{d\theta}{dt} = \frac{\rho}{\gamma} i^2 \tag{64}$$

The temperature rise thus follows the same law as it does with ordinary linear conductors. If resistivity ρ and specific heat γ are constant, the temperature rise is linear with time, and thus the admissible current density is

$$i = \sqrt{\frac{\gamma}{\rho}\,\frac{\theta}{t}} \tag{65}$$

Because the length dimension x does not appear in equation 64, this differential equation is true for every volume element of the soil, independent of the pattern of flow of the current. Hence equation 65 is valid for any shape of the ground electrode.

For example, an electrode in moist soil and for a temperature rise not exceeding 60 degrees centigrade is permitted to be loaded over a period of 100 seconds by a current density

$$i = \sqrt{\frac{1.75 \cdot 10^6}{10^2} \cdot \frac{60}{100}} = 100 \text{ amperes per square meter}$$

which is a fairly high value. For sandy soil with ten times higher resistivity a short-time loading at the same current density is permitted for ten seconds only. Higher current density or longer loading period would evaporate the moisture of the soil in a very short time, and an explosion of the space surrounding the electrode might follow, and is sometimes experienced.

The voltage at the electrode with short-time loading may attain a very high value. In the example for moist soil, the voltage is higher than in the steady-state example by the ratio of the current densities, and therefore it reaches here the value

$$E = 120 \cdot \frac{100}{1.2} = 10{,}000 \text{ volts}$$

The initial rise of temperature over a short time as given by equation 64 is shown in Figure 18. The final temperature rise, with fixed constants of the soil, would be given by equation 60, and also is indicated in Figure 18. The intermediate curve follows the complete differential equation 57 but is difficult to derive analytically. However, we can easily determine the time constant τ, defined as the time in which the linear initial rise would reach the steady-state temperature, see Figure 18. For this intersection we have, with use of equations 60 and 64

Fig 18
Change with Time of Ground Temperature

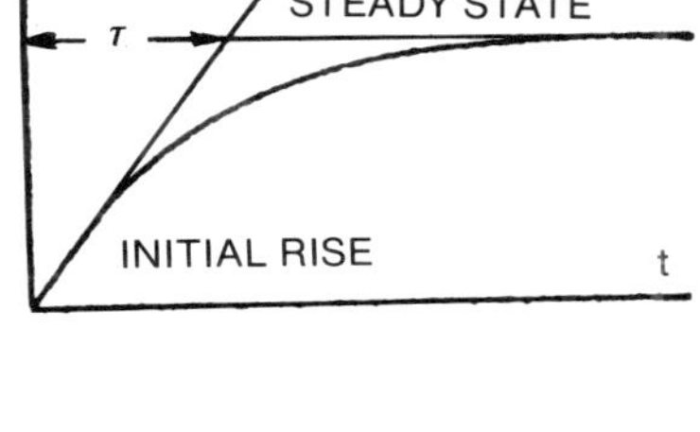

and substitution in the former the current density for the current

$$\theta_{\text{steady}} = \frac{1}{2}\frac{\rho}{\lambda} B^2 i^2 = \frac{\rho}{\gamma} i^2 \tau = \theta_{\text{rise}} \tag{66}$$

Hence the time constant of heating for a spherical ground electrode and its surrounding soil is

$$\tau = \frac{1}{2}\frac{\gamma}{\lambda} B^2 \tag{67}$$

In addition to the thermal constants of the soil, τ is dependent only on the square of the radius B. For other forms of electrodes the time-constant follows a similar form except that instead of B another characteristic dimension of the electrode must be used, being somewhere between the maximum and the minimum dimension of the electrode.

For example, for a sphere with radius B = 1 meter the time constant is

$$\tau = \frac{1}{2}\frac{1.75 \cdot 10^6}{1.2} 1^2 = 0.73 \cdot 10^6 \text{ sec} = 8.5 \text{ days}$$

This is a very long period, due to the very low heat conductivity of the soil, but it is completely in accord with test results.

Linearly Extended Fields

Buried Long Electrodes. In the foregoing discussions there have been considered only electrodes with constant voltage over their length. However, with the use of long electrodes such as wires or strips, cable sheaths, railroad or tramway tracks, pipe lines, or similar ground electrodes of considerable length, the current may suffer an ohmic drop due to the internal resistance r of the electrode before it spreads out in the resistance of R of the ground. Figure 19 shows how the total current I_0 enters a ground wire, follows its length, gradually being attenuated, and leaving the wire as spatial current i. There are measured now the wire current I in amperes, the ground current i in amperes per meter, the resistance r of the electrode in ohms per meter, and the resistance R of the ground in ohm-meters. The resistance R may vary slightly over the length of x of the wire due to the concentration of the ground-current density i near the end points. However, taking a mean value over the length is sufficiently accurate for this problem.

According to Kirchoff's laws, at any point x of the long wire the sum of the inflowing and outflowing current within the wire and the current flowing to ground is zero. On the other hand, the voltage over a narrow rectangle formed by a wire resistance element $r \cdot dx$ and the two adjacent ground resistances R is zero. Therefore

$$\left.\begin{aligned} i + \frac{dI}{dx} &= 0 \\ R\frac{di}{dx} + rI &= 0 \end{aligned}\right\} \tag{68}$$

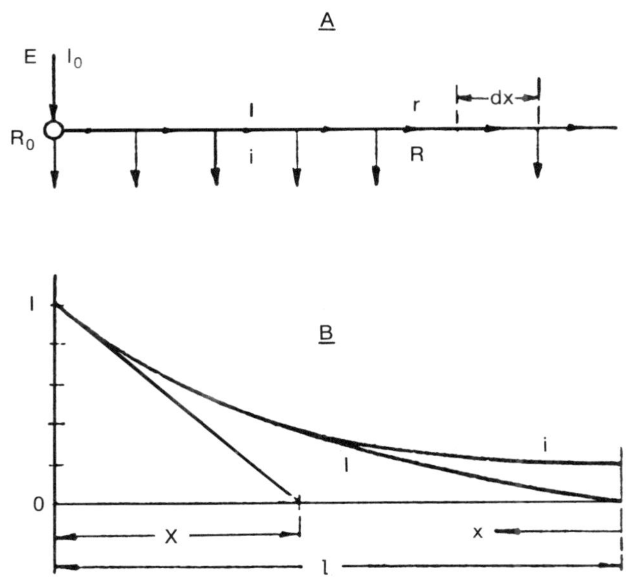

Fig 19
Current Distribution Along an Extended Electrode

from which can be obtained the differential equation, for the wire current I, for example

$$\frac{d^2I}{dx^2} - \frac{R}{r} I = 0 \tag{69}$$

If we reckon x from the far end of the wire of length l, and write for abbreviation

$$\delta = \sqrt{\frac{r}{R}} \tag{70}$$

the solution of equation 69 gives the current distribution in the wire as

$$I = \frac{\sinh \delta x}{\sinh \delta l} I_0 \tag{71}$$

Correspondingly, the ground-current distribution results from the first equation 68 as

$$i = \delta \cdot \frac{\cosh \delta x}{\sinh \delta l} I_0 \tag{72}$$

The variation of both currents along the length x of the electrode is shown in Figure 19b.

Near the terminal of a long conductor the hyperbolic distributions have a space constant X defined by the subtangent as

$$X = \frac{1}{\delta} = \sqrt{\frac{R}{r}} \tag{73}$$

If the electrode has a length of more than about $3X$, the ground current will have attenuated to an insignificant remainder over that length. Thus the effect of buried long electrodes is limited by gradual attenuation, determined by the ratio of wire to ground resistance, as seen by equation 73.

For low values of wire resistance r or length l the product δl is small and there is

$$\left.\begin{array}{l} \sinh \ \delta l \approx \delta l \\ \cosh \ \delta l \approx 1 \end{array}\right\} \tag{74}$$

Therefore

$$\left.\begin{array}{l} I = \dfrac{x}{l} \ I_0 \\ i = I_0/l \end{array}\right\} \tag{75}$$

Thus the wire current decreases linearly with x and the ground current is uniformly distributed over l, as in the case with constant-voltage electrodes.

For large values of wire resistance r or length l, excluding, however, the point $x = 0$ and its neighborhood, there is

$$\sinh \delta x \approx \cosh \delta \mathrm{x} \approx e^{\delta \mathrm{x}}/2 \tag{76}$$

Therefore, according to equations 71 and 72, both currents I and i attenuate exponentially from their original values, and the ground current is related to the wire current by

$$i = \sqrt{\frac{r}{R}} \ I \tag{77}$$

The resistance R_0 of the entire electrode to ground is given by the ratio of voltage E to terminal current I_0, the voltage being equal to the ohmic drop of the initial ground current i_0 across its ground resistance R. Thus

$$R_0 = \frac{E}{I_0} = \frac{R i_0}{I_0} \tag{78}$$

Using equation 72 at the terminal $x = l$, and substituting equation 70, the ground resistance of an extended buried electrode is generally

$$R_0 = \sqrt{rR} \cdot \coth \delta l \tag{79}$$

and for large δl, where coth δl becomes 1, is simply

$$R_0 = \sqrt{rR} \tag{80}$$

Hence the ground resistance of a long electrode depends equally on the electrode's metallic resistance electrode and the soil's resistance, the geometric mean value being effective.

For example, for the lead sheath of a cable of radius a = 2 centimeters and length l = 1 kilometer buried in moist soil close to the surface of the earth, the ground resistance of each meter of cable, using equation 30, is

$$R = \frac{\rho}{2\pi} \log_e\left(\frac{l}{a}\right) = \frac{10^2}{2\pi} \log_e\left(\frac{10^2}{2 \cdot 10^{-2}}\right)$$

$$= 170 \text{ ohm-meters}$$

This would give a total resistance of the lead sheath to ground of R/l = 0.17 ohm, if the voltage of the electrode were kept constant over the entire length. The resistance of the lead sheath is about $r = 1.5 \cdot 10^{-3}$ ohms per meter, giving a space constant

$$X = \sqrt{\frac{170}{1.5 \cdot 10^{-3}}} = 337 \text{ meters}$$

and since coth 1,000/337 = 1, a resultant resistance of lead sheath and ground

$$R_0 = \sqrt{1.5 \cdot 10^{-3} \cdot 170} = 0.5 \text{ ohm}$$

Hence the voltage on the sheath will decrease over the cable length to nearly zero and the effective resistance R_0 is three times as great as were this voltage constant.

Separate Conductors Buried in the Earth. If cross section and conductance of the electrode are considerable, the currents may be led to much greater distances. Such conductors sometimes play a tricky role, even if they are not directly connected to any electric system. They collect the streamlines of current in one region and carry them into a distant region of the ground.

For a long pipe in an extended ground field, as in Figure 20, an equilibrium is established between the parallel current density i_1 in the ground and i_2 in the pipe extending in the direction of the streamlines. The field strengths e in soil and conductor are the same, and since the current density is field strength over resistivity, the ratio of the current densities in electrode and soil is

$$\frac{i_2}{i_1} = \frac{e_2/\rho_2}{e_1/\rho_1} = \frac{\rho_1}{\rho_2} \tag{81}$$

Thus the current densities are inversely as the resistivities.

For a steel conductor buried in moist ground, there must be expected

$$\frac{i_{\text{steel}}}{i_{\text{soil}}} = \frac{10^2}{10^{-7}} = 10^9$$

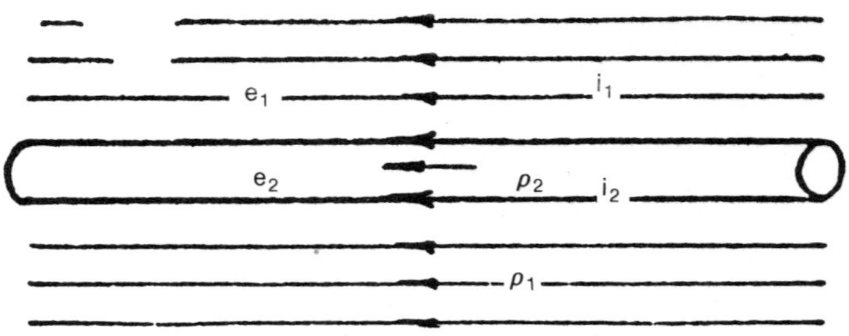

Fig 20
Long Pipe Embedded in an Extended Ground Field

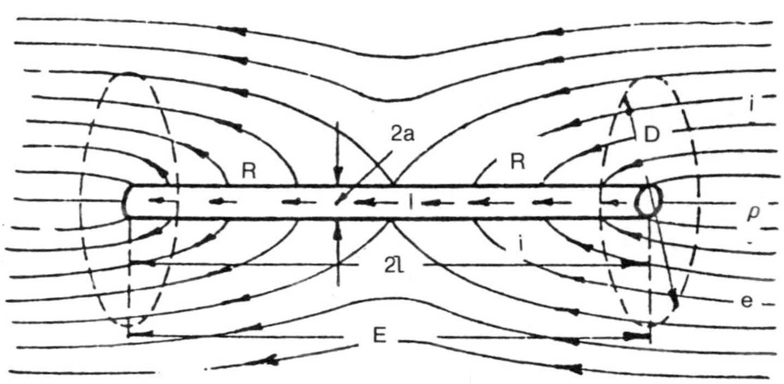

Fig 21
Short Pipe Buried in an Extended Ground Field

and a still greater ratio if the ground is dry. Therefore a steel pipe of ten square centimeters cross section can collect current from a ground area of

$$10^9 \frac{10}{10^4} = 10^6 \text{ square meters} = 1 \text{ square kilometer}$$

of cross section. Hence such a pipe is able to suck ground current from wide areas and carry it to distant points.

For a short pipe in an extended ground field, as in Figure 21, streamlines of current will enter at one end and leave from the other end. The voltage over the length $2l$ of the pipe in the undisturbed ground field is

$$E = 2le = 2l\rho i \tag{82}$$

This voltage works on the entrance and exit ground resistance of each end of the pipe and produces a current I in the center of the pipe, the metallic resistance being insignificant. Thus, with the use of equation 30,

$$E = 2RI = \frac{2\rho I}{2\pi l} \log_e \left(\frac{2l}{a}\right) \tag{83}$$

Equating equations 82 and 83 gives a pipe current

$$I = \frac{2\pi l^2}{\log_e\left(\frac{2l}{a}\right)} i \tag{84}$$

related to the current density i in the undisturbed underground field and independent of resistivity ρ. If groups of parallel pipes are present instead of a single pipe, their geometric mean distance A rather than the radius a of a single pipe is to be used, see equation 50.

For example, a pipe of half length l = 100 meters and radius a = 10 centimeters will collect from a field of current density $i = 10^{-4}$ amperes per square meter a maximum current

$$I = \frac{2\pi 100^2}{\log_e\left(\frac{2 \cdot 100}{0.1}\right)} 10^{-4} = 0.83 \text{ ampere}$$

Figure 21 shows that with such a pipe the current lines over a circular area are collected, the diameter D of which can be found from

$$I = \frac{\pi}{4} D^2 i = \frac{2\pi l^2}{\log_e\left(\frac{2l}{a}\right)} i \tag{85}$$

Thus the diameter of collection is

$$D = 2l \sqrt{\frac{2}{\log_e\left(\frac{2l}{a}\right)}} = 2l \sqrt{\frac{2}{6 \rightarrow 10}} \approx l \tag{86}$$

if an average value of the logarithm of about 8 is used.

Such pipes may suffer electrolytic corrosion if the d-c density exceeds a value of the order of 0.1 ampere per square meter. The average entrance or exit current density over the surface of the pipe in Figure 21 is

$$i = \frac{I}{2\pi a l} = \frac{l/a}{\log_e\left(\frac{2l}{a}\right)} i \tag{87}$$

showing a high current concentration over the density i in the free field, and this may even be doubled at the pipe ends. For the example the mean current density along the pipe is

$$i = \frac{100/0.1}{7.6} i = 132\, i = 132 \cdot 10^{-4} = 1.32 \cdot 10^{-2} \text{ ampere per square meter}$$

which is under the corrosion limit.

For a long pipe passing through a limited ground field, as in Figure 22, the problem of the development of the current lines in space is highly complex. Ollendorf found an approximate solution for the maximum current I_p developed in a pipe of radius a which passes at distance S an electrode from which a current I enters the ground, namely,

$$I_p = \frac{\sinh^{-1} \sqrt{\frac{S}{a}}}{\log_e \left(4 \frac{S}{a}\right)} I \tag{88}$$

The current collected in this way, as shown by Figure 22, may spread out over vast distances depending on the resistivities of ground and pipe. The maximum surface density of the current entering the pipe, according to Ollendorf's solution is

$$i = \frac{l}{\pi a S} \frac{1}{\log_e \left(4 \frac{S}{a}\right)} \tag{89}$$

With current I = 100 amperes entering the ground through an electrode at distance S = 30 meters from a pipe the current density entering the pipe of radius a = 10 centimeters will be

$$i = \frac{100}{\pi \cdot 0.1 \cdot 30} \frac{1}{\log_e \left(4 \frac{30}{0.1}\right)} = 1.5 \text{ amperes per square meter}$$

a value which will produce severe electrolytic corrosion, if it is direct current.

Fig 22
Long Pipe Running Through a Limited Ground Field

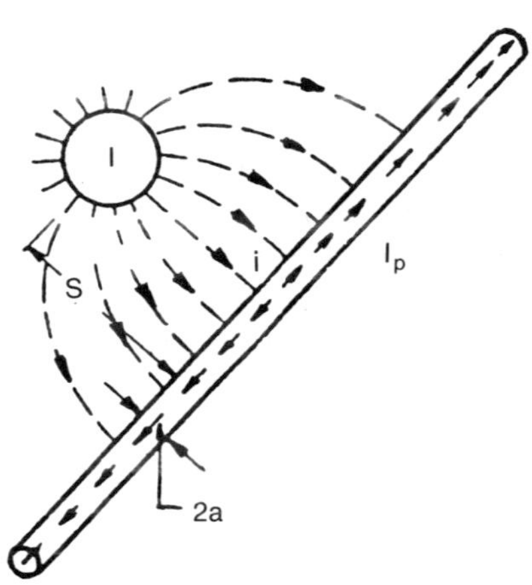

Alternating Return Current Under Transmission Lines. This article has dealt up to now with the effects of the resistivity of the ground without considering any inductive action of the currents. These derivations are true for direct current flowing through the ground. However, they are equally valid for alternating current as far as the proximity of the electrodes is concerned, for here the inductive action of the changing magnetic fields is negligible compared with the effect of the high resistivity of the ground. Only for extremely rapid change with time this may be modified, as will be shown later.

If ground current flows over a long distance between entrance and exit electrode, the current lines in the d-c case spread out over such a broad transverse area in the earth, as seen by Figure 4a, that the resistance is negligibly small except for the proximity of the electrodes, as previously shown. For alternating current, however, with decreasing effect of the resistance the self-inductive action of the magnetic field becomes preponderant, and thus the distribution of the current lines in the ground under a-c conditions will be determined mainly by the inductive effects. It is well known that the currents in such a case are so distributed that the energy of the magnetic field and thereby the self-inductance tend to become a minimum.

Therefore an alternating return current in the ground under a long transmission line will not spread to a great distance, as direct current does, but will concentrate on paths in proximity to the transmission line, as shown in Figure 23. Hence the distribution of alternating current in the ground is governed by the laws of skin effect, laterally as well as in depth. Because of the high resistivity of the soil, however, much larger dimensions come into consideration than with metallic conductors.

A single transmission line, as in Figure 23, will develop concentric magnetic lines of force of nearly circular form, at least in the neighborhood of the line, where the magnetic effect of the return current in the ground is of lesser importance because of its low density. Therefore the phenomena will be dependent primarily on the distance y from the center of the wire, carrying the total current I, to the point considered at the earth. The error caused by this assumption is usually small as compared to that caused by subsoil heterogeneity. With angular frequency ω, the differential equation of the distribution over the ground of the current density i then is

$$\frac{1}{y}\ \frac{\mathrm{d}}{dy}\left(y\ \frac{di}{dy}\right) - j \cdot 2\pi\mu\ \frac{\omega}{\rho}\ i = 0 \tag{90}$$

a complex relation similar to that for other skin-effect problems. The value $\mu = 10^{-7}$ is the permeability of nonmagnetic material in the meter-kilogram-second system.

The solution of this equation is given by a Bessel function H_0 of zero order. With abbreviation

$$x = \sqrt{2\pi\mu\ \frac{\omega}{\rho}} \tag{91}$$

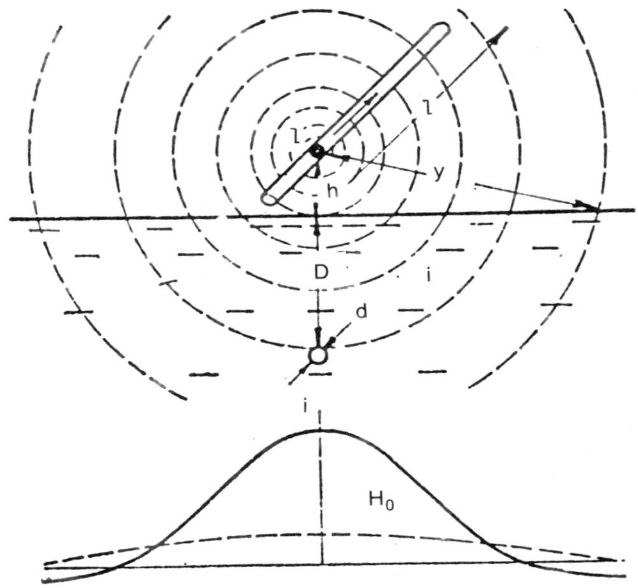

Fig 23
Distribution of Ground-Return Currents under AC Transmission Lines

the current density, up to frequencies of about 5,000 cycles per second, is

$$i = \frac{x^2}{2} \; I \cdot H_0 \left(\sqrt{j}\;\kappa y\right) \tag{92}$$

where the argument of the Bessel function depending on distance y is complex, indicating that the distribution of the return current through the ground occurs in waves.

For two instants, 90 degrees displaced in time, the values of the Bessel function H_0 are shown in Figure 23, determining the current density against distance. From numerical tables of this Bessel function it can be derived that directly under an over-head power line of height $y = h = 10$ meters the current density at 60 cycles per second in moist soil of $\rho = 10^2$ ohm-meters is nearly $i = 3 \cdot I$ in amperes per square kilometer.

Thus for every ampere of line current there develops a current density of three amperes per square kilometer in the ground under the line, and naturally a still smaller current density at greater distances.

The current density in the ground decreases to an insignificant magnitude at a distance of about $y = 3/\kappa$, as seen by a discussion of the function H_0. Hence for moist soil the zone of extinction is

For current of 60 cycles per second about 2,000 meters,
For current of 500 cycles per second about 700 meters,

and a still smaller distance for higher frequencies. Thus a return current under a transmission line flows only in a moderately broad zone through the ground under and to both sides of the line. This zone follows any trace of the line, it may be straight or curved, since the distribution of the return current always seeks to reduce the self-inductance to a minimum and therefore to avoid any open loop between line current and return path.

The field strength in the ground, with use of equations 91 and 92, is

$$e = \rho i = \pi \mu \omega I \cdot H_0 \left(\sqrt{j}\, \kappa y\right) \tag{93}$$

Therefore the voltage over a distance l on the ground surface immediately under the transmission line, namely for $y = h$, is

$$E = \pi \mu \omega I l \cdot H_0 \left(\sqrt{j}\, \kappa h\right) \approx \omega I l \left[\frac{\pi}{2} + j2 \log_e \left(\frac{1.12}{\kappa h}\right)\right] 10^{-7} \tag{94}$$

Herein an asymptotic approximation of the Bessel function is used for small values of the complex argument, resulting in a form which has distinct real and imaginary components. The voltage over the ground, therefore, is complex, consisting of two perpendicular vector components, where the real part gives the effective resistance of the return current

$$R = \frac{\pi}{2} \mu \omega l = \pi^2 f l \cdot 10^{-7} \tag{95}$$

while the imaginary part determines the effective self-inductance

$$L = 2l \cdot \log_e \left(\frac{0.178}{h} \sqrt{\frac{\rho}{f}}\right) \cdot 10^{-7} \tag{96}$$

In the last two equations the frequency f per second is used instead of the angular frequency ω.

Hence alternating return current through the ground experiences a resistance which is proportional to length l and to frequency f but independent of resistivity ρ of the soil. Table IV gives the return resistance R in ohms per kilometer for various frequencies in moist soil of $\rho = 10^2$ ohm-meters. The values are not unimportant for heavier currents or for frequencies above the power range. The same return resistance would be attained by a fictitious copper wire of an equivalent diameter d as given in the third line of Table IV.

The self-inductance of the return currents as given by equation 96 is also proportional to length l, depends slightly on the height h of the transmission line and,

Table IV
Resistance and Dimensions of Equivalent Ground-Return Conductor

f =	50	150	500	5000	c
R =	0.05	0.15	0.5	5	Ω/km
d =	2.1	1.3	0.7	0.2	cm
D =	800	460	250	80	m

to an even lesser degree, on resistivity and frequency, these being under the logarithm of a square root. Thus for a large range of such data the self-inductance is very nearly

$L \;=\; 0.9$ millihenry per kilometer

or

$\omega L \;=\; 0.34$ ohm per kilometer

This self-inductance of the ground current adds to the self-inductance of the line current, which is also determined by a well-known logarithmic relation. So is also the self-inductance of a fictitious return conductor of diameter d and depth D under the surface, the ohmic and inductive effects of which may be taken equivalent to those of the ground return. By comparison of equation 96 with the self-inductance of such an ideal return system, the equivalent depth can be determined. We see from the last two lines in Table IV that the equivalent diameter d of a copper return wire would be of the order of two centimeters for 50 cycles per second but of only 0.2 centimeters for 5,000 cycles per second. On the other hand, the equivalent depth D would vary from 800 meters to 80 meters over the same frequency range.

Inductive Interference of Parallel Lines. If at a distance y a neighboring line runs parallel to the transmission line with current I returning through the ground, as shown in Figure 24, and the two ends of the secondary line are directly or indirectly grounded, then a voltage develops over a length l of this line according to equation 93

$$E \;=\; \pi \mu \omega I l \cdot H_0 \left(\sqrt{j}\;\kappa y\right) \tag{97}$$

Fig 24
Mutual Inductance Between Lines with Ground Return

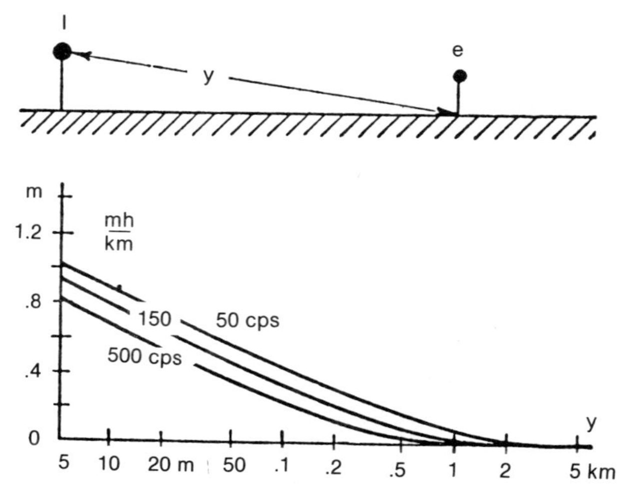

This is a voltage of mutual effect between primary and secondary line caused mainly by action of the ground-return currents and their magnetic field.

Hence the coefficient of mutual induction for the unit length of both lines is

$$m = \pi \cdot H_0 \left(\sqrt{j}\ \kappa y\right) \cdot 10^{-7} \qquad (98)$$

Since the value of the Bessel function H_0 is complex, this mutual inductance has amplitude and phase angle in contrast to the straight inductance of two linear circuits alone. This is effected by the alternating currents produced in a spatially extended third conductor, namely, the ground. Figure 24 shows the amplitude of the mutual inductance in millihenrys per kilometer dependent on distance y between primary and secondary circuit for various frequencies, if the resistivity of the intermediate ground is $\rho = 10^2$ ohm-meters, as for moist soil. In all practical cases the resistivity will not be uniform over surface and depth of the ground between and around such transmission lines, and thus an average resistivity of the entire field should be taken into account. High resistivity of the underground increases the action to a greater distance.

If, for example, a short-circuit current $I = 500$ amperes in a power line flows through an arcing ground to earth and returns through it to the grounded neutral of the power station, then, according to equation 97, a telephone line at a distance $y = 1$ kilometer, running parallel to the power line over a length $l = 10$ kilometers and using the ground for return, will suffer a voltage of

$$E = \pi \cdot 10^{-7} \cdot 377 \cdot 500 \cdot 10 \cdot 10^3 \cdot 0.20 = 118 \text{ volts}$$

a value which would annihilate any telephone conversation.

Impulse Characteristics

Performance of Driven Rods. It may be asked how closely the current in an electrode and the surrounding ground can follow a rapid variation of the voltage. The best criterion for this is given by the values of the electric time-constants and the natural periods of oscillation.

In Figure 25 the current I impinges on the rod electrode and enters the ground, which in addition to its resistivity has a dielectric constant k. Thus in parallel to the conductive current in the ground there develops a capacitive or displacement current in case the electrode voltage changes with time. The displacement current follows exactly the lines of the conductive current, and thus the ground electrode will have a capacitance, as reciprocal to the resistance of equation 30,

$$C = \frac{kl}{2 \cdot \log_e \left(\dfrac{2l}{a}\right)} \cdot \frac{10^{-9}}{9} \text{ farad} \qquad (99)$$

The currents in electrode and ground, furthermore, form a magnetic field indicated by the dashed lines in Figure 25. This field is highest where the current is most concentrated, namely around the rod electrode. Therefore the inductance of the ground currents is mainly given by the distribution of the currents in the rod,

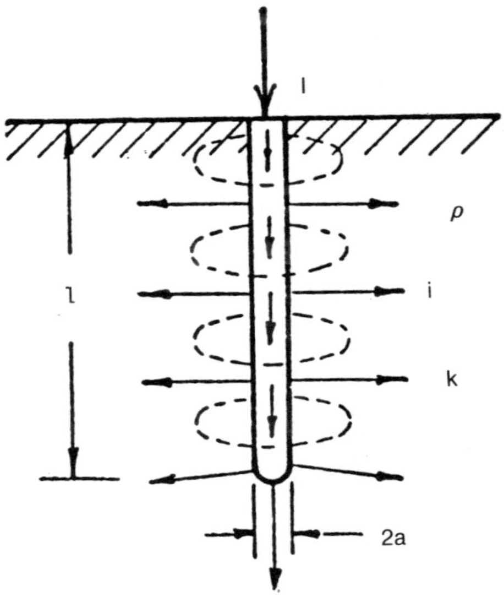

Fig 25
Impulse Current Spreading from Driven Rod

having full value at the top and decreasing to zero at the bottom. The inductance of such a rod is

$$L = 2l \cdot \log_e \left(\frac{2l}{a}\right) \cdot 10^{-7} \text{ henry} \tag{100}$$

The capacitive time constant of any circuit element is given by the product of capacitance and resistance, and thus a rod electrode has a time constant

$$\tau_C = CR = \frac{\rho k}{4\pi \cdot 9 \cdot 10^9} \tag{101}$$

It depends only on resistivity and dielectric constant k, the value of which for ordinary soil may be of the order of 9, in view of the high value for water.

For example, for moist soil

$$\tau_C = \frac{10^2 \cdot 9}{4\pi \cdot 9 \cdot 10^9} = 8 \cdot 10^{-9} \text{ second}$$

which is an extremely small value. For rock as underground the value is about $8 \cdot 10^{-7}$ second.

The inductive time constant as quotient of self-inductance and resistance is

$$\tau_L = \frac{L}{R} = 4\pi \frac{l^2}{\rho} \cdot 10^{-7} \tag{102}$$

It is proportional to the square of the length l of the electrode, and therefore greatly dependent on this dimension. For example, for a rod of l = 6 meters in moist soil

$$\tau_L = 4\pi \frac{6^2}{10^2} \cdot 10^{-7} = 4.5 \cdot 10^{-7} \text{ second}$$

For the same electrode in rock the value is $4.5 \cdot 10^{-9}$ second. Both these values are likewise very small.

Since all these numerical values are less than the time usually assumed for the shortest possible lightning impulse impinging on a driven rod, current and voltage in the rod will follow such impressed impulse of the order of 10^{-6} second without any significant time delay and will behave as in the steady state. Hence capacitance and inductance of a driven rod of moderate length play no significant part even with rapid lightning phenomena.

The natural oscillations of which such a rod is capable have a period

$$T = 2\pi\sqrt{LC} = \frac{2\pi\sqrt{k}}{3 \cdot 10^8} \cdot l \qquad (103)$$

depending only on dielectric constant and length. For the same example, the natural period is

$$T = \frac{2\pi\sqrt{9}}{3 \cdot 10^8} \cdot 6 = 3.8 \cdot 10^{-7} \text{ seconds}$$

This also is beyond the duration usually assumed for lightning strokes, and indicates the frequency up to which the rod behaves as an ohmic resistance.

The damping of the free oscillations is determined by the value of the quotient of resistance and surge impedance in comparison to the value one half, namely,

$$\frac{R}{\sqrt{L/C}} = \frac{\rho\sqrt{k}}{120\pi l} \quad \begin{cases} > \tfrac{1}{2} \text{ periodic} \\ < \tfrac{1}{2} \text{ aperiodic} \end{cases} \qquad (104)$$

For the same rod in moist soil

$$\frac{R}{\sqrt{L/C}} = \frac{10^2\sqrt{9}}{120\pi \cdot 6} = \frac{1}{7.5}$$

so that any natural oscillation would be aperiodically damped. For rock, however, the quotient assumes the value 13, indicating that here the driven rod can oscillate periodically.

An equivalent diagram is often convenient to represent the properties of a circuit element in a simplified form. In view of the foregoing numerical values there is no doubt that for low frequencies a driven rod and all its derived electrodes can be represented by a lumped resistance, as in Figure 26a. This remains true for a ground electrode of moderate length up to high and even very high frequencies. However, it should be noticed that a tower or a lead, through which a lightning discharge reaches the electrode may have considerable self-inductance because of its height. This is indicated in Figure 26b. If the voltage at the rod electrode is very high, as is frequently the case with lightning discharges, the field strength at the rod becomes so great that the electric strength of the ground is broken down and sparks discharge from the rod into the neighboring ground thereby considerably decreasing the entire ground resistance. This performance is indicated in the equivalent diagram of Figure 26b by a spark by-pass to a part of the ground resistance.

If the discharge to the ground electrode is extremely rapid or the equivalent frequency is extremely high, of an order of more than 10^6 cycles per second, a considerable displacement current in the capacitance of the ground will parallel the conductive current in the resistance, and together with the self-inductance of both of these currents through the rod, will effect an equivalent diagram as in Figure 26c. The time-constants, the natural oscillations, and the damping of the electrode, as in equations 101 to 104, now will be determinative.

Performance of Buried Wires. In ground electrodes of considerable length, as with buried long wires, the self-inductance will play a larger part with respect to impulse currents, since the inductive time constant, see equation 102, increases as the square of the length. The capacitive effect of the ground, see equation 101, will not change, however, and may be neglected. Since buried wires usually have substantial cross section, the internal resistance of the electrode also may be neglected in comparison to the self-inductance. Figure 27a shows the remaining data of the problem, namely, the wire current I in amperes, the ground current i in amperes per meter, the self-inductance L of the wire in henrys per meter, and the ground resistance R in ohm-meters.

Corresponding to the case where the wire resistance is taken into account, see Figure 18 and equation 68, the relations between the currents in the present example are

$$\left.\begin{aligned} i + \frac{dI}{dx} &= 0 \\ R\,\frac{di}{dx} + L\,\frac{dI}{dt} &= 0 \end{aligned}\right\} \tag{105}$$

Their combination gives the differential equation for the current in the wire

$$\frac{d^2I}{dx^2} - \frac{L}{R}\,\frac{dI}{dt} = 0 \tag{106}$$

and a corresponding equation for the ground current i. Equation 106 is the well-known differential equation of heat diffusion. A solution for a short impulse is

$$I = \frac{K}{\sqrt{t}}\, e^{-\frac{L}{R}\frac{x^2}{t}} \tag{107}$$

where the amplitude is still free as determined by a constant K of integration, while the exponent depends, in addition to L and R, on the square of distance x and on time t. This function is plotted in Figure 27b against distance x and in

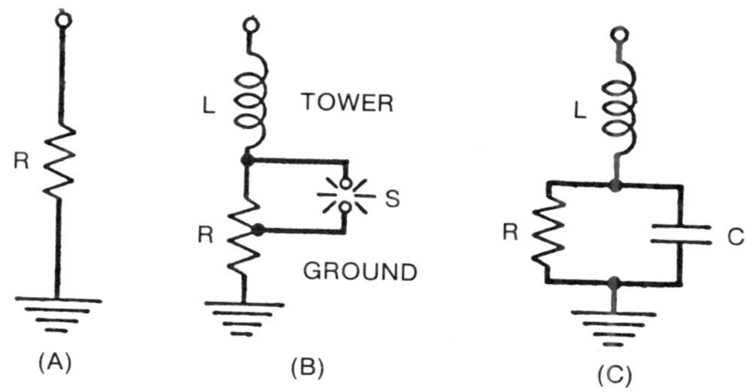

(A) LOW VOLTAGE, LOW FREQUENCY
(B) HIGH VOLTAGE, HIGH FREQUENCY
(C) ULTRAHIGH FREQUENCY

Fig 26
Equivalent Circuits of Ground Electrodes Over Wide Range of Voltage and Frequency

Figure 28a against time t. Equation 107 represents an impulse curve over the time axis, which increases at first slowly and then steeply, reaches a maximum peak, and decreases gradually toward zero. Such a curve might well be used to approximate the time characteristic of a lightning discharge to ground which is known from many experimental investigations.

Figure 27b shows the manner in which the current in the wire, from a narrow zone in the beginning, spreads gradually over the length of the wire. It is not necessary to identify the origins of x and t in equation 107 with the origin of the lightning current. Figures 27 and 28 show that displacements in space x_0 and in time t_0 can be used for constants of the solution (107) and by proper selection of these every actual case can be approximated closely.

The points at which the spatial curves of Figure 27b are steepest determine the front zone of the attenuated propagating wave of current. Their location can be derived from equation 107 and is given by

$$\frac{L}{R}\,\frac{x^2}{t} = \frac{1}{2} \tag{108}$$

These points travel with a velocity

$$v = \frac{dx}{dt} = \frac{1}{4}\,\frac{R/L}{x} = \frac{10^7}{4\pi}\,\frac{\rho}{x} \tag{109}$$

where R/L is taken from equation 102, both data, however, being here related to the unit length. Hence the velocity of the front is not constant but decreases rapidly with distance x.

Buried long wires are often used as electric counterpoises additional to the tower footing conductance of high-voltage transmission lines, in case the resistiv-

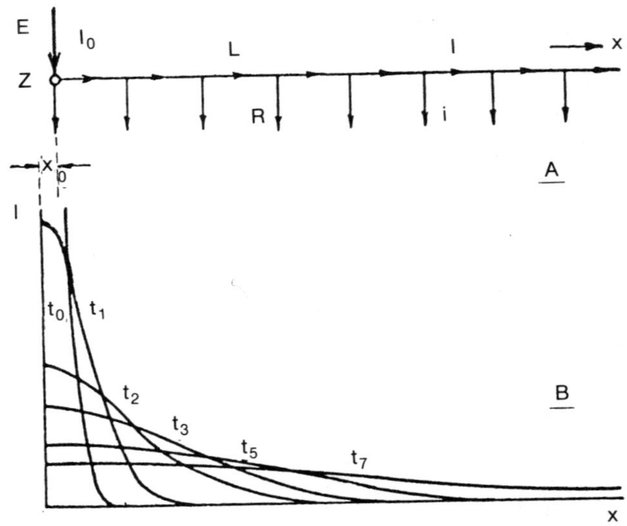

Fig 27
Distribution of Impulse Current Along a Buried Wire

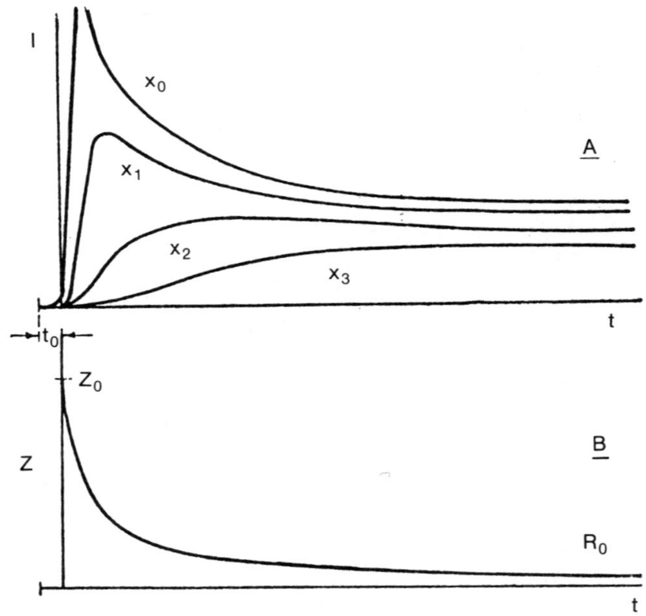

Fig 28
Change with Time of Current and Surge Impedance in Buried Wire

ity of the soil is so high that it is difficult otherwise to obtain a low ground resistance. With a resistivity, therefore, of $\rho = 10^3$ ohm-meters, the front velocity of the current wave will be, for example, at distance $x = 10$ meters on the wire,

$$v = \frac{10^7}{4\pi} \frac{10^3}{10} = 0.8 \cdot 10^8 \text{ meters per second}$$

This is about one-fourth the velocity of light, a velocity reached in straight conductors in air. In ground wires of considerable length, therefore, the joint effect of self-inductance and ground resistance leads to a lower and ever decreasing wave velocity.

The voltage at every point of the wire is determined by the ohmic drop of the ground current i in the ground resistance R. By using the first equation 105 and differentiating equation 107, the voltage is

$$E = R_i = -R \frac{dI}{dx} = 2L \frac{x}{t} I \tag{110}$$

whereby E is reduced for every distance and time to equation 107 for the current I. Since E is proportional to I, a definite value of the impedance is obtained, which, however, is dependent on distance and time. The surge impedance is

$$Z = \frac{E}{I} = 2L \frac{x}{t} \tag{111}$$

and is proportional to the self-inductance L per unit length of the wire. For any point x of the wire, Z is inversely proportional to time.

There may be correlated to the beginning of the steep ascent of the current with time, as shown by t_0 in Figure 28a, a location x_0 corresponding to about half the value of equation 108, so that

$$\frac{x_0^2}{t_0} = \frac{1}{4} \frac{R}{L} \tag{112}$$

By substitution of this value of x in equation 111 and use for R and L of values per unit length derived from equations 30 and 100 for a finite wire, the surge impedance for the terminal of the wire becomes

$$Z_0 = \frac{t_0}{t} \sqrt{\frac{RL}{t_0}} = \frac{t_0}{t} \sqrt{\frac{2\rho \cdot 10^{-7}}{\pi t_0}} \log_e \left(\frac{2i}{a}\right) \tag{113}$$

Since buried wires usually are embedded near the surface of the ground, extending to both sides from the terminal, the value of equation 30 is doubled for the resistance here in question.

We consider for an example a wire of length $l = 75$ meters and of radius $a = 0.5$ centimeter, buried in ground of high resistivity $\rho = 10^3$ ohm-meters, and we estimate the time until the steep ascent of the lightning current occurs as $t_0 = 0.2 \cdot 10^{-6}$ second. Then the surge impedance for the practical start of the current at $t = t_0$ will be

$$Z_0 = \sqrt{\frac{2 \cdot 10^3 \cdot 10^{-7}}{\pi \cdot 0.2 \cdot 10^{-6}}} \log_e \left(\frac{2.75}{0.5 \cdot 10^{-2}}\right) = 184 \text{ ohms}$$

and this value decreases hyperbolically with increasing time as shown in Figure 28b.

The impedance would decrease gradually to zero, see equation 113, due to the omission of ohmic resistance of the infinite wire in this calculation. Actually the impedance cannot drop lower than the d-c ground resistance of the finite buried wire, which in this example is

$$R_0 = \frac{10^3}{\pi \cdot 75} \log_e (3 \cdot 10^4) = 44 \text{ ohms}$$

thus limiting the decrease of Z, as shown in Figure 28b by the dashed line.

Conclusion

By using for the development of the current distribution in the ground such mathematical analyses as are best suited to each special problem, rather than any general method of rigorous calculation, the treatment does not depart far from the physical aspect, and the solutions become fairly simple and easy to survey, even in complicated examples. In view of the great heterogeneity of the ground under the surface of the earth, the effect of which can never be taken into account by any rigorous deduction, such a simplified foundation of the principles of ground currents appears appropriate from the viewpoint of engineering as well as from the viewpoint of the economic use of science.

Appendix I
General Fundamentals of Electrical Grounding Techniques

(Les Bases Générales de la Technique des Mises à la Terre Dans les Installations Electriques)

by Pierre Laurent
from *Le Bulletin de la Société Française des Electriciens* *
July 1951

I. Classification and General Performance of Earth Electrodes

I-1. The Earth Electrode Problem

The earthing of electric installations plays an important part as regards the behavior of the network and personal safety when there are disturbances on the lines. There are, however, few fields of electrical engineering in which empirical methods are more largely employed.

In spite of important and impartial study devoted to the subject, the opinion is still sometimes expressed that "earthed", even if only by the use of a few inches of rod pushed into the ground, immediately becomes incapable of being the seat of a dangerous voltage. Several people lose their lives every year as a result of this simple belief. In one place the same type of earthing will be used on all cases, although its effectiveness may vary in the ratio of 100 to 1, depending on the type of ground. In another place considerable sums will be spent on earth circuits, whereas the foundations themselves, or the buried links between the different structures forming the plant, would be quite sufficient to give an excellent connection with the ground. The methods used for checking earth connections are, too, often used without the correct conditions for carrying out satisfactory measurements being properly appreciated.

It must be recognized that earthing problems are often of a very complex nature for a number of reasons. The ground is a poor conductor (the conductivity

*Appreciation is due to the Central Electricity Generating Board of Great Britain for making available an English translation of this article, and to the author, M. Laurent, for reviewing and making some detail revisions in the English text. Section VI of the original article (dealing with heating of the ground around electrodes), has been omitted since the author believes it outdated by more recent investigations.

of a mass of good soil with a cross-section of 2.47 acres is comparable to that of a copper wire of one square millimeter; it is not homogeneous and has characteristics of which little knowledge is available). Conductors and electrodes buried in the soil are out of sight, often have a complicated shape and can only be examined, and their properties measured, with difficulty. Personal safety introduces questions of the probability of contact which are difficult to formulate in equations. Finally, installing good earthing systems often involves serious financial difficulties, especially under poor ground conditions.

The aim of the present paper is to lay down, in the simplest manner possible, the basis governing the determination and behavior of earthing systems. It is divided into five sections. The first deals with earthing arrangements in general; the second, with soil characteristics; the third, with the most usual types of earth electrodes; the fourth, with multielement connections and interconnected earths; and the fifth, with the special problems of large substations and currents through the human body.

Our chief purpose has been to give the basic principles. The more complex equations are put forward for reference purposes only or for illustration, and the working has been omitted except in very simple cases. A number of these equations have been taken from the excellent paper by Dwight in the December 1946 issue of "Electrical Engineering".

We have not considered the question of the behavior of earth electrodes when subjected to very steep waves, as this would require a special paper.

I-2. General Purpose and Classification of Earth Electrodes

The purposes of earthing electric installations are protection of the installation, improvement in the quality of the service and safety of personnel.

These general aims by no means determine the methods to be used, and widely different techniques can lead to the same average degree of perfection. Absolute perfection in these fields is impossible, since electricity, however much it may be controlled, remains intrinsically the source of incidents to plant, breakdowns in service, and accidents to people, which it is impossible to avoid completely without condemning the principle of application altogether.

It will be seen to what extent earthing of certain units can contribute to better operation and greater safety on networks. Distinction should first be made between two types of earth electrodes — viz. the "system" earth and the "protective" earth.

The system earth forms an integral part of the network and forms links, permanent or otherwise, between the conductors either directly or through suitable impedances, the effect of which is to influence the behavior of the network in case of an earth fault in the plant or on the line.

The protective earth originates rather with the consumer than with the supplier. It is not connected to the conductors of the network, but to objects which risk accidental contact with the conductors and which would be dangerous if maintained at a high potential over a period.

System earths are, for instance, the neutral earths or voltage limiter earths. Protective earths apply to housing and frames or line supports. (Figure 1).

It may happen that the same earth electrode fulfills both functions. Further, the impedance of the earth electrode through which the fault current flows obviously has an effect on the electrical behavior of the network similar to that of the system earth. The distinction between the two is thus not always completely clear.

It should further be mentioned that earth faults can occur at points which have not intentionally been provided with protective earths, for instance, in the case of a wire which has broken and fallen to earth, or a person with his feet on the ground who makes direct contact with a live bare conductor. In this case, it is a quite accidental resistance, with a value difficult to determine, which plays the part of protective earth (Figure 2).

Fig 1
System Earths (A) and Protective Earths (B)

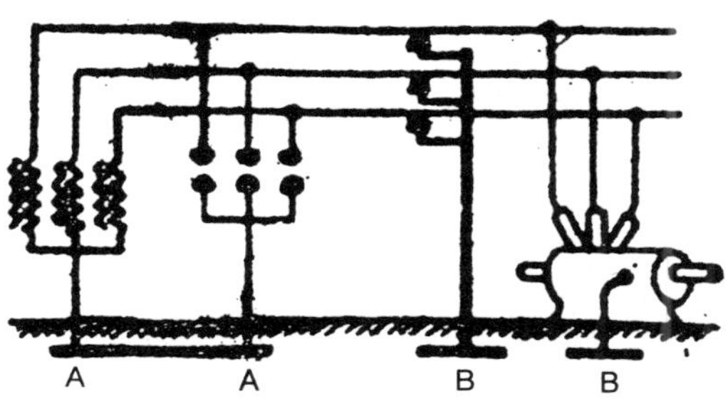

Fig 2
Protective Earths (A) and Accidental Earths (B)

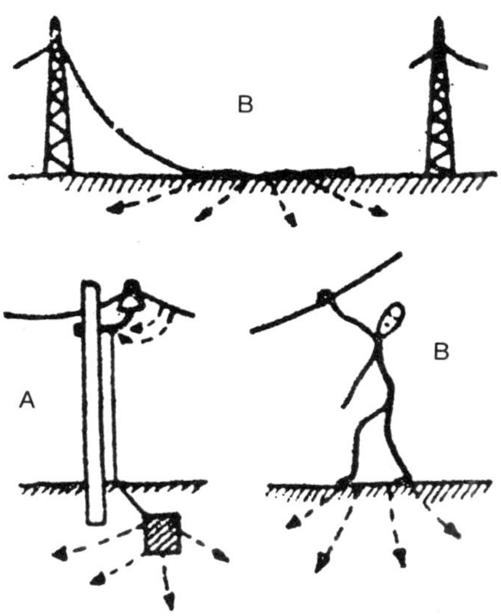

I-3. Function of the Earth Electrode in High-Voltage Networks

We do not propose here to go into the question of the different high-voltage systems, that is, with neutral either insulated or more or less directly connected to earth. It will be sufficient to mention the different ways in which these systems contribute to the three aims set out above, that is, protection of plant, continuity of service and safety of personnel.

On high-voltage networks with a sufficiently short length and an insulated neutral or with Petersen coils, the absence of a direct earth contributes to good performance of the plant by limiting earth fault current to low values, so that the thermal and electrodynamic effects are less pronounced. This low fault current is of help in ensuring reliable service and in favoring self-extinction of arcs to earth. It also reduces the discharge voltage of these arcs through the protective earth and thus contributes to human safety.

On the other hand, a network with insulated neutral or with quenching coils induces relatively high dielectric stresses in the plant. Further, advantage is often taken of the low fault currents to allow discharge to earth to continue over a considerable period and this may give rise to a serious risk of accident when the protective earths or contacts taking their place have a certain resistance.

A high-voltage network with earthed neutral protects the plant at the cost of high overcurrents, first by reducing the overvoltages and then by limiting the duration of the faults, thanks to high-speed interruption circuit breaking caused by the overcurrent. This system improves the quality of the service by making it easier to cut out defective sectors selectively. Finally, it helps in ensuring human safety by high-speed clearing of faults.

On the other hand, the earthed neutral, unless combined with high-speed reclosing, does not have the advantage of self-extinction of transient faults. Further, automatic disconnection of defective units is compromised when the resistance of the protective earth or contact is so high that the fault current falls below the sensitivity limit of the protective devices.

This rough comparison only serves to show the two widely different ways in which protective earths act, either by reducing the fault current discharge voltages to earth to safe values, or by favoring the rapid disconnection of defective sectors.

Assume a potential of 125 volts as the danger limit and imagine that a person simultaneously touching the structure connected to the earth lead and the ground in proximity is subjected to half the total voltage on the earth connection. Assume further that the network has an insulated neutral and that the capacitive fault current to earth has a value of 10 amperes. Danger will arise if the earth electrode resistance exceeds 25 ohms, unless the sensitivity of the protective devices or the instructions given to the staff are such that a fault discharging at 10 amperes is rapidly eliminated.

Assume now a 15-kilovolt network with earthed neutral and a fault current reaching 1,000 amperes at the place where the fault occurs, and a sensitivity limit, for the earth current relay arrangements, of 50 amperes. Reduction of the local potential to a safe value of 125 volts would necessitate an earth resistance of 0.25 ohm, a value which is obviously impossibile to attain. It is then necessary to count exclusively on high-speed clearing in order to ensure safety, and this requires that the earth resistance should not exceed 173 ohms for the fault current to be higher than 50 amperes.

Adequate protection is thus not only a question of earths, but of coordination between these and other means of protection. The sensitivity of the relays should, in principle, be the greater the more difficult it is to make good earth connections on the network in question. Serious technical difficulties will often be met within the attempt.

It is necessary to point out that one cannot always expect protective earths to limit the voltages at the fault to safe values, especially on high-voltage networks with earthed neutral. Human safety is thus almost exclusively ensured on these networks by high-speed circuit-breaking, which reduces to an infinitesimal value the chances of a person finding himself at the spot where a fault occurs at the precise instant and in a hazardous attitude. The object of protective earths is simply to allow the passage of a sufficiently high current to ensure high-speed protection. Practice shows the safety thus offered is satisfactory. Accidents to people are practically non-existent on high-voltage networks with earth neutral and suitable protection. They are, indeed, certainly much rarer than on networks with insulated neutral, where the fault currents to earth are much smaller, but where they are allowed to continue for a considerable time.

I-4. Functions of Earths on Low-Voltage Networks

We will only deal briefly here with the familiar problem of the relationship between earthing systems and safety on low-voltage networks. In the case of low voltages, all public power distribution networks have the neutrals more or less well earthed. The effect of this earthing should be to maintain the neutral, in all circumstances, at a low voltage and the phase conductors at a voltage close to the rated phase-to-earth voltage, since this reduces the gravity of direct contact with them without altogether eliminating it. Actually, the neutral earthing of a low-voltage network is often poor and sometimes worse than that of the intentional or accidental protective earths which discharge the fault currents. In such cases it is the protective earth which imposes a low potential on the conductor concerned and, by a balancing effect, the voltages at the sound conductors are raised to the neighborhood of the voltage between phases. In this way to a certain extent, although somewhat erratically, we gain in safety as regards contact with an object which is the subject of a fault what we lose as regards direct contact with the neutral and the unaffected conductors.

It will be noted that it is very difficult to ensure simultaneously a very low voltage on the neutral conductor and on the earthed frame of the defective apparatus as long as a connection exists between the latter and the source. For this to be achieved it will be necessary for both the neutral earth and the frame earth to have sufficently low resistance to practically form a short circuit. Further, in actual circumstances it is scarcely possible to count on automatic clearance ensured by the apparatus damaged, as is the case with high voltages. The faults are often of too high a resistance to blow the fuses which, in practice, is the only form of protection used in France on the majority of equipment, since improved devices are considered rightly or wrongly as too costly and too delicate.

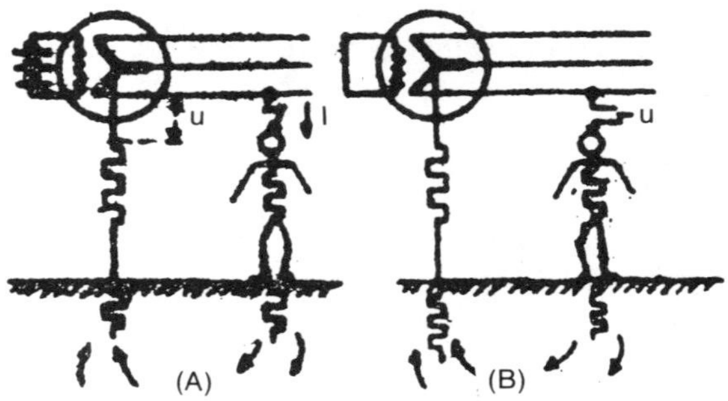

Fig 3
Application of Thevenin's Theorem to Currents Flowing in the Ground: A, Actual Circuit; B, Equivalent Circuit

One method used in Switzerland consists in connecting the frame of consumer apparatus to protective earths, which are intentionally made always better than neutral earth. Protection against contact with the earthed frame is then completely ensured and good insulation of the conductors is relied on to prevent direct contact with the latter. This system should give excellent results, but it means that all consumers must have very high-quality earths, the arrangement of which is often costly.

It will be seen that the problem of earthing on low-voltage networks is delicate. Earthing the neutral remains, however, desirable on public supply networks, particularly because of the risk of mixing with high voltages. The same is not necessarily the case with private networks, especially when it is possible to maintain permanent supervision of the insulation of the network in relation to earth. Every simple fault, or all direct contacts with the conductor, are then immediately signalled and can be eliminated before a double fault can occur. The simple fault is harmless because it only causes the passage of a very low leakage or capacitive current.

I-5. Current Path Through the Ground

Before considering each earth electrode separately, and in order to justify the liberty often taken of separating it from the complex circuit in which it is inserted, it is of advantage to scan all the elements met with sucessively by the current discharging into the ground at the fault. We would mention straight away that the current which will appear when there is contact between a conductor and the ground close to it is most usually, according to Thevenin's theorem, identical with that caused by a single-phase potential equal to the pre-existing voltage between these two points (that is to say, equal to the normal phase voltage of the network), injected into the accidental connection which joins them, and flowing between the earth and the network assumed to be completely de-energized (Figure 3).

For a current to penetrate from the conductor into the ground, it must obviously have a return circuit to the network. This return might take place either

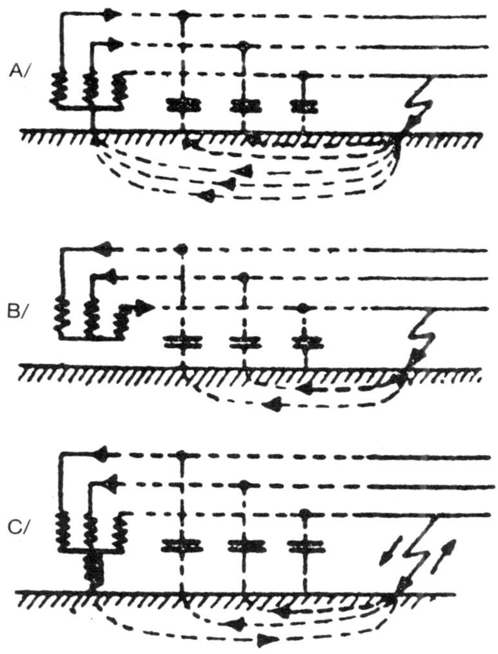

Fig 4
Return Path of a Fault Current to Earth:
A, Earthed Neutral System; B, Insulated Neutral System; C, Quenching Coil System

by one or more neutrals arranged for this purpose, or through the distributed capacitance between the ground and the unaffected phases of the network. The impedance of the return system is added to that of the input earth connection and to the internal longitudinal impedances of the network to limit the current discharged into the ground.

It often happens that the current has several parallel return paths available, some concentrated and some distributed. We know how this fact is taken advantage of in earthing through Petersen coils, which provide an inductive return circuit tuned to the capacitance of the lines in such a manner that the whole forms a very-high-impedance choke, limiting the current to a low value (Figure 4). It is also well to consider the case where the current discharged into the ground does not come from the ordinary sources on the network, but from a lightning stroke. The return of the current to the thundercloud must then take place by way of the distributed capacitance between the latter and the area of the ground surface covered by it. (Figure 5).

Current which enters the ground through a local earth electrode begins by spreading out around the latter. Then each thread of current bends toward the earth return connection or towards the zone which takes its place if the return operates through a distributed capacitance.

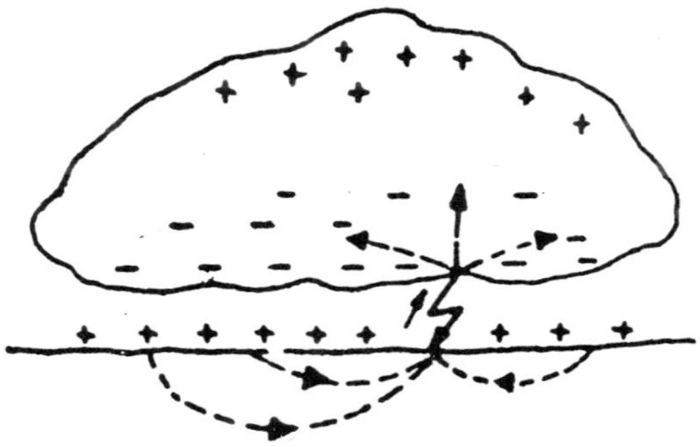

Fig 5
Lightning-Current Circuit

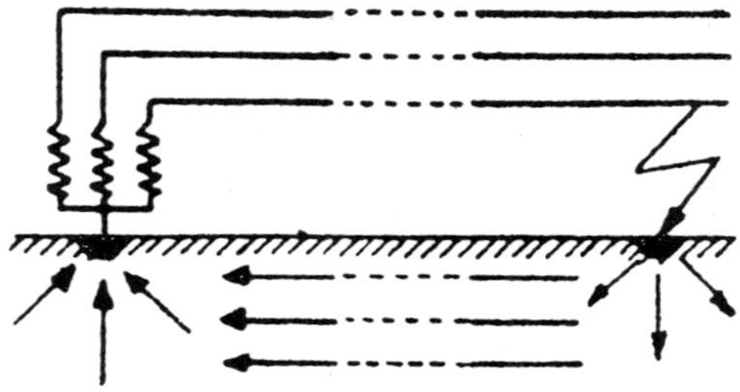

Fig 6
Path of a Fault Current in the Ground

In principle, the range of spread of such a current extends to the whole planet. In practice, the current densities and the potential gradients are obviously only perceptible within a restricted area. As a general rule, and this constitutes an important simplification, we can consider that the entry and exit earth electrodes for the current are suffciently far apart not to affect each other. The current entering by an earth electrode starts by spreading vaguely in all directions up to a considerable distance, as compared with the dimensions of the earth electrode, and it can be assumed that the latter provides the same impedance as if the return earth electrode was uniformly distributed to infinity in all directions. The second phase of the path consists of circulation in the ground and the current thus diluted from the area of entry up to the area of exit. A last phase corresponds to drainage of the current from the return earth electrode to the network if this earth electrode is itself concentrated (Figure 6). It is further obvious that the terms "entry" and "exit", "discharge" and "drainage" are only used as simple references and that the two earth connections operate in a completely symmetrical manner for low-frequency phenomena where the propagation time can be ignored.

These considerations lead to the convenient idea of the natural impedance of an earth electrode, considered independently of the return circuit, as if this latter occurred to infinity in all directions.

As to the passage of the current in the ground between the entry and exit areas, we will confine ourselves here to mentioning that it tends to take place by following the feeder lines at an average depth of some hundreds of meters. The resistance and the reactance along the line of this return path are dependent on the nature of the ground and are generally in the neighborhood of a rather small fraction of an ohm per kilometer.

In any case, there is usually no need to take this impedance into account because this has already been done in the calculations for determining the zero-sequence line impedance of the feeder lines. It should merely be noted that the tendency of return currents in the ground to follow the lines can be more or less falsified if they find other paths which are good conductors, such as metal conduits, cable sheathing, etc. Certain inconveniences may result from this, especially when the electrical continuity of these paths or their links with neighboring earth electrodes is badly arranged.

II. Electrical Characteristics of the Ground

II-1. Ground Resistance; Effect of Temperature; Humidity and Grain Size

The effect of the ground as regards the behavior of earth electrodes is based largely on its resisitivity. Now the resistivity of natural ground has the following peculiarities:

(a) It is extremely variable from one spot to another, according to the nature of the rocks and their degree of humidity. The extreme values met with in practice may vary in the ratio of 1000 to 1 and more.

(b) The ground at a given spot is often very nonuniform, both horizontally and vertically.

(c) The resistivity of the superficial layers of the ground shows considerable seasonal variations, due to the effect of frost and drought, which increase resistivity, or to humidity, which reduces resistivity. This effect is found to have a depth of approximately one or two meters. It is well to count on resistance variations of at least 1 or 2 and, exceptionally, 1 to 5 or more, for electrodes of average dimensions buried at a depth of 1 meter.

(d) It is rare for the resistivity of the ground at the surface, and still less at a depth, to be known before the installation of the plant, and it is often only the installation of the earth electrodes which gives the first indications as to the properties of the ground.

We will give a few values, which are not very accurate, of resistivity in ohm-meters. Resistivity in ohm-meters is mathematically equal to the resistance in ohms of a cube with one-meter edges. It is expressed by a figure one-hundredth the resistivity in ohm-centimeters. A resistivity of 100 ohm-meters corresponds to a conductivity of 10^{-13} cgs, a figure which is sometimes assumed for calculations relating to propagation in the ground.

Heavy soils, clays, marls, and damp chalky sands: Some multiples of 10 ohm-meters which may reach 100 to 200 for very compact clays.

Stony soils, covered with poor herbage: Often 300 to 400 ohm-meters.

Siliceous sands: 200 to 500 ohm-meters when damp; thousands when dry.

Damp peats, low dissolved-salt content: In certain mountainous areas up to 150 to 300 ohm-meters.

Calcareous rocks: Some multiples of 10 ohm-meters in cases of impregnation by underground water; 1000 to 3000 in a dry state.

Old rocks (granite, gneiss, schists, basalts): The older and the more compact the rock, the greater the resistivity. It may reach several tens of thousands ohm-meters for very healthy and very dry granites, but may often drop to values between 1000 and 5000. In the case of rocks which are greatly decomposed and mixed with clay, it may fall to around 100 ohm-meters.

The resistivity of *river waters* varies according to the place and season. It is 16 ohm-meters for the Seine in Paris, 40 for the Garonne in St. Gaudens, 20 to 40 for the Rhine, 30 to 60 for the Rhone, 80 to 300 for the Dordogne, and 5 to 80 for the Gave de Pau. It will be seen that it is approximately the same as for good soils.

Ballasts and dry concentrates are very good insulators. On the other hand concrete buried in a damp soil has a resistivity which scarcely exceeds some hundreds of ohm-meters. It would not be far wrong to assume that the resistivity of buried concrete is approximately the same as that of the ground around it.

The conductivity of the soil is essentially of an electrolytic nature. It is thus very low in a dry state and increases with humidity for a constant temperature, and increases with the temperature for a given degree of humidity.

Trials carried out in England on a light mould gave the following results:[9]

Temperature, C	Resistivity ohm-meters
-15	3300
- 5	790
0 (Ice)	300
0 (Water)	138
+10	99
+20	75

Degree of humidity (%)	Resistivity ohm-meters
0	More than 10^7
2.5	1500
5	430
10	185
15	105
20	63
30	42

Another important element in the ground formation is its grain size, which influences both its porosity and its capacity for retaining humidity and the degree of contact with the electrodes. Large-grain earths such as gravel, pebbles, etc.

make it difficult to establish satisfactory earth electrodes, and it is often necessary to remedy this state of affairs by surrounding the surface of the electrodes with a thickness of fine rich soil or with some other material which is a relatively good conductor.

II-2. Variation in Earth Electrode Resistance with Current

As a general rule the resistance of earth electrodes varies little in relation to the current. The values measured with the usual portable apparatus which involve relatively small currents, are often quite representative of those obtained by injecting heavy currents, if account is taken of the inaccuracies from which these instruments often suffer. The resistance frequently has a tendency, however, to fall as the current increases under the action of somewhat complex factors.

A first reason for this drop may be heating of the earth in the neighborhood of the electrodes, due to the passage of current. The heating, as we have seen, tends to reduce the resistivity of the ground.

A second factor which acts in the same sense is the sometimes defective contact between the metal of the electrode and the surrounding earth. In cases of earth shrinkage there may be a thin air space which offers a high resistance to small currents, but which is readily traversed by arcs caused by a higher voltage applied to the electrode. Similar bad contacts may also occur at the joints of the connections.

Again, with the very heavy currents, another phenomenon is encountered. The ground, considered an imperfect insulator, has a breakdown gradient in the neighborhood of 200 kilovolts per meter[10] for clayey soils, for instance. The result is that a higher gradient cannot exist in the soil in the neighborhood of the electrode without giving rise to internal discharges which increase the equivalent dimensions of the electrode and return the gradients to this limiting value.

This phenomenon is clearly demonstrated if a high voltage is applied to a very small electrode plunged into water for a short distance. Branching sparks will be seen to appear on the electrode, which they make longer in effect, their own length increasing with the voltage (Figure 7). The result is that, however small the original dimensions of the electrode, the resistance cannot exceed a certain value

Fig 7
Flow of a Heavy Current Through a Small Electrode Immersed in Water (Piece of Fine Insulated Wire); Equivalent Hemispherical Electrode

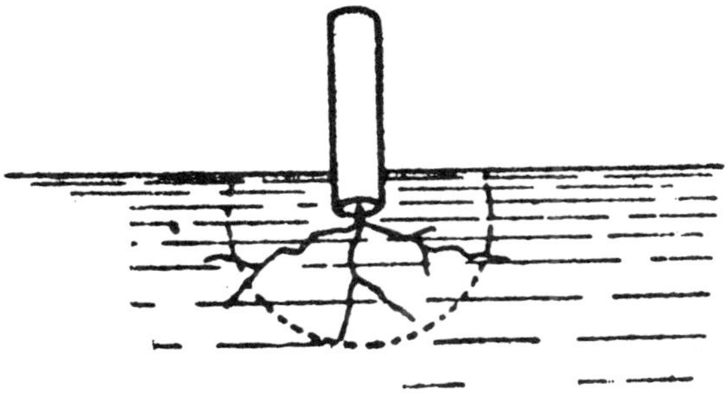

which is the lower, the higher the voltage. The resistance has an approximate value of

$$R = \frac{K}{U}$$

The constant K is in the neighborhood of 3000 ohm-kilovolts for a resistivity of 100 ohm-meters.

This result is easily interpreted from equations given later for hemispherical electrodes. It will be seen that the gradient at the surface of a hemispherical volume discharging current is equal to the quotient of the applied voltage over the distance to the center. Under these conditions the limitation of the gradient to 200 kilovolts per meter implies that the internal discharges in the ground bring the dimensions of the earth electrode (if it has not already exceeded them) up to approximately those of a hemispherical conducting medium, having a radius equal to the quotient by 200 of the voltage applied in kilovolts. The expression above for resistance is deduced if a value of some 100 ohm-meters is taken for the resistivity.

The table below gives the minimum radius r_m to which the electrode is thus enlarged when the full phase voltage is applied to it, as well as the maximum resistance R_M deduced from it and the corresponding fault current I, for a soil of this nature and for different network voltages:

U (kilovolts)	r_m (meters)	R_M (ohms)	I (amperes)
15	0.043	350	250
60	0.17	87	1000
150	0.43	35	2500
220	0.65	23	5600

A comparison of the figures in the second column with the usual dimensions of the earthing arrangements of overhead line supports shows that soil breakdown will rarely occur in good ground at normal operating voltages. It will occur more frequently when lightning strokes are being discharged, since these involve much heavier currents.

Data are lacking regarding the limiting gradients for the ground in the case of poor soils, but the passage of arcs certainly results more frequently.

II-3. Effect of Soil Capacitance

It is easy to verify the fact that the equipotential surfaces caused by the flow of a current through an electrode in a homogeneous conducting medium have the same form as the equipotential surfaces of the electrostatic field developed by the same electrode in a homogeneous insulating medium ure 8).

In the case of a spherical electrode, for instance, the equipotential surfaces are concentric spheres in both cases. A very narrow and very simple relationship thus results between any electrode in a homogeneous conducting medium and the capacitance of the same electrode in a homogeneous insulating medium. This relationship can be obtained readily from any elementary case, for instance, by considering a medium limited by two surfaces S, parallel and removed by L, between which flows a current or an electrostatic field develops.

If ρ is the resistivity of the conducting medium in ohm-meters and K is the dielectric constant of the insulating medium referred to vacuum, the resistance in the first case and the capacitance in the second, respectively, are expressed by:

$$R = \frac{\rho L}{S} \text{ ohms} \qquad C = \frac{KS}{4\pi L} \text{ meters}$$

from which, noting that a microfarad represents 9000 meters, we can deduce directly:

$$CR = \frac{K\rho}{36000\,\pi} \text{ ohm-microfarads}$$

or approximately:

$$CR = \frac{K\rho}{100{,}000} \text{ ohm-microfarads}$$

It will be noted further that the product CR is homogeneous at a time expressed in microseconds in the preceding equation. These considerations allow of linking calculations for the resistance of earth electrodes to those for conductor capacity in an insulating medium. Another important consequence can be deduced from this.

No substance is a perfect conductor or insulator. Every material can be considered as possessing both a finite resistivity ρ and a dielectric constant K, so that each small volume between two sections of equipotential surface can be considered as a combination of a resistance and of a capacitance in parallel.

For a highly conductive material, the resistance will be very small and will completely shunt the capacity of the element. For a good insulating material on the contrary, the effect of the capacity will be preponderant.

In all cases the product $CR = \dfrac{K\rho}{100{,}000}$ will represent (in microseconds) the time constant for the charging and discharging of the capacitance of the elementary volume. The following rule can be directly deduced from this:

A material in which $\dfrac{K\rho}{100{,}000}$ is very small as compared with the duration of the variation in the applied voltages will behave as a pure resistance with regard to these variations. The capacitance charge will not materially restrict the establishment of the voltages. On the other hand, a material in which $\dfrac{K\rho}{100{,}000}$ is large in relation to the duration mentioned will behave like a low-loss capacitance.

With alternating currents, the capacitance does not begin to intervene to an appreciable extent until a certain frequency threshold has been passed, this threshold being very high in practice.

Assume 10 and 4 as average dielectric constants of slightly humid and very dry grounds. The charge and discharge time constants for the first ground will only be 0.01 microsecond for a resistivity of 100 ohm-meters, and that of the second

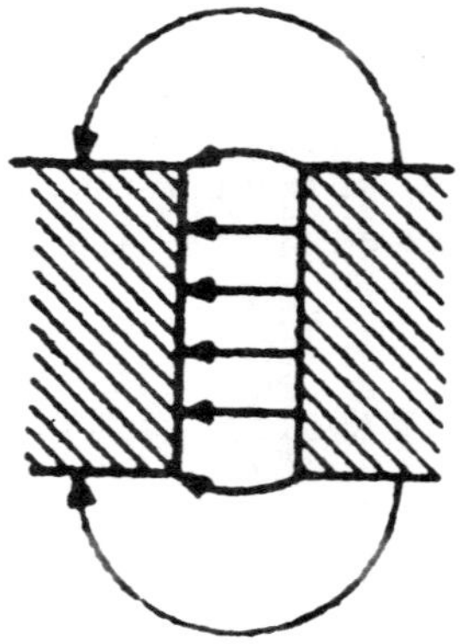

Fig 8
Flow of Current Between Two Electrodes, by Displacement in an Insulating Medium or by Conduction in a Conducting Medium

0.4 microsecond for a resistivity of 10,000 ohm-meters. We can thus check the fact that the ground behaves like a quasi-pure resistance in regard to nearly all electric phenomena. It is only when there is a combination of an extremely high resistance with a very-high frequency or with extremely steep-fronted surges that, if our argument is accurate, the capacitance should begin to have an effect.

III. Review of Earthing Electrodes According to Shape

III-1. Inaccuracy of Practical Earthing Electrode Calculations

Uncertainties as to the resistivity of the ground and its seasonal variations, as the heterogeneous nature of the ground and the more or less good contacts provided, result in it being impossible to claim in practice that earth electrodes can be calculated. It is only possible at the most, when there is an idea as to what the resistivity is, to predetermine the order of magnitude of their resistance and of the potential gradients in their vicinity, or inversely, to estimate the dimensions they should be given for the purpose of maintaining resistances and gradients within the limits desired.

The apparent accuracy of the equations we are going to give should not, therefore, give rise to any illusions. These equations are of interest because they permit comparisons and because of the information they give as to the relative effect of one factor or other. Strict accuracy must not be expected in the final mathematical result. It is thus permissible to simplify without inconvenience and to reduce them to a few broad rules which can be remembered readily and which are quite sufficient in practice, and so avoid the need for referring to curves and nomographs.

This is the essential purpose of this chapter.

III-2. Resistance, Potentials and Local Gradients

The resistance of an earth electrode is not sufficient in itself to characterize performance. It only defines the over-all potential $U_0 = RI$, which the passage of a

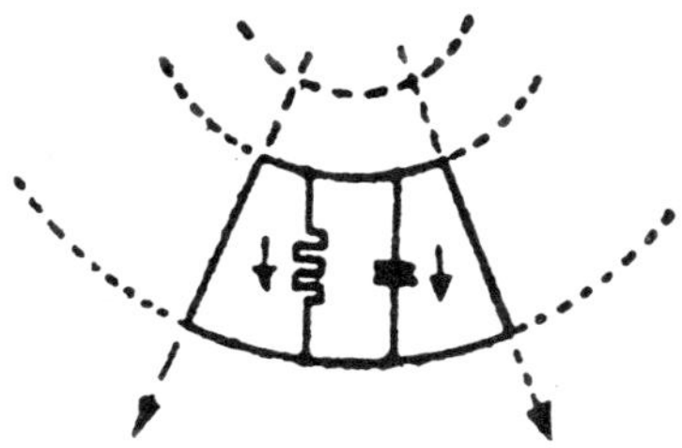

Fig 9
Resistance and Capacitance of a Small Unit of Ground

current I produces in an electrode in relation to the earth at a great distance. The potential U_0 is the integral of all the elemental increases which are met with when following any path from infinity to the electrode. Practical interest attaches to the potential between specific points along this path which may not be the electrode and infinity. In order to characterize an electrode completely, it is necessary in practice to know the distribution, or the "map", of the potential throughout the whole of the surrounding ground, taking into account its natural and artificial lack of homogeneity.

It should be noticed that potentials which have to be considered in nearly all practical problems are between points which are physically close. For instance, the flow of a current through the earth of a substation raises the earthed metal frames of machines to a certain potential. Does this potential risk cause a "reverse flashover" between these frames and another circuit in the substation? This will depend on the potential between the top and bottom of the insulators which carry the circuit. Or again, the danger for anyone passing will be a function of the potential which he picks up across the feet (some 75 centimeters apart), or between feet and hand resting on a structure connected to the frame (Figure 10).

This applies when checking earths after the installation has been completed. To investigate the danger presented (both to persons and to equipment) by an earth electrode, it is advisable to pass a current and to carry a suitable voltmeter for measuring the potentials at suspected points close together, such as insulator terminals, places where people pass frequently and neighboring earthed frames, etc. Data are thus obtained which are much more concrete than those provided when simply measuring the earthed frame potential with respect to a distant reference point.

Another interesting point is that these physically close points, as shown in Figure 10, may be at very different distances electrically. In the case of contact between the feet, the potential U_2 picked up will usually be fairly closely related to the local potential gradient at the ground surface. With contact between feet and hand, on the other hand, the potential U_3 picked up may integrate the gradients covering a path which may be fairly long or complex because the structure touched will occasionally be earthed at a considerable distance from the place of contact.

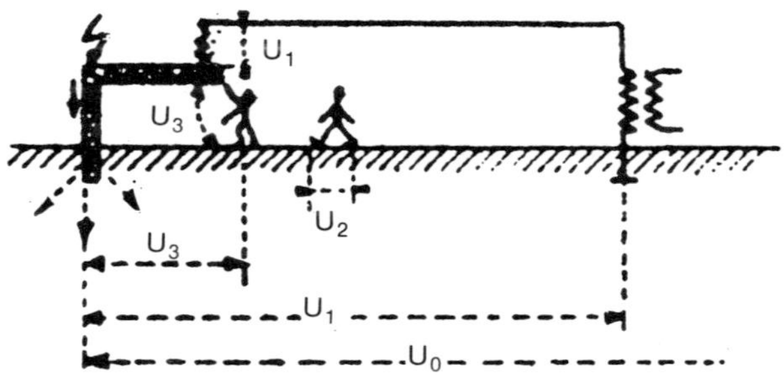

Fig 10
Voltages Which May Result from the Flow of a Current in the Ground

If, for instance, we consider the case of an electrode buried at a certain depth and connected to the ground by an insulated connection, it is obvious that increasing the depth at which it is buried will lower the potential gradients on the surface and will thus reduce the step potential. Although the result is thus to reduce the absolute potential of the surrounding earth, it will increase to some extent the potential between the latter and the protruding structures connected to earth, as well as the danger of placing the hand on them. Burying an earth electrode is not, therefore, a panacea against danger to life.

If we now examine the danger of return flashovers between the earthed frames carrying a current and a safe electric circuit entering the substation, we shall have to ask in what manner the average potential of this circuit is determined. In many cases it will be determined by a very distant earth electrode, or by the distributed capacities of lines which are also at a distance. The circuit will then bring into the substation the potential of the distant ground, and the overall potential $U_0 = RI$, corresponding to the resistance of the frame earths, will appear at U_1 across the ends of the insulators which separate this circuit from the earthed frames.

There will, on the contrary, be only a fraction of this total potential if the earth electrode for the circuit is within the bounds of the substation and shares more or less, as a result, the potential of the earthed frames. At the limit, the potential would be nil if this earth is common with the frame earth. On the other hand, in this latter case, the potential RI at the earthed frames will be propogated by the circuit to distant areas and it will be found there between the terminals of the insulators. Under intermediate conditions, the potential RI will be distributed, in proportions which are a function of the coupling of circuit earth and earthed frames, between the insulators in the substation and those at a distance. This question of coupling will be defined later.

It will be seen that an estimate of the quality of an actual earth electrode introduces local considerations which escape general investigation. The simple factors which will be retained below are, in addition to the over-all resistance R, the potential U_x and the gradient G_x at a distance X from the electrode.

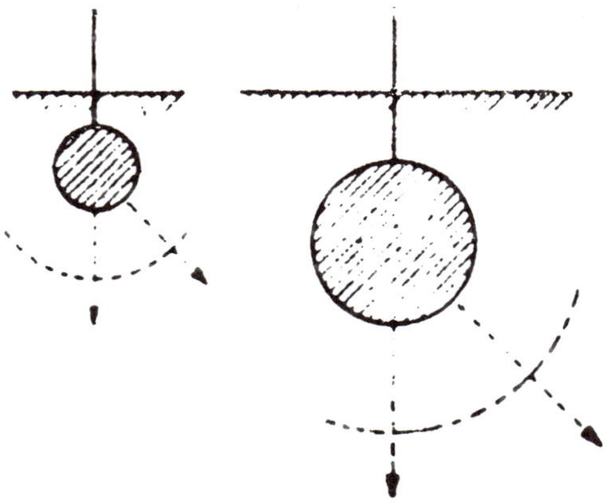

Fig 11
Alternation in Resistance According to Size of Electrode

III-3. General Remarks

(a) Consider a given earth electrode in homogeneous ground and assume that the resistivity of the ground is doubled. The path of the threads of current will not be modified, but all the potential drops will be doubled:

The resistance of an earth electrode of given shape and dimensions is proportional to the resistivity of the ground.

(b) Assume, now, that we expand all the dimensions concerned with the electrode to twice the original figure without altering its form or the resistivity of the ground. The equipotential surfaces will have their surface area quadrupled, but the path of the current between two equipotential surfaces will be doubled. The resistance will be consequently reduced by a half (Figure 11):

The resistance of an earth electrode of given shape in homogeneous ground is inversely proportional to its linear dimensions.

(c) A consequence of this expansion of the whole of the equipotential surfaces along with the dimensions of the electrode is that the resistance of the earthing arrangements, when these are very large, brings in the effect of the distant layers of ground to a relatively greater extent.

This is so because to the resistance corresponding to the distant layers (the only resistance involved in the case of large electrodes), there is added (in the case of small electrodes) the resistance introduced by the intermediate layers, which is greater and thus reduces the relative importance of the former.

The potential on an earth plate or rod has already been absorbed to the extent of approximately half at a distance of one pace. On the other hand, the rise in potential of the ground area of a large substation has a substantial effect up to several hundred meters away.

(d) It would result from the two first laws that, in order to obtain a given resistance in ground with a resistivity 100 times greater, an installation with a volume one million times larger would be required. Luckily, it is possible to escape this consequence, as we shall explain later by showing that the "small dimensions" of very elongated electrodes only come into the equations giving the resistance as logarithmic terms, which variables change slowly.

> The resistance of earth electrodes of very elongated form is only slightly dependent on the "small dimensions" and varies approximately in inverse proportion to the "large dimension".

With very elongated electrodes a ground resistivity 100 times greater only increases by some 100 times the volume of installation work necessary for obtaining a given resistance.

(e) The metal is thus used to better advantage in electrodes of elongated shape and, further, the earthing installations become more costly as the earth resistance rises. The result is that:

> The poorer the conductivity of the ground, the more desirable it is to use electrodes of elongated form.

(f) We will now consider the effect of the depth to which the electrode is buried, assuming that the connection linking it with the surface is insulated.

If the electrode is symmetrical, it is obvious that the current has half the area for discharge when only half buried, as when buried at infinite depth (Figure 12):

> The resistance of an electrode with an insulated connection thus varies in the ratio of 2 to 1 for half and complete immersion respectively.

At intermediate depths an evaluation by rule of thumb, based on these two limits, will usually allow a sufficiently close estimate without further calculation.

III-4. Hemispherical Earth Electrode in Homogeneous Ground

In homogeneous ground the simplest electrode consists of a conducting hemisphere with the flat side level with the surface. The current paths are then extensions of the radii of the electrode and the equipotential surfaces are concentric half-spheres (Figure 13).

Between two equipotential surfaces, separated by X and X + dX from the center of the electrode, the current I has a cross-section $2\pi X^2$ through which to pass.

The corresponding element of resistance is thus $\frac{\rho dX}{2\pi X^2}$.

The voltage drop in the element is $\frac{\rho I dX}{2\pi X^2}$, and the local gradient $G_x = \frac{\rho I}{2\pi X^2}$

The resistance R and the potential U_0 at the surface of the electrode are readily deduced from this by integration, together with the potential U_x at a distance X from the center (Figure 14).

$$R = \frac{\rho}{2\pi r} = \rho \frac{\frac{r}{2}}{\pi r^2}, \quad U_0 = \frac{\rho I}{2\pi r}, \quad U_x = \frac{\rho I}{2\pi X} = U_0 \frac{r}{X} = rGx$$

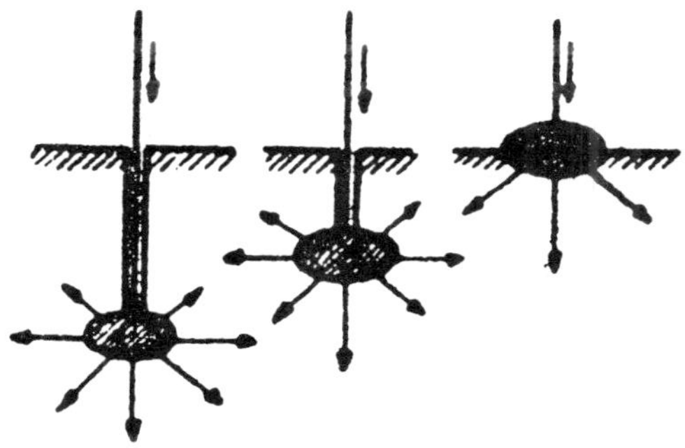

Fig 12
Effect of Depth of Immersion

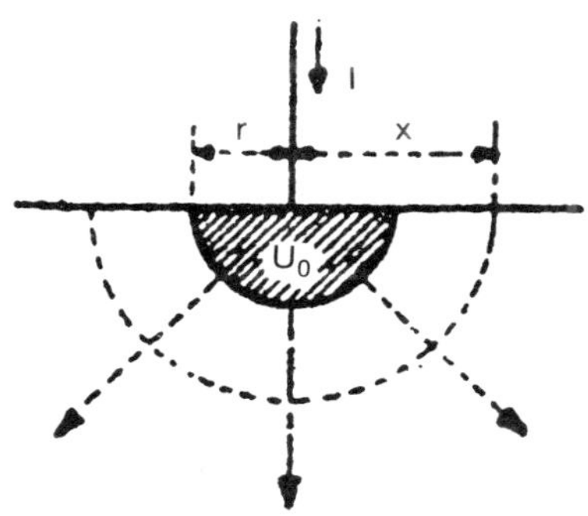

Fig 13
Hemispherical Earth Electrode in Homogenous Ground

It will be seen that:

The resistance offered by the ground to the flow of current from a hemispherical electrode is equal to that of the cylindrical unit of ground having the diameter cross-section of the electrode with a length half the radius. The resistance varies as the resistivity of the ground and as the inverse of the radius of the electrode.

The absolute potential and the gradient at any external point are dependent on the distance of the point from the center of the electrode, but not on the dimensions of the latter. The potential varies as the inverse of the distance to the center and the gradient as the inverse of the square of this distance. The gradient falls relatively quickly. Half the total potential is absorbed at a distance from the periphery of the electrode equal to its radius.

Hemispherical electrodes are scarcely ever used in practice because the concentrated shape results in bad utilization of the metal. It is preferable to spread this over a large stretch of ground.

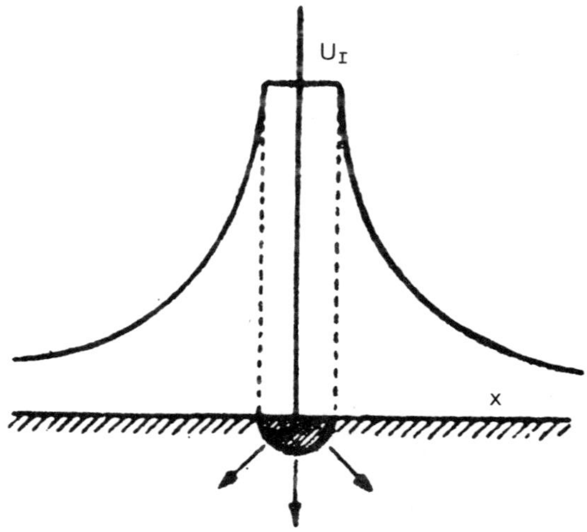

Fig 14
Potential Around a Hemispherical Earth Electrode

It is important to mention that, however irregular the shape of an earth electrode, the equipotential surfaces always tend to approach a hemispherical shape, with increasing distance from the electrode. The absolute potential and the gradient at even a short distance, by comparison with the dimensions of the electrode, are thus dependent neither on the shape nor on the dimensions of the latter and are expressed by the equations given above as a function of the distance and of the current.

Further, as an equivalent hemispherical electrode, which has the same resistance and which the flow of the same current would thus raise to the same potential, can be made to correspond to any kind of electrode.

The radius of this equivalent hemispherical electrode will evidently be intermediate between the three dimensions of the actual electrode. We shall see, for instance, that a vertical rod is equivalent to a half-sphere of radius something like a sixth of the length and that a superficial plate to a half-sphere of radius some 65 percent that of the plate, etc. It is also clear that the resistance of any electrode will be higher than that of any circumscribed hemispherical electrode.

The simple rules given above are thus valid for all earth electrodes, even when not hemispherical, insofar as the irregularities are not too great or as the distance is sufficient for these local effects and irregularities to be inappreciable.

III-5. Buried Spherical Earth Electrode with Insulated Connection

In the case of a spherical electrode of radius r buried at a great depth in homogenous ground, the surface offered for the discharge of current at a distance X is $4\,\pi\,X^2$, that is, twice as large as with a hemispherical electrode with its flat top level with the ground. Whence resistance, potentials and gradients are halved:

$$U_x = \frac{\rho I}{4\pi X} \qquad G_x = \frac{\rho I}{4\pi X^2}$$

$$U_0 = \frac{\rho I}{4\pi r} \qquad R = \frac{\rho}{4\pi r}$$

When the depth h to which the center of the electrode is buried is finite while remaining large in relation to thė radius (Figure 15), the non-uniformity due to the surface of the ground can be eliminated by super-imposing on the electrode and ground systems their image system in relation to this surface and by making a similar current I flow from the image electrode. Considerations of symmetry would readily show that the field in the ground remains unchanged. Further, the individual field of one of the electrodes is not noticeably modified by the presence of the second electrode which is assumed to be of fairly small dimensions. The result is that the field in the over-all area is practically the pure and simple superposition of the individual effects of the two electrodes (Figure 16).

For the potential at a point, distant by X and X′ from the centers of the two electrodes, we can immediately deduce from this

$$U_x = \frac{\rho I}{4\pi X} + \frac{\rho I}{4\pi X'}$$

In particular, if we choose the point at the surface of one of the electrodes, and insofar as the distance 2h is large in relation to the radius r:

$$U_0 = \frac{\rho I}{4\pi}\left(\frac{1}{r} + \frac{1}{2h}\right),$$

whence the resistance

$$R = \frac{\rho}{4\pi r}\left(1 + \frac{r}{2h}\right).$$

The resistance is thus raised in the ratio $\frac{r}{2h}$ in relation to an infinite buried depth. It will be noticed that the magnification factor is only 25 percent for a depth equal to the diameter. It would be 100 percent as we have seen for half emergence, the previous equation then ceasing to apply.

The potential and the gradient at the surface of the ground at a (horizontal) distance X from a point directly above the sphere are (Figure 17):

$$U_x = \frac{\rho I}{4\pi}\frac{2}{\sqrt{X^2 + h^2}}, \quad G_x = \frac{\rho I}{4\pi}\frac{2X}{(X^2 + h^2)^{\frac{3}{2}}},$$

In particular, vertically above the center

$$U_v = \frac{\rho I}{2\pi h}, \quad G_v = 0, \quad U_0 - U_v = \frac{\rho I}{4\pi}\left(\frac{1}{r} - \frac{3}{2h}\right).$$

The potential vertically above the center is the lower, the greater the buried depth. On the other hand, the difference of potential between this area of the

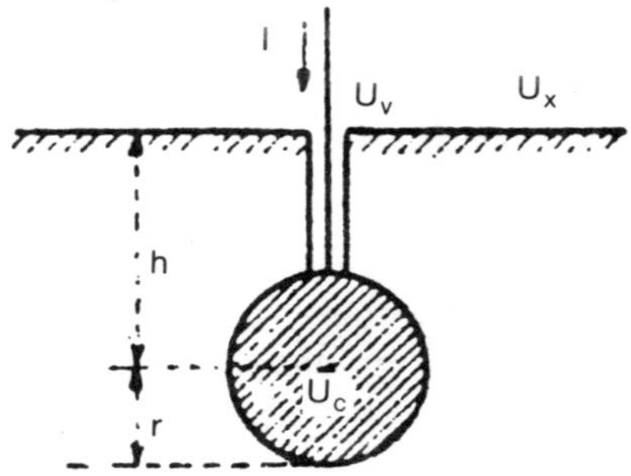

Fig 15
Buried Spherical Electrode

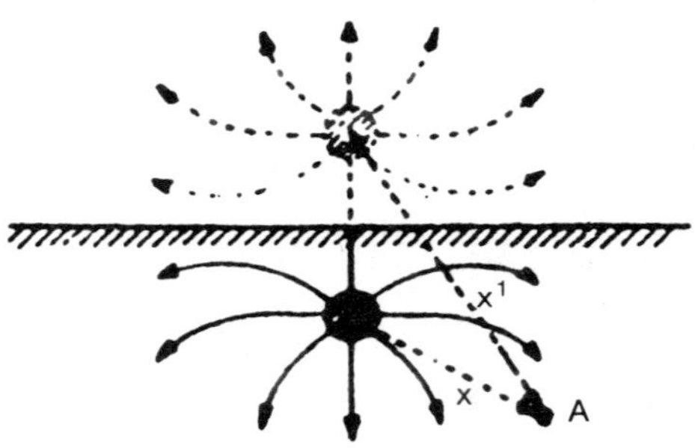

Fig 16
Introduction of Electrode "Image" in Relation to Earth

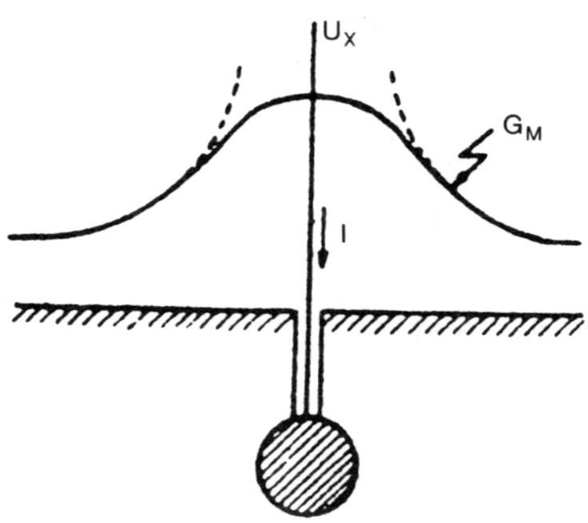

Fig 17
Potential Around a Buried Spherical Electrode

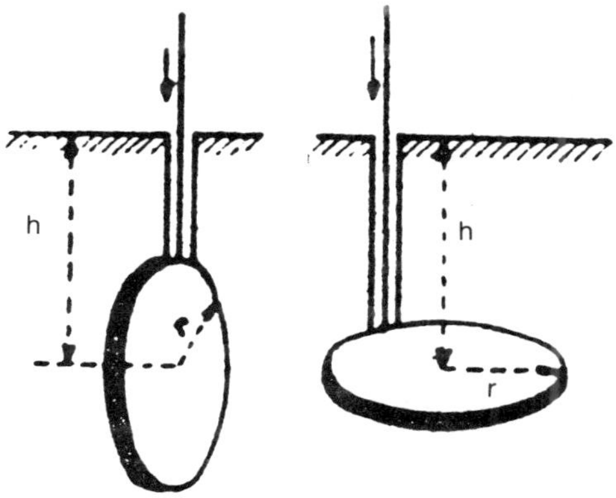

Fig 18
Plate Electrode

ground and the metal structures connected to the electrode increases somewhat with the depth. Burying an electrode is thus no protection against contact with a structure connected to it.

The potential gradient on the ground surface, which is zero vertically above the center, passes through a maximum at a distance from this point equal to 0.7 times the buried depth. The maximum gradient can be written:

$$G_M = \frac{\rho I}{2\pi(1.6h)^2}$$

Comparing this expression with that for a gradient around a hemispherical earth at the surface of the ground, it is found that burying a more or less spherical electrode with an insulated connection to a depth h reduces the maximum gradient to the same extent as an enclosure surrounding it at a distance 1.6h from the center of the surface-electrode.

III-6. Plate Electrode

A horizontal or vertical circular plate of radius r, the center of which is buried to a depth h, with the connection insulated, has the approximate resistance (Figure 18):

$$R = \frac{\rho}{8r}\left(1 + \frac{r}{2.5h + r}\right).$$

This purely empirical equation gives results which diverge very little from the theory. In practice, it is sufficient to remember that

$$R = \frac{\rho}{8r} \text{ for infinite depth,}$$

$$R = \frac{\rho}{4r} \text{ for zero depth,}$$

applying a suitable interpolation factor empirically.

Comparison of these values with those corresponding to spherical electrodes shows that a plate electrode has substantially the same resistance as a sphere of radius 1.5 times smaller.

The thickness of the plate does not seriously affect its resistance. Further, it will be seen later that a plate electrode is not much more effective than a simple ring of the same diameter. Finally, it will be noted that a square plate is comparable to a circular one with a radius equal to 0.6 times the side, or again that a plate one meter by one meter buried at a depth of one m represents a resistance 0.25 ρ, which is close to that of a four-meter rod.

The distribution of the potential on the ground near the plate, buried at a considerable depth, differs only slightly, to a first approximation, from what it would be with an equivalent spherical electrode.

III-7. Loop-Shaped Electrode

The resistance of a horizontal circular loop of radius r, buried at a depth h and consisting of a conductor of diameter d can be expressed, according to Dwight by:

$$R = 0.366 \frac{\rho}{2\pi r} \left(\log \frac{16r}{d} + \log \frac{4r}{h} \right).$$

The second term is approximate and does not apply to extreme values of h. It should be made to tend towards zero for infinite depth and towards the first term for a depth of nil.

As an example, the resistance of a ring with a radius of 50 centimeters, formed of a conductor 1 centimeter diameter and buried 1 meter down, is 0.374 ρ. The resistance of a solid disc with the same diameter would be 0.288 ρ, which is only 23 percent less than that of the ring.

A solid plate is not much more effective than a simple ring of the same diameter.

The distribution of earth potential in the neighborhood of a loop of small dimensions buried at an appreciable depth is approximately the same as with a plate of the same diameter or the equivalent sphere. If, on the other hand, it is no longer a question of a compact electrode, but of a loop with a diameter which is large as compared with the buried depth, the ground potential in the neighborhood of the electrode should be calculated by use of the same equation as in the case (investigated below) of a buried rectilinear electrode.

It is of interest to note that the absolute potential at the center of a loop, not buried or only sunk to a slight depth, and carrying a current i per unit length, has the very simple expression $U_c = \rho i$

A circular loop buried at a shallow depth and carrying one ampere per meter in homogeneous ground has its center at a potential numerically equal to the resistivity of the ground.

III-8. Vertical Rods in Homogeneous Ground

(a) *Note on the calculation of rectilinear electrodes.* The characteristics of earth electrodes of more or less rectilinear shape can be quite simply calculated by breaking them down into small sectional elements between which it can be assumed, to a first approximation, that the current is equally distributed. The discontinuity introduced by the surface of the ground is also avoided by superimposing its image system on the original system, as was done in connection with buried spheres.

To arrive at the potential at a given point, the sum of the effects of all the sectional elements at this point is found (considered as a chain of small separate spheres). To obtain the potential of the electrode and its resistance, it is sufficient to move the reference point to the surface of one of the sections.

This is the same method that will be given later for calculating multiple electrodes. The approximations come, on the one hand, from the assumption of a uniform distribution of the currents carried by the electrode, as if the different sections did not react on each other; and on the other hand from taking a circular section as being equivalent to a sphere, a process which is valid if the reference point is only a short distance away, but which becomes risky when it is actually brought into contact with the electrode. Some authors, such as Dwight, have investigated the problem more closely, but the approximation of the elementary equation remains sufficient in practice.

(b) The *resistance* of a vertical rod with a length L and a diameter d has the following fairly accurate expression (Figure 19):

$$R = 0.366 \frac{\rho}{L} \log \frac{3L}{d} \text{ ohms.}$$

The flow of a current $I = L_i$ thus raises the rod to a potential

$$U_0 = 0.366\rho \frac{I}{L} \log \frac{3L}{d} = 0.366\rho\, i \log \frac{3L}{d} \text{ volts.}$$

Fig 19
Vertical Rod in Homogenous Soil

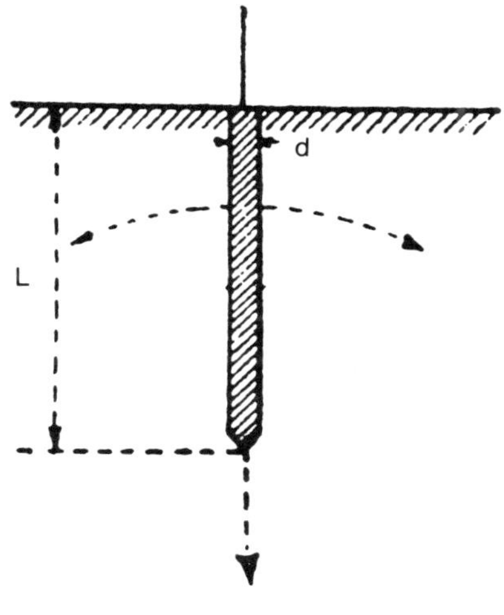

The simplified method of calculation referred to above would give the coefficient 4, instead of 3, for the logarithmic term, and this coefficient 4 has been adopted by many authors. The resultant error is usually of no importance and we have ourselves kept to the rough form of equation for the simplified calculations of the gradients and potentials which will be gone into later.

A particularly interesting value is the earthing resistance RL per each meter of rod, which is also the potential $\frac{U_0}{i}$ to which the rod is raised when it carries one ampere per meter.

$$RL = \frac{U_0}{i} = 0.366\,\rho \log \frac{3L}{d} \quad \text{ohms or volts per ampere per meter.}$$

The resistance per meter increases a little with the length, and as a consequence long rods are a little less effectively used than short ones; resistance falls a little as the diameter increases and enlarging the diameter is thus not without some advantage.

But it should be mentioned that L and d appear in this expression in a logarithmic term which only changes very slowly when L and d have the usual values. This term is almost always comprised between 2 and 3, and doubling the length or diameter only alters it to the extent of 0.3, that is, merely 10 to 15 percent.

The following simple rules can be deduced from this:

> The resistance per meter of rod, or the voltage of the rod when it carries one ampere per meter, is only slightly dependent on the dimensions of the rod and is approximately equal to the resistivity of the soil.

For instance, in a soil with a resistivity of 100 ohm-meters, a one meter rod will have a resistance in the neighborhood of 100 ohms, and will be raised to some 100 volts by the flow of one ampere. A 20 meter rod, again, will have a resistance close to 5 ohms and will be raised to some 100 volts by a flow of 20 amperes.

> Doubling the length of a rod reduces the effectiveness of the meter length by 10 to 15 percent; doubling its diameter increases it by the same amount

It is therefore pointless to attempt to reduce the resistance of an earthing arrangement by increasing the size of the rods. It is more effective to increase their length or number.

As an example, here are overall resistances and the resistances per meter (or potentials for 1 ampere per meter) of 3 centimeters diameter rods of various lengths:

L (meters)	R (ohms)	RL (ohm-meter or volts per ampere per meter)
1	0.73 ρ	0.73 ρ
2	0.42	0.85
4	0.24	0.95
8	0.13	1.07
16	0.07	1.17
32	0.04	1.28

We would mention again that a hemispherical earth, equivalent to a rod, has a radius close to a sixth of the length of the rod. A hemisphere with a resistance of 0.24 ρ has, for instance, a radius of 0.66 meters as compared with a length of 4 meters for the corresponding rod.

(c) *The potential on the ground* at a distance X from the rod (Figure 20) has the expression (which is slightly inaccurate for distances approximating the diameter of the rod):

$$U_x = 0.366\,\rho\,\frac{I}{\mathrm{L}}\,\log\left(\frac{\mathrm{L}}{\mathrm{X}} + \sqrt{1 + \frac{\mathrm{L}^2}{\mathrm{X}^2}}\right).$$

The gradient on the ground at a distance X is:

$$G_x = \frac{\rho I}{\left(2\pi\mathrm{X}\sqrt{\mathrm{L}^2 + \mathrm{X}^2}\right)}\,.$$

These equations can be simplified for distances X which are long or short as compared with L.

For long distances, the gradient evidently tends towards the general law $\dfrac{\rho\,I}{2\pi\mathrm{X}^2}$ which corresponds to hemispherical flow.

For distances which are short as compared with the length of the rod, we get:

$$U_x = 0.366\,\rho\,\frac{I}{\mathrm{L}}\,\log\frac{2\mathrm{L}}{\mathrm{X}} = 0.366\,\rho\mathrm{i}\,\log\frac{2\mathrm{L}}{\mathrm{X}}\,,$$

$$G_x = \frac{\rho I}{2\pi\mathrm{XL}} = \frac{\rho\mathrm{i}}{2\pi\mathrm{X}}\,.$$

The potential found from these equations for the surface of the rod is:

$$U_0 = 0.366\,\rho\,\frac{I}{\mathrm{L}}\,\log\frac{4\mathrm{L}}{\mathrm{d}}$$

(a value which is slightly inaccurate as mentioned above in section (b)). The error is, however, so slight that this expression can be retained in making practical calculations. The difference of potential between the rod and the ground at a distance X, which is short as compared with L, is then:

$$U_0 - U_x = 0.366\,\rho\,\frac{I}{\mathrm{L}}\,\log\frac{2\mathrm{X}}{\mathrm{d}}$$

The mathematical application of this equation shows that:

The voltage between a rod of the usual dimensions and the ground at the distance of a pace is of the order of 50 to 80 percent of the total voltage on the rod, that is, 50 percent for very long rods and 80 percent for very short rods (Figure 20).

III-9. Vertical Rod with Top Buried in Homogeneous Soil

Burying the top of the rod reduces the potentials and gradients at the surface of the ground. For the same bare useful length, burying up to a depth h gives approximately the same reduction in the maximum gradient as an enclosure surrounding the non-insulated rod at a distance of 3 or 4 h (Figure 21).

The resistance of the rod for a given bare length is not affected by burying the head for all practical purposes.

III-10. Rectilinear Electrode Buried Horizontally

(a) *The resistance* of a rectilinear electrode with a length L and a diameter d, buried at a depth h, is expressed by:

$$R = 0.366 \frac{\rho}{L} \left(\log \frac{3L}{2d} + \log \frac{3L}{8h} \right)$$

The second term in brackets is suitable for moderate depths, but should be replaced by zero for infinite depth and by the value of the first term for semi-emergence.

The flow of a current $I = L_i$ raises the electrode, under the same conditions, to a potential

$$U_0 = 0.366 \rho i \left(\log \frac{3L}{2d} + \log \frac{3L}{8h} \right) .$$

The resistance per unit length RL, which is equal to the voltage for one ampere flowing per meter, is usually a little larger for horizontal electrodes than for vertical rods because of usually greater lengths, smaller diameters and shallower depths.

With an electrode one centimeter diameter buried at a depth of one meter or 25 centimeters, the transverse resistance per unit length and the voltage taken by the electrode for one ampere carried per meter have, as common value:

L (meters)	RL (Ohm-meters or volts per ampere per meter) (h = 1 meter)	(h = 25 centimeters)
10	1.37 ρ	1.59 ρ
20	1.59	1.81
50	1.89	2.11
100	2.11	2.33

It is sufficient to remember that:

> The earthing resistance per meter of rectilinear electrodes buried horizontally is of the order of 1.5 to 2 ρ ohm-meters and the flow of one ampere per meter raises the electrode to about 1.5 to 2 ρ volts. The buried depth has no great effect in practice on the resistance.

(b) *The potential difference* between the electrode carrying one ampere per meter and a point on the ground situated at a distance X (which is small with respect to L) from the vertical plane of the electrode is expressed by (Figure 23):

$$U_0 - U_x = 0.366 \rho i \log \frac{X^2 + h^2}{dh} .$$

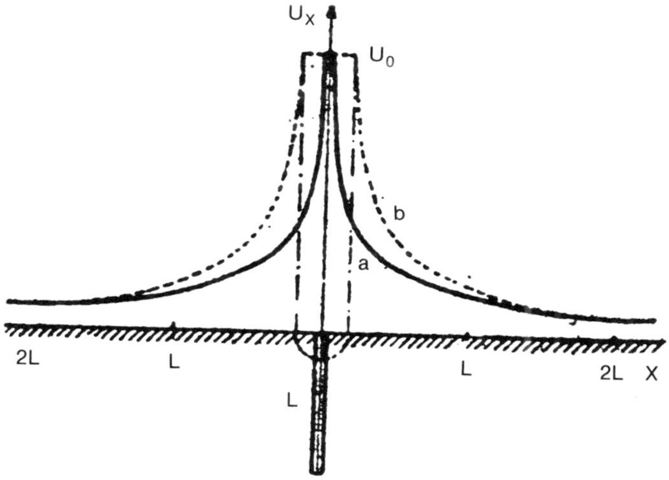

Fig 20
Potential Around a Rod (A) and Around the Equivalent Hemispherical Electrode (B)

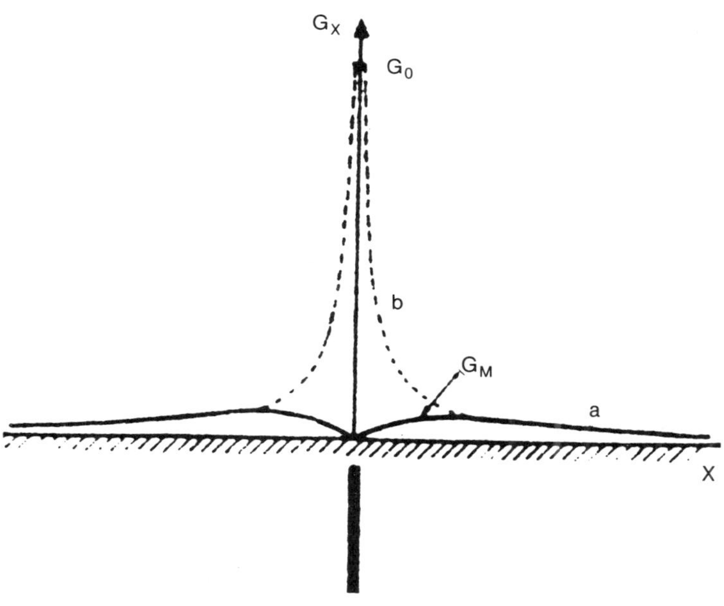

Fig 21
Earth Gradient Around an Earth Rod with Head Buried (A), or Level with Ground (B)

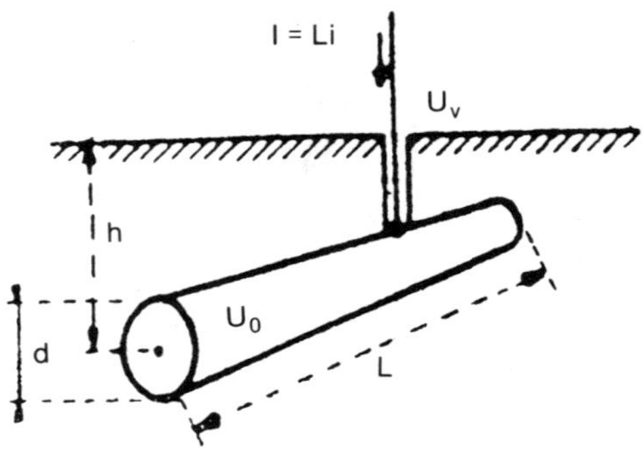

Fig 22
Straight Conductor Buried Horizontally

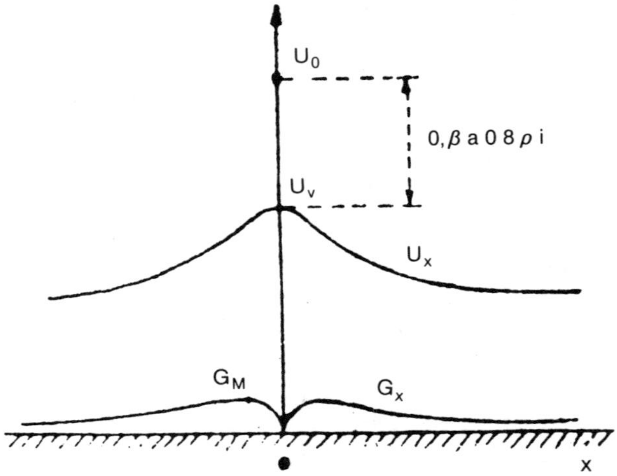

Fig 23
Potential and Gradient in Neighborhood of a Buried Horizontal Conductor

The maximum gradient on the ground is found at a horizontal distance from a point vertically above the electrode equal to the buried depth and has a value of

$$G_M = 0.16 \frac{\rho i}{h} .$$

The application of these principles to usual conditions leads to the following rules:

The potential difference between the buried electrode and the soil around it is of the order of 60 to 80 percent of the product of the resistivity of the soil times the current carried away per meter. It increases slightly with the buried depth.

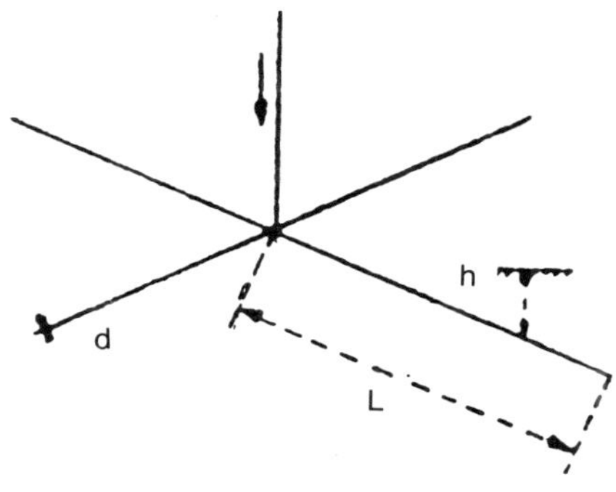

Fig 24
Earth Electrode with Radiating Arms

The potential gradient on the ground falls as the buried depth increases. For depth of the order of a pace, the maximum step voltage is about 10 to 15 percent of the product of the resistivity times the current carried away per meter.

III-11. Earthing Electrode with Radiating Arms

We can apply the following equations given by Dwight, in which L is the length of each arm, d the diameter of the electrode and h the depth at which it is buried (Figure 24):

Two arms at 180 degrees:

$$R = 0.366 \frac{\rho}{2L}\left(\log\frac{4L}{d} + \log\frac{L}{h} - 0.250\right),$$

Two arms at 90 degrees:

$$R = 0.366 \frac{\rho}{2L}\left(\log\frac{4L}{d} + \log\frac{L}{h} - 0.103 + 0.19\frac{h}{L}\right),$$

Three-branched star:

$$R = 0.366 \frac{\rho}{3L}\left(\log\frac{4L}{d} + \log\frac{L}{h} + 0.465 - 0.18\frac{h}{L}\right),$$

Four-branched star:

$$R = 0.366 \frac{\rho}{4L}\left(\log\frac{4L}{d} + \log\frac{L}{h} + 1.265 - 0.93\frac{h}{L}\right),$$

Six-branched star:

$$R = 0.366 \frac{\rho}{6L}\left(\log\frac{4L}{d} + \log\frac{L}{h} + 2.98 - 1.36\frac{h}{L}\right).$$

These equations are only suitable for average buried depths. For zero depth the second logarithmic term should be made equal to the first, and for an infinite depth the resistance should be reduced by half as compared with zero depth.

These equations bring out the fact that the effectiveness of radiating arms decreases when their number is increased.

Assuming arms 10 meters long and one centimeter diameter buried at a depth of one meter, we have the following results:

Number of branches	Resistance (ohm)	Increase of length (percentage)	Gain in conductivity (percentage)
1	0.137 ρ	—	—
2	0.082	100	68
3	0.062	50	32
4	0.053	33	17
6	0.046	50	15
Ring, r = 10 meters	0.034	4.7	35
Plate, r = 10 meters	0.023	—	48

The gains would be slightly better with longer arms. The star with several arms has the further advantage over the simple type in that it facilitates the discharge of the very steep-fronted surges. However, it is usually completely useless to exceed three or four arms.

III-12. Effect of Non-Insulated Connections

Bare buried connections linking the earth electrode with the structures to be earthed behave like extensions of the electrodes. Their diameter is generally small, but it has been seen that this fact is of little importance provided that contact with the soil is well arranged. On the other hand, account must be taken of possible drying out of the superficial layers of the ground in which they are set.

The effect of bare connections on resistance may be of secondary or predominant importance, according to their lengths as compared with that of the electrodes.

Consider a plate with an area one square meter buried at a depth of 1.5 meters in homogeneous ground and connected by a bare vertical connection. The plate along would have a resistance of approximately 0.25 ρ and the connection alone, considered as a rod, a distinctly higher resistance of about 0.70 ρ. Further, the plate will interfere with the diffusion of the current by the connection. In actual fact, the presence of the connection will not modify the overall resistance to any great extent.

Consider now four 3-meter rods at the angles of a 6-meter square and linked by cross-shaped connections. The resistance of the 4 rods alone will be in the neighborhood of 0.10 ρ and that of the connections some 0.11 ρ. The over-all resistance will be of the order of 0.08 ρ. The placing of the rods, provided the ground is homogeneous and that no drying-out effect of the superficial layers is introduced, will not change the value of the resistance by comparison with that which would correspond to the connections along, to any great extent.

It will be seen later that, for substations covering a large area, the connections between the structures to be earthed are usually sufficient to form an excellent earth electrode by themselves.

III-13. Use of Non-Circular Section Earthing Conductors

The circular form of earth conductor gives the smallest perimeter for a given section and thus has the lowest capacity for discharging current.

It will be noted in particular that a long thin strip with a width W is approximately equivalent to a round conductor with diameter $\frac{W}{2}$.

The resistance of a long electrode depends only to a slight extent on its diameter, to such an extent that the gain to be expected from non-circular conductors is generally illusory. Compare, for instance, a one centimeter round conductor and a strip 4 centimeters wide and 0.2 centimeters thick. They have approximately the same cross-section. The strip is equivalent to a two centimeter diameter round conductor, double that of the first one, but for a same buried length of ten meters, the resistance will only differ by 6 percent. The difference would be even more insignificant for longer buried lengths.

It is therefore useless to depart from the circular type of conductor which is the sturdiest mechanically.

III-14. Earth Electrode in a Heterogeneous Subsoil

We will only discuss here the simple case in which the ground consists of a superficial horizontal layer with a thickness H and a resistivity ρ_1 and of an homogeneous subsoil with a different resistivity ρ_2.

We will confine ourselves to indicating briefly how much non-uniformity of the soil is able to react on the resistance of earth electrodes and the gradients in their vicinity without going into the question of methods used for exploring the resistivity of the soil in depth.

(a) *Resistance of a small electrode on non-uniform subsoil.*

If the electrode can be compared to a hemispherical electrode having a radius r, sufficiently small as compared with the thickness H of the superficial layer, its resistance has the approximate value:

$$R = \frac{\rho_1}{2\pi r} + 0.366 \frac{\rho_1}{H} \log \frac{\rho_1 + \rho_2}{2\rho_1} .$$

The first term will be the resistance of the hemisphere in a homogeneous medium having a resistivity ρ_1. The second term which is independent of the radius r, provided that it is small, is additive or subtractive according as ρ_2 is higher or lower than ρ_1. It tends toward infinity for a subsoil with infinite resistance.

To give an example, for an electrode with a radius half that of the thickness H of the superficial layer, the resistance takes (as a function of the relationship

$\frac{\rho_2}{\rho_1}$), the following relative values:

$\frac{\rho_2}{\rho_1}$	0	0.1	1	10	100	1000
R	0.65	0.7	1	1.85	3.1	4.1

Figure 25 gives the general course of the flow of current in the case of a point electrode for different values of $\frac{\rho_1}{\rho_2}$. The same paths would be suitable for all electrodes which conform approximately to the contour of an equipotential surface, and in particular for small hemispherical electrodes.

The potential gradients are practically the same, in the close vicinity of small electrodes, as if the soil were homogeneous and had a resistivity ρ_1. The effect of the subsoil becomes more and more pronounced the greater the distance. It tends to reduce the gradients if the subsoil is more conductive than the surface and to increase them under the opposite conditions by bringing them, very quickly in the first case and more slowly in the second, close to the value corresponding to a homogeneous soil with a resistivity ρ_2.

(b) *Subsoil more conductive than the surface*

The equation given above allows of calculating the resistance of small electrodes. It is generally possible to get an idea of the behavior of more complex electrodes without complicated calculations. If, for instance, the electrode is a long vertical rod having a length H in soil of resistivity ρ_1, and a length L-H in subsoil or resistivity ρ_2, it can be assumed that the ground is homogeneous and has a resistivity ρ_2, by simply reducing the length H situated in the resistant layer in the proportion $\frac{\rho_2}{\rho_1}$.

If the electrode can be compared to a superficial plate of larger dimensions as compared with the thickness H, it may be assumed that the plate is placed directly on subsoil having a resistivity ρ_2, and to the resistance thus found should be added that of a cylinder having a resistivity ρ_2, and to the resistance thus found should be added that of cylinder having a resistivity ρ_1 with a height H and a cross-section equal to the surface of the plate.

As mentioned above, the potential and gradient round an electrode implanted on a subsoil which is more conductive than the surface, very quickly reach (at a distance of approximately the thickness of the superficial resistant layer) values which would correspond to homogeneous ground with the same resistivity as that of the subsoil.

(c) *Subsoil more resistant than the surface*

With subsoil having a very high resistivity ρ_2, the threads of current have difficulty in spreading in depth and run superficially to distances which become all the greater as ρ_2 gets higher. The result, as we have seen, is an increase in the resistance and a less rapid fall in the potential gradients around the electrode as compared with homogeneous soil with a resistivity ρ_1.

A particularly simple limiting case is that of a subsoil which is a perfect insulator. The equation above would then give the resistance an infinite value. An

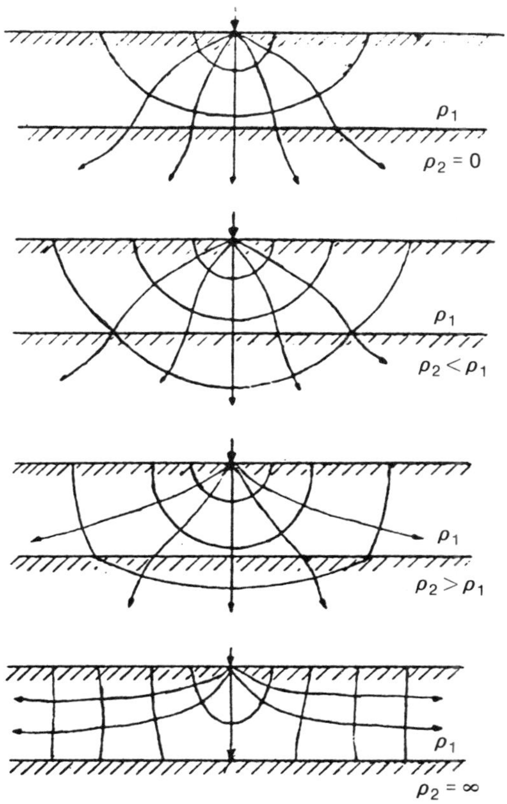

Fig 25
Field Around a Point Electrode, in Nonuniform Ground, for Different Values of Subsoil Resistivity

infinite potential is, in fact, necessary to force the current to flow as far as infinity when it cannot spread out in depth. But it would still be possible to define the potential difference between the electrode having a small radius r and a point situated at a distance X. It will be approximately

$$U_0 - U_x = \frac{\rho_1 I}{2\pi}\left(\frac{1}{r} - \frac{1}{X}\right) + 0.366\,\rho_1\, I \log \frac{1 + \sqrt{1 + \frac{X^2}{H^2}}}{1 + \sqrt{1 + \frac{r2}{H^2}}}$$

Further, in such a case the most simple electrode is no longer hemispherical, but a vertical cylinder having height H and diameter d, and passing through the whole of the conducting layer (Figure 26). The threads of current then spread radially in horizontal planes and the equipotential surfaces are cylinders concentric with the electrode. The current I = Hi has available for its passage, at a distance X from the electrode, a cross-section 2 π HX, whence a resulting local gradient:

$$G_x = -\frac{\rho I}{2\pi HX} = -\frac{\rho i}{2\pi X}$$

and, by integration, a potential difference between two points X and X′:

$$U_x' - U_x = 0.366\ \rho\ i \log \frac{X'}{X},$$

whence, by taking one of the points at the surface of the electrode

$$U_0 - U_x = 0.366\ \rho\ i \log \frac{2X}{d}.$$

We thus find again the same expression as for gradient and potential in the vicinity of a rod in homogeneous ground. This was to be expected because, in this case, too, the current lines extend radially before spreading in depth.

Each multiplication by 10 of the distance X increases the voltage drop $U_0 - U_x$ by an identical amount 0.366 ρ i.

These simple equations can be applied to electrodes of any shape on condition that they are made to correspond to equivalent cylinders having a length H and that the point X is chosen sufficiently far from the irregularities of the surface.

They also apply for moderate distances from the electrode in the usual case in which the soil has not an infinite resistance. At greater distances account must be taken of the diminution in current which continues to follow the superficial layer, since an ever larger fraction of the total current leaves it progressively and spreads in depth.

The mean path of the current in the superficial layer, that is, the distance from the electrode at which half the current has left this layer, can be defined. This distance is approximately

$$X = \frac{\rho_2}{\rho_1} H$$

in the case of a small electrode.

Thus, with a subsoil having ten times the resistance of the surface, the average path of the currents in the more conductive superficial layer is of the order of 10 times the thickness of this layer. The current which discharges through the superficial layer at this distance is only half the total current. The gradient is half what would be computed, assuming the subsoil to have an infinite resistance, or, again, half that which would correspond to a homogeneous soil with the same resistivity ρ_2 as the subsoil (Figure 27).

At definitely greater distances the superficial layer contributes less and less to the current discharge and the potential and gradient can be calculated in practice as if the soil were homogeneous with a resistivity ρ_2.

With an earth electrode of very large dimensions as compared with the thickness of the superficial conducting layer, the latter has practically no effect on the

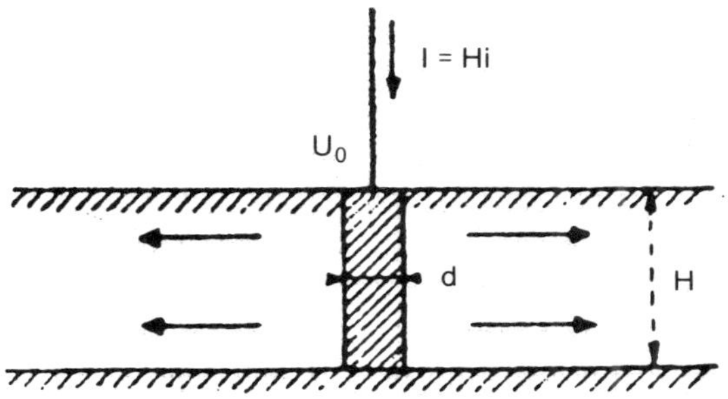

Fig 26
Cylindrical Electrode with Insulating Subsoil

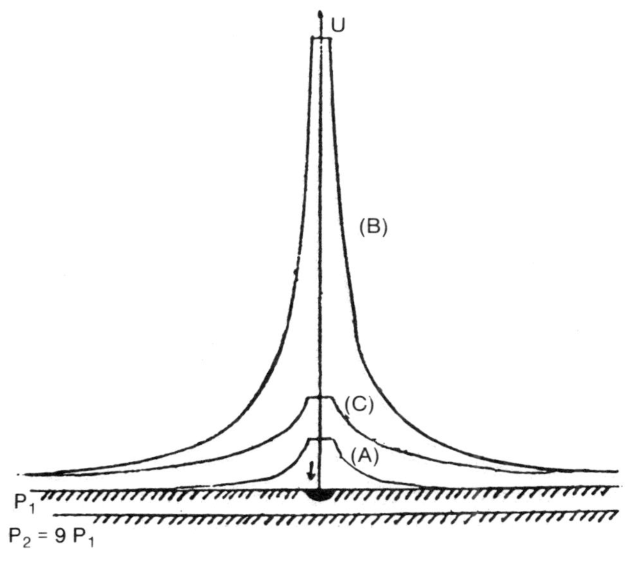

Fig 27
Potential Around an Earth Electrode Having a Radius r:
A, Homogenous Medium with Resistivity P_1;
B, Homogenous Medium with Resistivity $P_2 = 9\ P_1$;
C, Layer with Resistivity P_1 and Thickness 2r on Subsoil with Resistivity $P_2 = 9\ P_1$

value of the resistance. Its only effect is to reduce the local potential gradients in the close vicinity of the boundaries of the electrode.

IV. Multiple Earth Electrodes. Interaction Between Electrodes.

IV-1. Electrodes in Parallel in Homogeneous Soil

(a) *General considerations*

A multiplicity of electrodes in the ground interferes with the diffusion of their partial currents in the soil and thus increases their individual resistances. (Figure 28).

The central electrodes are more affected than the peripheral ones and the current will tend to be displaced towards the latter.

These effects are small when there is a small number of electrodes and the separations are large compared with their dimensions. They are considerable when the electrodes are numerous and close together.

> Above a certain degree of occupation of a ground area of a given size there is practically no gain, as regards resistance, by adding additional electrodes. However, even beyond this limit, increasing the number of electrodes reduces the current discharged by each and thus reduces the local potential gradients.

A basic statement, linked with the fact that the gradient decreases rapidly as the distance from the electrode increases, runs as follows:

> The potential gradients in close proximity to a particular electrode depend solely on the current discharged by this electrode and are only very slightly affected by the more distant electrodes.

In consequence, in the neighborhood of a horizontally buried electrode carrying IA/m, the potential between electrode and ground and the maximum gradient will be approximately the same as if the electrode were single, that 0.50 to 0.80 ρI and 0.10 to 0.15 ρI respectively, however complex may be the whole of the buried circuits to which the electrode considered belongs.

(b) *Review of methods of computation*

The computation of multiple or complex electrodes is based on two statements made above regarding hemispherical and spherical earth electrodes.

First, the effect of irregularities in the shape of an electrode becomes less as the distance from it increases, and the ground potential at a distance X from the electrode by the flow of a current I tends to approach the simple law

$$U_x = \frac{\rho I}{2\pi X} .$$

In the second place, the presence of electrodes which are of small dimensions as compared with the distances causes only a slight degree of non-uniformity in the ground and thus produces little change in the field of an individual electrode with the consequence that the resulting potential, at a point situated at distances X_1, X_2 ... from electrodes discharging currents, I_1, I_2 ... in a homogenous medium, is substantially the sum of the individual effects of each electrode assumed to exist singly (Figure 29).

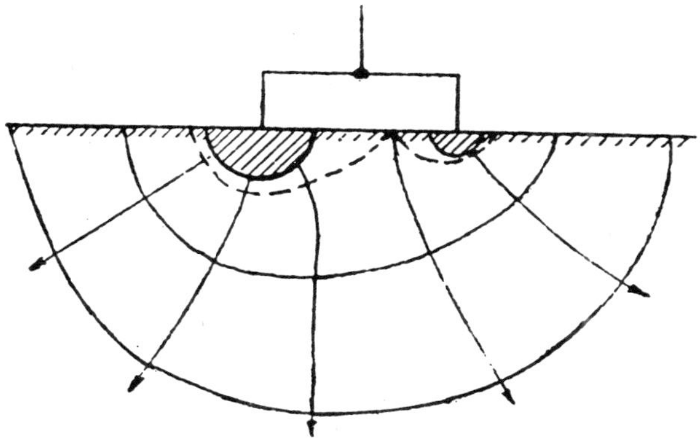

Fig 28
Distribution of Ground Between Two Electrodes in Parallel

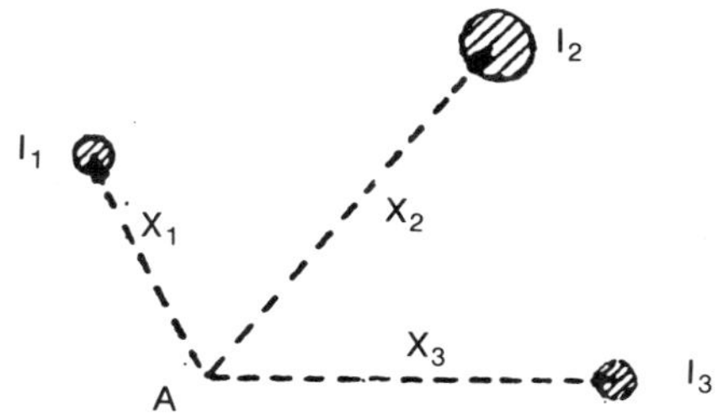

Fig 29
Point A in Relation to Several Electrodes

$$U = \frac{\rho I_1}{2\pi X_1} + \frac{\rho I_2}{2\pi X_2} + \ldots$$

This equation, in which neither the shape nor the dimensions of the electrodes are introduced, is sufficient for taking into account all the electrodes distant from the point considered. When this point is very close to some particular electrode, the more accurate equations given in preceding paragraphs should naturally be used for the term corresponding to this electrode.

This method permits treatment of problems of multiple or complex electrodes with fairly close approximation and we have already referred to it in connection with the calculation of rods and other rectilinear electrodes.

The question arises as to how the total current is distributed between individual electrodes in parallel. The calculation can be carried out by giving arbitrary values $I_1, I_2 \ldots$ to the individual currents and examining the potentials resulting at the surface of each electrode and stating that all the potentials are equal. This is tedious when the number of electrodes is large.

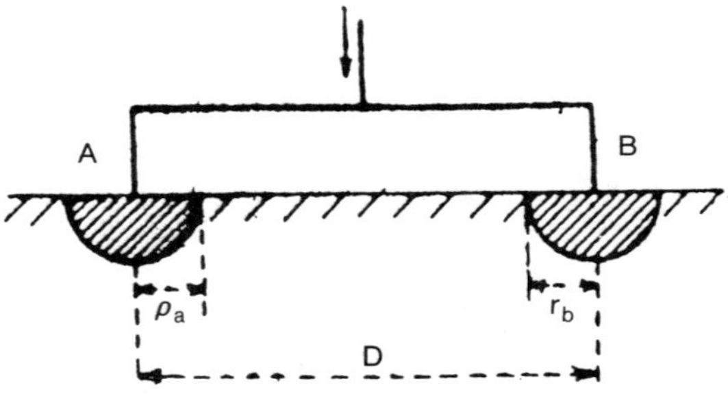

Fig 30
Two Hemispherical Electrodes in Parallel

As a rule, only permissible errors occur with regard to the over-all resistance when assuming that the current is distributed between the electrodes as the inverse of the individual resistances. When calculating the local gradients, the results obtained can be made more accurate by somewhat increasing the contribution of the peripheral electrodes to the disadvantage of the central ones, or in the case of a buried electrode, by raising towards the extremities the current flowing per unit of length. Such corrections can, without objection, be made empirically in view of the low accuracy of practical data as to the nature of soils.

(c) *Two electrodes not overlapping*

As an example, in the very simple case in which two earth electrodes A and B are sufficiently far apart to be considered as two equivalent hemispheres of radii r_a and r_b spaced D apart, we get

$$U_0 = \frac{\rho}{2\pi}\left(\frac{I_a}{r_a} + \frac{I_b}{D}\right)$$

$$U_0 = \frac{\rho}{2\pi}\left(\frac{I_a}{D} + \frac{I_b}{r_b}\right),$$

by taking the reference point successively on A and B (Figure 30), and further

$$I_a + I_b = I$$

The currents I_a and I_b can readily be deduced from these equations as well as the overall resistance $\frac{U_0}{I}$.

In particular, if the two electrodes are identical:

$$I_a = I_b = \frac{I}{2}$$

$$R = \frac{1}{2}\ \frac{\rho}{2\pi r}\left(1 + \frac{r}{D}\right).$$

The coefficient of increase in resistance due to proximity of the electrodes is $\frac{r}{D}$.

We will come back to these calculations in paragraph 4, which deals with coupling between neighboring earth electrodes.

(d) *Vertical rods forming a regular pattern of squares*

The rate of increase of the individual resistances is about:

15% for 2 rods spaced at once their length
15% for 3 rods spaced at twice their length
15% for 6 rods spaced at four times their length

50% for 6 rods spaced at once their length
50% for 13 rods spaced at twice their length
50% for 40 rods spaced at four times their length

100% for 10 rods spaced at once their length
100% for 28 rods spaced at twice their length
100% for 100 rods spaced at four times their length

This table allows for rapid interpolation so that intermediate cases can be dealt with to a considerable degree of accuracy without the use of nomographs. For electrodes other than rods it should be noted that a four-meter rod is approximately equivalent to a plate with one-meter sides or to a hemisphere with a radius of 0.65 meters.

From another standpoint, if R is the resistance of a solid plate covering 1 hectare (100 meters · 100 meters), 30 three-meter rods distributed over this surface will have a resistance 3 R, 56 rods a resistance 2 R and 100 rods a resistance 1.5 R. It will be seen that the ground is almost saturated with a total of 300 meter length of rods.

These evaluations assume that the rods are linked by non-buried connections, an arrangement in considerable favor in certain foreign countries, combined with small-diameter rods driven in with a pneumatic hammer. In this way the work of digging and boring is greatly reduced. If, on the other hand, the connections are buried they will, since their length is usually considerably greater than that of the rods, often be sufficient to saturate the ground without it being necessary to use the rods.

(e) *Electrodes forming horizontal meshes or loops*

As an example we will compare the relative values of conductances and local gradients obtained by burying two or three regularly spaced concentric loops formed from a one-centimeter-diameter conductor in a circular area of ground 80 meters in diameter (Figure 31).

The gradients in the vicinity of the loops, which are proportionate to the current per meter, are designated, starting from the peripheral to the center loop:

Number of Loops	Buried Length, Percent	Conductance, Percent	Local Gradients, Percent		
1	100	100	100	—	—
2	150	123	70	59	—
3	200	135	56	44	41
Solid plate	—	148	—	—	—

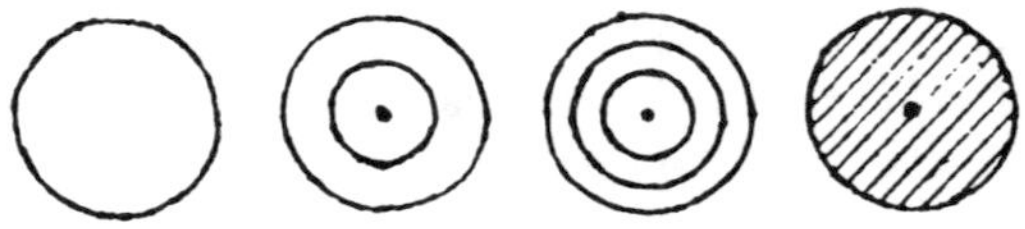

Fig 31
Concentric Rings and Solid Plate

It will be seen that one or two loops are almost sufficient to saturate the ground as far as resistance is concerned. As regards the gradients, on the other hand, increasing the lengths buried secures a gain approximately proportional to the lengths. The fact is confirmed that the currents discharged per meter are greater at the periphery than at the center, but this effect is not very pronounced. Regular distribution would have given a uniform gradient of 50 percent for the three loops, which is not very far removed from the extreme value of 56 percent found for the outside loop.

It will be noted, too, that obtaining a given resistance requires greater length in continuous conductors than in spaced rods. But the cost per meter may be different and each case needs special investigation from the financial viewpoint.

The case for earthing arrangements for very large stations where the length of buried conductors becomes very great will be considered in chapter V.

IV-2. Multiple Electrodes on a Non-Uniform Subsoil

(a) Neighboring electrodes in the superficial layer of the ground interfere with each other relatively less than the subsoil as a better conductor than the surface. The mutual interference between the current paths is confined more especially to the portion situated in the more conductive subsoil and thus only influences a relatively small part of the resistance. If, however, the electrodes penetrate far into the conducting subsoil, the superficial layer can be more or less ignored and the disturbance is approximately the same as with homogeneous ground.

(b) When the subsoil, on the other hand, has a much higher resistance than the surface, it has been seen that the current has difficulty in spreading out in depth and that its effects show much further away from the electrodes.

The obvious result is that electrodes in parallel cause more interference and fewer electrodes are needed to saturate a given area.

In the extreme case of a subsoil which is a perfect insulator, the sum of the individual effects of the electrodes would immediately give

$$U - U' = 0.366\, \rho_1 \left(i_1 \log \frac{X'_1}{X_1} + i_2 \log \frac{X'_2}{X_2} + \ldots \right)$$

for the potential difference between any two points at distances X_1 X_2, ... and X'_1, X'_2, ... from electrodes discharging currents i_1, i_2 ... per meter of conductive thickness.

The distribution of the total current between electrodes in parallel would be obtained by taking the two points of reference at the surface of different pairs of electrodes and by stating that all the corresponding left-hand sides of the equations are zero.

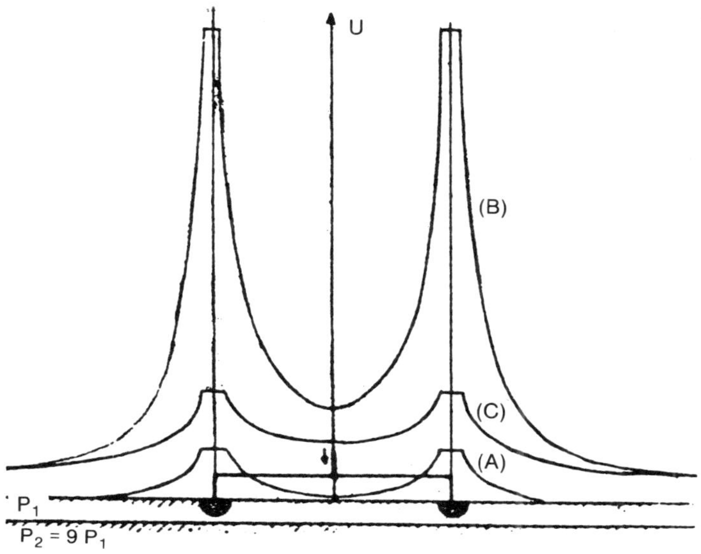

Fig 32
Potential Around Two Earth Electrodes Having a Radius r and Spaced 20r:
A, Homogenous Medium with Resistivity P_1,
B, Homogenous Medium with Resistivity $P_2 = 9\ P_1$,
C, Layer with Resistivity P_1 and Thickness 2r on Subsoil with Resistivity $P_2 = 9\ P_1$

This equation also applies approximately to cases in which the resistivity of the subsoil is finite, on condition that the electrodes are not too far from each other. It gives lower values for the local potential differences than if the high resistivity of the subsoil prevailed throughout the whole ground area. These local potential differences depend essentially on the resistivity of the surface, whereas the overall potential of the area occupied by the earth electrodes depends rather on that of the subsoil (Figure 32).

On a low-voltage rural network run over a granite subsoil and earthed at a number of points separated by several hundred meters, it was found that the discharge of current through the neutral conductor raised the whole of the area supplied to a fairly high fraction of the potential of this conductor.

IV-3. Effect of Proximity of Earth Electrodes Carrying Return Current; Application to Measurement of Earth Electrode Resistances.

The general equations so far given for calculating the earth potential in the vicinity of multiple electrodes, in homogeneous soil or with a subsoil which is an insulator, obviously apply also to cases in which one or more of the electrodes act as return current paths. It is only necessary to change the sign of the currents they carry.

The case of return-current electrodes at a short distance is met with, more especially, in the measurement of earthing resistances when the distance of the

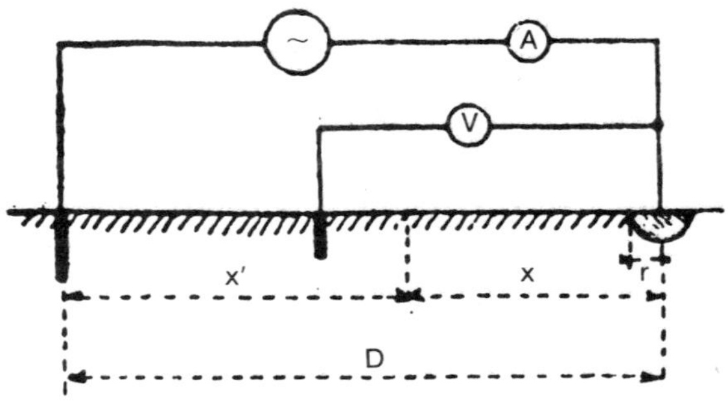

Fig 33
Resistance Measurement of an Earth Electrode

auxiliary electrodes is not great as compared with the dimensions of the earth to be investigated. The error resulting may then be large. Consider the case of an earth to be measured and a return-current earth separated by D in homogeneous ground (Figure 33). The potential of a point at a distance X and X′ from the two earth electrodes will be

$$U = \frac{\rho I}{2\pi X} - \frac{\rho I}{2\pi X'} = \frac{\rho I}{2\pi}\left(\frac{1}{X} - \frac{1}{X'}\right)$$

to the extent that irregularities in the shape of the electrodes have no effect as far as this point. The potential is zero when X = X′, and the additional electrode for measuring the potential should be placed at an equal distance from the centers of the two electrodes whatever the dimensions of the electrodes.

Assume that the electrode to be measured is equivalent to a hemisphere electrode with radius r. Its potential can be deduced from the equation above by taking the reference point on the surface of the hemisphere:

$$U_0 = \frac{\rho_1}{2\pi}\left(\frac{1}{r} - \frac{1}{D}\right).$$

Assuming that the connection for measuring potential has been properly attached at a point of zero potential, it is this voltage U_0 which measurement will show, and

$$R' = \frac{\rho}{2\pi r}\left(1 - \frac{r}{D}\right)$$

will be deduced for the resistance of the earth electrode instead of the theoretical value with the return-current earth distributed to infinity:

$$R = \frac{\rho}{2\pi r}$$

The relative error due to the proximity of the return-current electrode is thus $\frac{r}{D}$, that is 10 percent if the distance from the earth electrodes is ten times the equivalent radius of the tested earth and 25 percent if it is only four times this radius, etc. Since high accuracy is not essential, distances of 10 to 30 meters are sufficient for measuring the resistance of a plate or rod of average value.

On the other hand, in case of a station 200 by 200 meters, equivalent to a half sphere of some 75 meters radius, the return-current earth must be placed at more than 300 meters from the center to ensure that the error does not exceed 25 percent.

Of course the error thus committed is added to that caused by the fact that the connection for ascertaining potential is never placed exactly at a point of zero potential. In the calculations, too, the ground is assumed to be homogeneous and fairly large errors may otherwise be introduced, in particular, when buried cables or ducts pass close to auxiliary earth electrodes.

We would mention, however, that there are cases in which the true value of the resistance has only a platonic interest, as regards safety, and in which it is much more important to measure the potential differences in reference to the different points at which one may stand while touching structures connected to the earth electrode.

IV-4. Coupling Effect Between Neighboring Earth Electrodes

Consider an elementary volume of ground B close to an earth electrode A, which the flow of current raises to a potential U_o. The ground unit B will itself be raised to a potential U equal to a fraction of U_o, and this fraction $\frac{U}{U_o}$ may be called the "coefficient of coupling" of B with A.

In cases where the unit B is small and distant by D from the center of the hemisphere of radius r_a, equivalent to the electrode A, the coefficient of coupling of B with A in homogeneous soil has the simple expression:

$$k_{ba} = \frac{r_a}{D} .$$

When B is also an earth electrode (Figure 30), it is possible similarly to define a coefficient of coupling of A with B, this being equal to the fraction of the potential of B to which the flow of a current through the latter raises A. If B is equivalent to a hemisphere of radius r_b, we now have:

$$k_{ab} = \frac{r_b}{D} .$$

Again, it may be necessary to deal with the ratio K of the potentials which one of the electrodes A or B acquires according as it is inactive or as it is placed in parallel with the other electrode to discharge the current.

The calculation of the over-all resistance of the two electrodes has been mentioned in IV-1. The new coefficient K is deduced from this without difficulty and the fact that it has the same value for A and B can be checked.

It is preferable to deal with the problem in a more general way in noting that the unit comprising any two earth electrodes A and B and the return-current electrode (consisting here of the distant ground) forms a passive system with three terminals. But it is known that the exchanges between the terminals of such a system, however complicated this latter may be, are completely defined by three

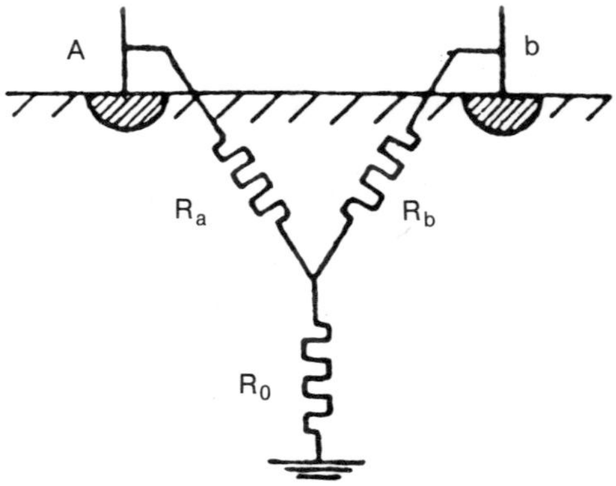

Fig 34
Equivalent Diagram of Two Coupled Earth Electrodes

internal parameters only, for which we can choose, for instance, three resistances R_a, R_b, and R_o joining the three terminals in a Y-circuit (Figure 34).

$R_o + R_a$ is the total resistance offered to the flow of current via A if B is not concerned.

$R_o + R_b$ is again the resistance which determines the potential of B when A is not concerned.

R_o is the part common to these resistances and it thus determines the potential at one of the terminals when the other is the one which discharges the current.

In particular, if it is possible to consider the earth electrodes as equivalent hemispherical electrodes, we have:

$$R_o + R_a = \frac{\rho}{2\pi r_a} \quad , \quad R_o + R_b = \frac{\rho}{2\pi r_b} \quad , \quad R_o = \frac{\rho}{2\pi D} \; .$$

The three coefficients of coupling defined above can be expressed as a function of the three resistances by the general relationships:

$$k_{ba} = \frac{R_o}{R_o + R_a} \quad , \quad k_{ab} = \frac{R_o}{R_o + R_b}$$

$$K = \frac{R_o}{R_o + \dfrac{R_a \quad R_b}{R_a + R_b}}$$

Further, the three coefficients are linked by the relationship

$$K = \frac{k_{ba} + k_{ab} - 2k_{ba}k_{ab}}{1 - k_{ba}k_{ab}} ,$$

which becomes

$$K = k_{ba} + k_{ab} = \frac{r_a + r_b}{D}$$

when the coupling coefficients are small, that is, when the distance from the electrodes is great as compared with their equivalent diameters.

From these expressions it is possible to immediately deduce the following rules are valid in homogeneous ground:

> If the potential of a passive electrode is to be less than 10 percent of that of the active (live) electrode, the former must be spaced from the latter by at least ten times the radius of the active element. If the potential of a passive electrode is to be less than 10 percent of the potential it would acquire if placed in parallel with the active electrode, the two electrodes must be spaced by at least ten times the sum of their equivalent radii.

A subsoil which is a better conductor than the superficial layers tends to reduce the coupling and a more resistant subsoil to increase it.

In cases in which one of the earth electrodes is of very large dimensions, it is practically impossible to obtain low values for the coupling. This point will be taken up again in the next chapter in connection with the question of earthing arrangements for large stations.

IV-5. Advantages and Limitations of Electrode Interconnection

Interconnection of all the safety earthing arrangements within an installation is always desirable. It is the best method of obtaining an earth having the lowest resistance at the minimum cost. It is also the method of reducing the potential differences within the installation to the greatest extent possible.

Interconnecting the earths is also the simplest system to realize, as well as the one which requires the minimum of investigation and thought. Apart from the fact that effective separation of the earths may be, as we have seen, difficult to effect it is often arduous to provide for all the phenomena to which the appearance of considerable local potential differences might give rise in complex installations, the subsoil of which is often crossed by many conducting wires and ducts laid with no general layout in view by workmen from different sections of the undertaking.

Although the same arguments apply to system earths, the solution is not so obvious for the neutrals of high or low potential circuits which extend outside the boundary of the plant.

(a) Where there is a high-potential circuit leaving the station, every earth fault *on the line* causes a flow of current, part of which at least re-enters by the local neutral and raises its potential. If the protective earth is also used for earthing the neutral, the whole of earthed frames and structures of the station is raised to this potential. A separate neutral earth situated in an inaccessible spot avoids in principle inconveniences which might result for personnel. (Figure 35, a and c).

But it may be objected that stations in which the neutral of high-potential networks is earthed are often large ones where the earth for the frames is frequently good enough to carry heavy currents without excessive potential gradients and where an effective separation of the earth electrodes would be very difficult to carry out. It was seen, too, at the beginning of this paper that safety is primarily assured on these networks by high-speed circuit-breaking and not by limiting the potentials on the ground.

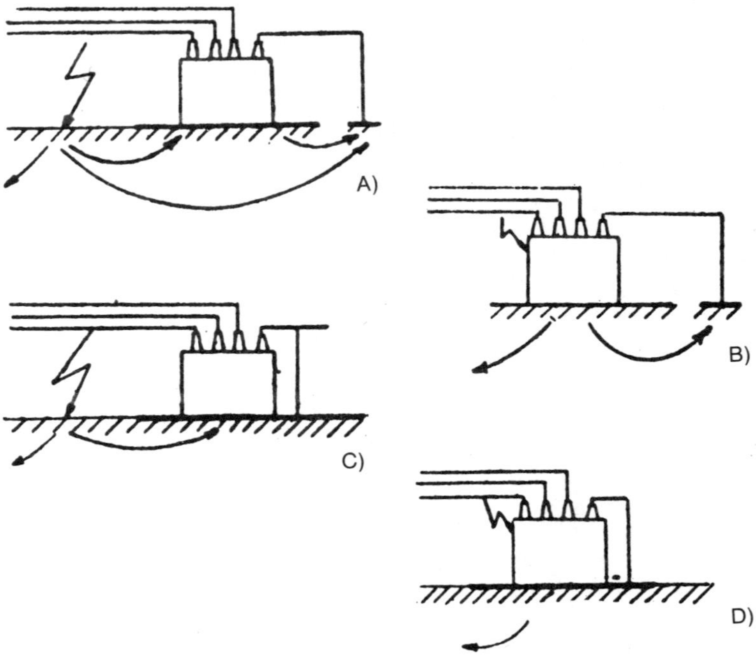

Fig 35
A, Separate Earths: Line Fault;
B, Separate Earths: Fault In Station;
C, Interconnected Earths: Fault On The Line;
D, Interconnected Earths: Fault In Station

Again, if the fault occurs in the plant itself and not on the line, separating the earths compels the frame earthing arrangement to carry the whole of the fault current, whereas their interconnection would offer a metal path to the part of the current which completes the circuit through the local neutral (Figure 35, b and d). The circulation in this case of large currents from earth to earth in a restricted area of ground is, too, not without various risks.

It can be considered that separation of the earthing arrangements for high-potential neutrals is not to be recommended except in exceptional cases.

The question should then be the subject of careful investigation into the positioning of the separate electrodes and the course of the buried conduits and cables.

(b) With neutrals of low-potential circuits extending outside the station, the problem is no longer to protect the staff and the material in the plant, but to protect distant consumers against chance rises in potential of the station earth, a potential which low-potential overhead lines connected to this earth may propagate to localities in which the local potential is greatly different.

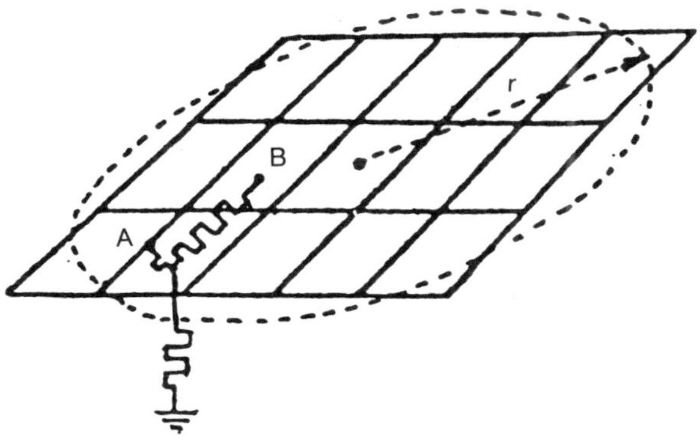

Fig 36
Earth Circuits A of a Large Station; Equivalent Plate (Shown Dotted) and Coupling With a Small Local Electrode B

It may be argued against separation that the earthing arrangements for the low-potential neutral conductor are often very good and that the fault currents to be discharged by the station earth are sometimes very small, for instance, when high-potential networks have an insulated neutral or Petersen coils. Again, potentials will still exist in the station between circuits and frames, from which the low-potential circuit is to be safeguarded. In the case of these potentials being high, flashover may result (apart from the danger to personnel), thus making violent connection.

All these arguments need to be weighed impartially in each particular case. It must be recognized, however, that with the poor earth electrodes still to be found in some small substations, the separation of the low-potential earth is often justified there.

V. Review of Some Special Problems

V-1. Earth Resistance and Potential Gradients in Large Stations

The length of the conductors it is necessary to bury at large stations in order to effect the required connections between frames, structures, etc., is so great that the ground can be considered to be practically saturated and almost equivalent to a solid plate from the point of view of resistance, if not always as regards local gradients.

(a) *Homogeneous ground.* The resistance of a substation with an average radius r, erected on homogeneous ground and having a length L of buried wiring (Figure 36) can be expressed approximately by the following equation:

$$R = \frac{\rho}{4r} + \frac{\rho}{L} ,$$

or still more simply:

$$R = \frac{2\rho}{\text{perimeter}} .$$

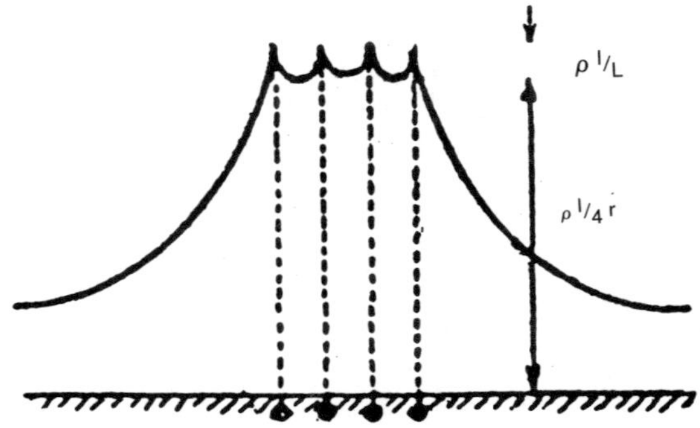

Fig 37
Potential in Neighborhood of Earth Circuits of a Large Station

The resistance of the earth circuits of a large station in homogeneous ground differs only slightly from the quotient of the resistivity of the soil over the half-perimeter of the installation.

In the first expression for R, the first term is the resistance of a superficial plate having a radius r. The second term, which is usually much smaller, expresses the divergence of the real earth as compared with this solid plate. The corresponding voltage $\frac{\rho I}{L}$, which is proportional to the average current flowing per meter of conductor, represents to some extent the potential drop between the buried electrodes and the central area of the more or less regular meshes which they form. (Figure 37).

The difference in potential between the soil in the middle of the meshwork and the structures connected to the earth circuits will thus be of the order of ρi.

The step-potential in the neighborhood of the buried electrodes will be close to 0.10 to 0.15 ρi, being somewhat higher towards the periphery and somewhat lower toward the center.

These rough equations yield only approximations, since they take no account of the size of the meshwork, but the approximate interpretation they provide for the phenomena involved is not without interest.

Consider a substation 250 meters · 150 meters, erected on ground with a resistivity of 200 ohm-meters and having 4000 meters of buried electrodes which have to discharge 4000 amperes in case of fault in the installation. The resistance of the earth circuits will be of the order of:

$$R = \frac{200}{250 + 150} = 0.50 \text{ ohm.}$$

The discharge of 4000 amperes will raise the earth circuits of the station to some 2000 volts in relation to the distant ground.

The part of the resistance which is a function of the buried length will be about

$$\frac{\rho}{L} = \frac{200}{4000} = 0.05 \text{ ohm.}$$

The differences in local potentials in the substation may reach approximately:

$$\rho i = \frac{\rho I}{L} = 200 \text{ volts}$$

and the step potentials in the neighborhood of the buried electrodes will be approximately

$$0.15\ \rho i = 30 \text{ volts}$$

This calculation, though approximate, shows to what extent large stations are usually favorably placed as regards the potentials which may appear, in the case of an earth fault, within the area covered by the earth circuits.

(b) *When the subsoil is a better conductor than the surface*

Assume that the subsoil has a resistivity ρ_2 which is less than the resistivity ρ_1 at the depth at which the earth circuits are buried.

In the resistance equation, ρ_2 should be introduced for the term corresponding to the effect of the plate, and ρ_1 for that corresponding to the local potential drops in the neighborhood of the electrodes. The equation then becomes:

$$R = \frac{\rho_2}{4r} + \frac{\rho_1}{L}$$

In the previous example, if drying of the ground in summer raises ρ_1 from 200 to 1000 ohm-meters, the second term will change from 0.05 to 0.25 ohm, and the total resistance will only vary from 0.67 to 0.87 ohm.

The earths of stations covering a large area will, as a rule, only show small seasonal variations.

The gradients at the surface will be increased to a greater extent than the resistance by drying of the soil, but the very high resistivity of the latter in contact with the feet will then, by itself, often constitute a form of protection for the staff.

The very rudimentary form of the equation above may be criticized since it does not take into account the thickness of the resistive layer. It assumes by implication that this thickness is about half the dimension of the meshes at most. But it may easily be seen that 10 meters additional thickness of a soil with a resistivity of 1000, for a station covering two hectares (20,000 square meters), only introduce a resistance of some 0.5 ohm. It is also easy to adapt the equation to particular cases when measurements of resistivity at different depths are available. It is only given as a guide and does not dispense with the need for a minimum of inference in each specific case.

Similar considerations show that it is generally useless to add deep earthing "wells" to the earth circuits of large stations, even when the subsoil is a better conductor than the surface. A very large total length of rods would be necessary if their effectiveness were to be made greater than the normal earth circuits of the station by themselves.

These conclusions, however, only apply to large stations and it would be unwise to apply them to substations erected on confined areas.

c) *Highly resistive subsoil*

With a subsoil having a resistivity ρ_2, which is much higher than the superficial resistivity ρ_1, the term $\frac{\rho_2}{4r}$ corresponding to the effect of the solid plate becomes much larger and the term $\frac{\rho_1}{L}$, corresponding to the total voltage drops in the neighborhood of the buried conductors, is even more negligible than before as a relative value.

These local potential drops, however, retain approximately the same values in the absolute sense as if the ground were homogeneous throughout and had a resistivity ρ_1. The resistance itself is, practically speaking, dependent on the resistivity ρ_2 only.

It is even more useless to carry out deep boring in such cases. The buried connections are amply sufficient to saturate the ground covered by the station, and any substantial reduction in resistance could only be obtained by a considerable increase in the superficial dimensions of the area occupied by the earth circuits.

On a subsoil with a resistivity of 1000 ohm-meters, a station with a more or less square-shaped area will have a resistance of some four ohms for one hectare (10,000 square meters) covered and about three ohms for two hectares.

V-2. Coupling Between Earth Electrodes in Large Stations

We have seen that the resistance of the earthing network at a large station is some $\frac{2\,\rho_2}{P}$, where P is the perimeter of the installation and ρ_2 the resistivity of the subsoil. Again, the part of this resistance which corresponds to a potential drop between the earth circuits and the center of the mesh is some $\frac{\rho_1 I}{L}$, where L is the total length of buried conductors and ρ_1 the superficial resistivity.

The consequence is that a small independent earth electrode B, placed in the center of a mesh (Figure 36), will have a coupling of some

$$k_{BA} = 1 - \frac{\frac{\rho_1}{L}}{\frac{2\,\rho_2}{P}},$$

with the frame earthing system A, or in homogeneous ground:

$$k_{BA} = 1 - \frac{P}{2_L}.$$

As placing the small electrode B in parallel with A would not substantially modify the latter's resistance, this coefficient also represents the ratio of the potential assumed by one of the electrodes depending on whether it is separated from the active electrode or connected to it.

With homogeneous ground and a station having a half-parameter of 400 meters embracing 4000 meters of buried conductors, the coefficient of coupling will be some 90 percent, and the separation of the electrodes will only reduce the potential on the inactive electrode by 10 percent.

When the subsoil is a better conductor than the superficial layers, the coefficient of coupling will be less, but will remain considerable.

The above calculations only give a rough value, but one confirmed by experience. It is generally found that the coupling between earths situated within the area of large stations is of the order of 50 to 100 percent.

To obtain reduced coupling factors, the electrodes to be separated must be removed well outside the area of the earthing circuits.

V-3. Flow of Earth Along Overhead Earth Wires

An earth wire connected to earth at many points behaves as a conductor having a certain longitudinal line impedance Z_1 and a certain transverse line conductance $\frac{1}{R_2}$. If the line is sufficiently long, calculation shows that it behaves towards the source like an equivalent impedance independent of this length having the value:

$$Z = \sqrt{Z_1 R_2}$$

A 70 square millimeter steel earth wire of the type generally used on overhead lines has a longitudinal impedance of four ohms per kilometer. Assume that it has three earth connections per kilometer with an impedance of 30 ohms each. The transverse resistance per kilometer R_2 will be some 10 ohms and the wire, seen from the source, will appear as an impedance of 6.3 ohms, which is slightly inductive and will act in parallel with the earth resistance proper of the station if the wire is connected to the structural framework of the latter.

If several lines terminating at a station have earth wires connected to structural frameworks, the conductance of the earth wires of the different lines will add up more or less arithmetically.

It will be noticed that the equivalent impedance of the earth wire in the calculation above is roughly equal to the longitudinal impedance of 1500 meters of cable only. This shows that the currents do not travel far along these very resistive wires and flow rapidly to earth (Figure 38).

V-4. Discharge of Earth Currents Through Buried Cables

The sheathing of a large cross-section armored cable, if it is connected to the frame earthing of a station, and leaves the area of the latter, constitutes an excellent earth electrode which tends to shunt an important part of the currents to be discharged into the ground.

The impedance of such a discharge path, viewed from the station end, can be calculated in the same way as is done for an earth wire by estimating the longitudinal line impedance and the transverse line conductance. This method of calculation is very inaccurate because neither is the longitudinal inductance (which is a function of the distance from the buried cable to the feeder lines) precisely known, nor is the transverse resistance, which is dependent on the resistivity of the ground, the quality of the contacts and the useful equivalent length of the sheath. However, as an approximate value is sufficient, the first of these factors can be estimated at 0.5 ohm per kilometer and the second at 4ρ ohms for one meter, that is, 0.004ρ ohm-kilometer. The longitudinal resistance can be readily calculated from the cross-section and resistivity of the sheath.

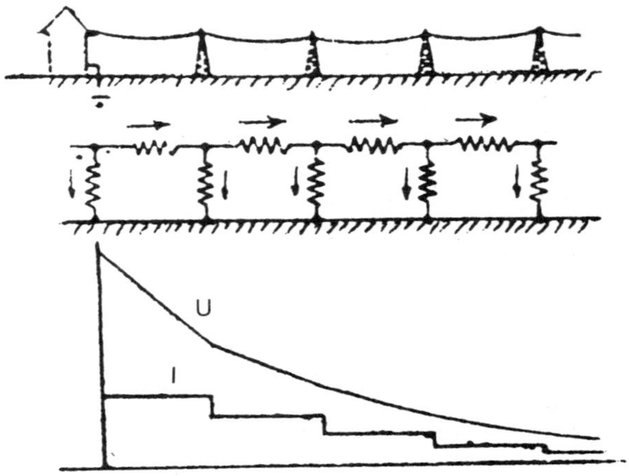

Fig 38
Flow of Earth Current Through an Earth Wire

The resulting impedance thus obtained will generally be about a fraction of one ohm, or of a small number of ohms, according to the cross-section of the sheath and the quality of the ground. In most cases the duration of the faults will be too short for heating of the lead covering by the diverted current to constitute a problem. In critical cases, the sheathing could have an alternative path provided over a certain distance by the use of a bare copper conductor connected to the junction boxes.

If the cable is sufficiently long the potential on the sheath will fall progressively from the station as a result of *longitudinal* potential drops until it blends with that of the distant ground. It is of interest to note that inductive potential drops have their full effect on the internal conductors in the cable, whose magnetic coupling with the sheath is perfect. In consequence, an internal conductor, if connected at one of its ends to the frame earthing of the station, will also arrive on its part at the other end of the cable with a potential close to the local potential, insofar as it is possible to ignore the longitudinal ohmic drops as compared with the inductive.

This self-adaptation of the potential of the internal cable conductors to the local potentials of the territory traversed is of particular interest as regards telephone cables. It obviously requires that the resistance of the sheath should be made low, as compared with the reactance, by appropriate means.

Apart from the cable sheathing all metal conduits (wiring, piping, etc.) leaving the station contribute towards discharging earth currents. This may cause incidents or constitute a danger, especially when the metallic continuity of the conduits is not perfect. If, however, they are properly arranged, they can play a useful part. Without stressing the point, it will merely be mentioned that it is generally much easier and safer to guide the currents by increasing the number of conducting connections than to attempt to bar their path by means of increased distances and barriers, the insulating value of which is often doubtful or insufficient.

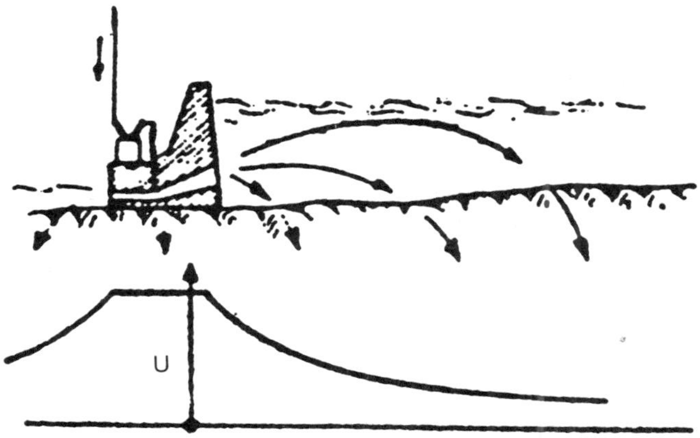

Fig 39
Flow of Earth Current Through Headwater at a Dam

V-5. Discharge of Earth Current by Sheets of Water

Geological faults having good conductivity and considerable thickness and sheets of surface or underground water may play the part of channels draining or discharging earth currents in cases in which the general ground is very resistant.

A case of particular interest is that of power stations adjoining dams built on granite. These frequently have a step-up substation, the earth electrodes of which often have to discharge heavy currents. The installation of good earthing systems in such ground is very difficult.

On the other hand, the earth circuits of the substation and power station are in good contact with the headwater via the metal conduits passing through the dam, and the latter's own conductivity which is high because of the dampness of the concrete.

The resistivity of the water in the most unfavorable cases rarely exceeds 100 to 200 ohm-meters and may be very much lower than that of the adjacent rocks. The headwater then behaves to a certain extent as a conducting channel which will divert a portion of the fault currents and finally diffuse them into the ground (Figure 39).

It is true that such a conductor has a very high longitudinal resistance which is about the same for a hectare section of water as for a one square millimeter copper wire. Calculations show, however, that the impedance of a large volume of headwater, seen from the dam, is moderate and in the neighborhood of only a few ohms. The portion of the fault currents flowing through the water may thus be considerable.

As a result of measurements made by M. Bodier at the Valentine station on a shallow canal, we have suggested improving the discharging capacity of the headwaters by burying them in considerable lengths of copper conductors connected to the frame earthing of the power station with the aim of reducing the longitudinal resistance to a very low value. It is often possible, even in very poor quality ground, to limit the resistance of the ground to values less than one ohm.

Fig 40
Restricted Area of Conductive Soil In a Resistant Mass

However, it may be of interest to check the fact that the flow of current does not raise the potential gradient at the surface of the water to excessive values at locations liable to be frequented.

V-6. Restrictive Area of Conductive Ground at a Very Resistant Site

It sometimes happens that a hollow is available for installing an earthing system which contains high conductivity sediment, isolated in the middle of a highly resistant mass.

The value of such an earthing arrangement should not be over-estimated. Its resistance comprises two terms, one corresponding to the potential drop across the conductive ground and the other to the potential drop in the resistant medium. (Figure 40).

It is possible to modify the first term, but not the second, which is dependent on the dimensions of the restricted spot of good conductivity only. Let r be the average radius for the conducting zone. In comparing this to an electrode with a shape intermediate between plate and hemisphere we arrive for the second term at a resistance of about $\frac{\rho}{5r}$, which represents the minimum resistance that can be obtained when using the conducting isolated zone.

If the resistivity of the rock is, say, 2000 ohm-meters, and if the conducting zone has an average radius of 20 meters, the resistance of the earthing arrangement could not fall below 20 ohms.

It remains to design the electrodes in such a way that the potential drop in the conducting soil does not increase this figure unduly. If we allot, say, 5 ohms for the passage through the conducting soil and if the resistivity is 100 ohm-meters, it would be necessary to bury about 20 meters of vertical earthing rods or 40 meters of horizontal electrodes.

V-7. Effect of Relatively High Conductivity Geological Faults

The appearance of such conductive faults is sometimes counted on for reducing the resistance of earth electrodes in rocky ground. Their importance should not, however, be over-valued when they are narrow and only spread in area after having reached a considerable distance.

All that has been said regarding superficial layers which are better conductors than the subsoil also applies to geological faults. They behave as diffusers, making it possible to connect up to the ground through electrodes of smaller sizes, but it is not wise to count on them to carry current to great distances while maintaining moderate voltage drops. It results from what has been said in paragraph III-14, that half the current will already have left the fault and spread out in the media which, though more resistant, are more extensive, at a distance from the electrode 10 to 100 times the half-thickness of the fault, according as this latter is 10 to 100 times more conductive than the neighboring media.

V-8. Improving Earth Electrodes by the Use of Conducting Salts, Coke or Wood Charcoal

It is possible to improve the conductivity of soil around earth electrodes locally by the injection of electrolytes such as common salt or sodium carbonate. The resistance of the connection first begins to fall as the salt diffuses into the ground and then finishes by increasing again after water has carried the salt away from the electrode.

The time the treatment is effective is variable according to the permeability of the soil. In many cases it may attain or exceed ten years. On the other hand, it is almost zero when there is an appreciable flow of underlying water or when the ground is very porous.

The diffusion of the salt can, however, be retarded by various means, for instance, by using insoluable gels as electrolytes (a system tested by M. Ducrot in 1944 and taken up again in Sweden by M. Sanick).

For the injection of salts to have an appreciable effect on the resistance, the area of ground treated must be large as compared with the principal dimension of the electrodes. Reduction in resistance will be considerable with a rod of small size or a plate, but the method will be of no benefit for improving the earthing arrangements of large stations.

Common salt accelerates the corrosion of electrodes, but to an extent which is not serious if copper or soft steel are used. The latter should not be employed with salt, however, in chalky soils.

Another means of improving earths consists of surrounding the electrodes with a bed of coke, or better still with wood charcoal, which is distinctly less corrosive.

It should be mentioned that these conducting beds do not, as a rule, increase the largest dimension of the electrodes to a great extent, so that their effect on resistance is often poor. They make it necessary, too, to increase the diameter of the excavations, when using rods, and it is then less expensive to increase the number of small cross-section rods. On the other hand, they have the advantage (according to experiments made in England) of reducing seasonal variations in resistance at the same time as they increase the current that the electrodes can carry without the ground being heated dangerously.

V-9. Currents Diverted Through the Human Body Between Two Points of Contact

The risk of electrocution is a function of many factors such as the value of the current, its duration and path. The parts of the body particularly sensitive to the passage of currents are: the heart, the beats of which may be stopped or disarranged; the ganglion controlling respiratory movements; and the hands, because of the muscular contraction which prevents release of a live conductor carrying alternating current. The contraction threshold is usually 10 to 15 milliamperes. The threshold of cardiac fibrillation appears to be higher than 50 milliamperes for periods of contact of one second or more. We have no accurate information as to currents through the ganglion causing respiration to cease.

The most dangerous contacts are, in fact, between head and feet and between hands and feet. Contacts between foot and foot are of little danger except from resultant falls which may of course change them into more dangerous contacts. Further, it is impossible to disassociate the idea of danger due to a current from that of the probability of contact. As seen at the beginning of this paper, supports for lines may be raised to a potential of thousands of volts without any appreciable danger resulting for the public, if they are situated in a not-too-frequented spot and if the damage to insulators is sufficiently rare and sufficiently, quickly eliminated to restrict danger to, say, a fraction of a second over several decades.

As, however, it is necessary to adopt a figure for purposes of comparison, we will assume that a current of 50 milliamperes through the body is a dangerous value which should not be reached, insofar as it is desired to ensure safety by limitation of potential only. The rules and equations given in preceding chapters permit calculation of the potential U, which separates two points A and B between which the body will establish a contact, and we must now determine the resultant current through the body. For this purpose we will apply (as in Chapter I-5) Thevenin's theorem (Figure 3):

> The current which will appear in the body between the point of contact A and B will be equal to that which the pre-existing potential U will cause to flow in the circuit constituted by the body closing on the impedance of the external network viewed between points A and B.

If the impedance of the external system, seen from AB, is low compared with the resistance R_c of the body, the current shunted by the latter will be substantially $\frac{U}{R_c}$. In another extreme case in which the impedance of the circuit external to the body is high compared with the resistance of the latter, the potential drops in this external impedance will determine the current diverted.

We can distinguish two conditions, according as to whether it is a question of direct contact with a live conductor, or simple shunting through the body of the whole or part of the path to earth of a fault current.

(a) *Direct contact with a live conductor*

In this case, the pre-existing potential is generally that of the network. The impedance of the external system, viewed between the hand and feet of the victim, comprises the impedance of the small earth electrode formed by shoes resting on the ground, then, in series, the impedance of the return circuit to the network which is a function of the position of the neutral (Figure 4), and finally the longitudinal impedance of the circuit.

These latter terms (except with low-voltage networks having an insulated neutral) will always be low as compared with the resistance of the body. On the other hand, the resistance of the earth electrode consisting of the feet may be high if they are resting on material which is a poor conductor. The resistance of the body is a value which is not clearly defined. Roughly, the internal tissues may be compared to an electrolyte having a resistivity of some 4 to 6 ohm-meters. These tissues are protected by the skin wall, which offers a resistance of several tens of thousands of ohms per square centimeter in a dry state to penetration of the current, but much lower when the skin is damp, in which state it is fairly sensitive to the value of the potential applied. The skin may also show gaps where there are wounds or scars, and is perforated by potentials of a few hundreds of volts only.

Actually, the resistance measured between a hand and the soles is, in the majority of cases, higher than 3000 ohms, but it may fall to values less than 1000 ohms, when the skin loses all protective power and the shoes are good conductors. Here again, we must confine ourselves to a typical value and will choose 3000 ohms for this purpose. The variables and uncertainties in these problems are so many that it is hopeless to attempt to deal with them more fully in this brief paper. The resistance constituted by the feet is calculated by considering them as two small neighboring earth plates. On ground with a resistivity ρ, each foot, flat down, will have a resistance of some $2.5\,\rho$, and the two feet in parallel a resistance of about $1.5\,\rho$. When there is direct contact with live conductor U; the current shunted through the body between a hand and the two feet resting on the ground will be (on these assumptions):

$$i = \frac{U}{3000 + 1.5\,\rho} \quad \text{amperes.}$$

For resistivities up to some 1000 ohm-meters, the body resistance will almost solely determine the current flowing. For resistivities of 10,000 ohm-meters or over, on the contrary, the body will only be subjected to a small part of the total potential. This fact shows the value of insulating coverings, such as wood or very dry concrete, ballast, bitumen, etc., in places where there is considerable risk of contact with live objects.

(b) *Diversion by the body of the path of a fault current*

We will now examine the case in which the body simply shunts the path to earth of a fault current (Figure 10). If the shunting occurs between one foot and the other when walking, with one of the feet not completely flat down, the impedance of the ground from one foot to the other will be about $6\,\rho$ and the shunted current will have the approximate value:

$$i = \frac{U}{3000 + 6\,\rho} \quad \text{amperes.}$$

U being the pre-existing potential.

If, on the other hand, the shunt takes place between the feet resting on the ground and a hand touching an object connected to the earthing system, the resistance of the return path as far as the hand must be introduced into the denominator in series with the resistance $1.5\,\rho$ at the entrance into the ground through the two feet. This return may take place through two paths in parallel, one direct, constituted by the earth electrode and its connection to the object touched, and the other via the network return-circuit system and the accidental

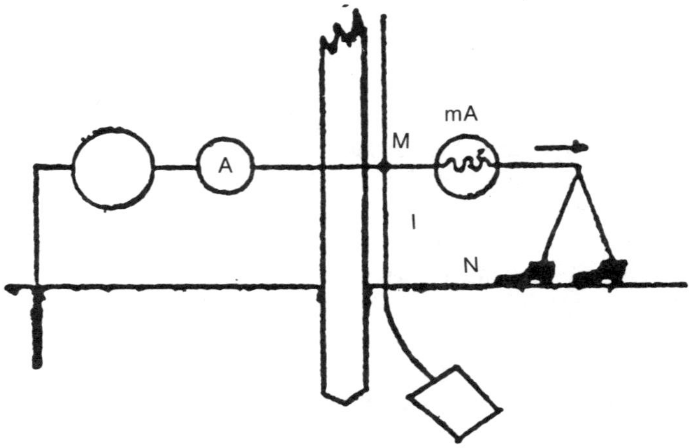

Fig 41
Measurement of Current Shunted by a Human Body Between M and N

contact between this and the object. In many cases, this additional term will be negligible and we shall get approximately:

$$i = \frac{U}{3000 + 1.5\rho} \text{ amperes.}$$

It will be seen that, in all cases, in order to determine directly the current shunted, without having to measure separately U and ρ, it is sufficient to substitute for the person a voltmeter having an internal resistance approximately that of the human body and in contact with the ground through soles representing the feet, and to read off the current i flowing through this voltmeter when a fault current I is made to flow to the earth electrode (Figure 41).

c) *Examples*

1) Assume a one-meter bare earthing rod carrying a current of 5 amperes in ground with a resistivity of 100 ohm-meters. A person touches the top of the rod. What current will flow through him? The potential at the rod in relation to distant earth will be about 400 volts. The potential between the top of the rod and the ground at a distance of one pace will be some 300 volts. The resistance of the two feet resting on the ground will be about 150 ohms. The resistance of the circuit comprising the two feet in series with the earth rod will be about 200 ohms. The current shunted will thus be approximately:

$$i = \frac{300}{3000 + 200} \quad 0.094 \text{ ampere}$$

assuming a resistance of 3000 ohms between hand and soles. This contact may be fatal if conditions are unfavorable.

2) Assume an electrode buried horizontally, and carrying one ampere per meter, in ground with a resistivity of 200 ohm-meters. What will be the current

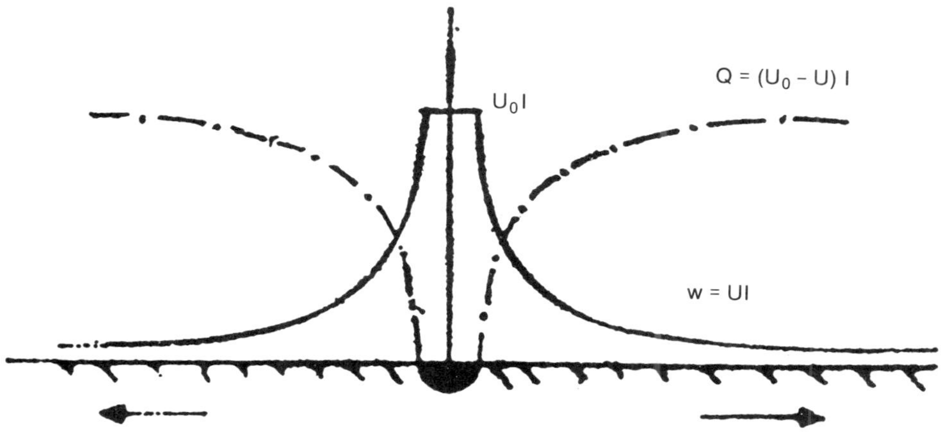

Fig 42
Radial Flow of Electric Energy and Heat

shunted from foot to foot or between neighboring ground and a structure connected to the electrode? The maximum voltage from one foot to the other in the neighborhood of the conductor will be 0.10 to 0.15 ρi, that is, 20 to 30 volts. The current from one foot to the other will have the maximum value:

$$i = \frac{30}{3000 + 1200} \quad 0.007 \text{ ampere.}$$

This is quite harmless.

The potential between the feet and hand might reach 0.6 to 0.8 ρi, that is, 120 to 160 volts. The corresponding current will be about

$$i = \frac{120 \text{ to } 160}{3000 + 150} \quad 0.038 \text{ to } 0.050 \text{ ampere.}$$

Contact in this case may occasionally be dangerous.

The instances met with in practice vary far too much for it to be possible to give general rules. The simplest method is often, as mentioned in paragraph III-2, to cause a current to flow to the earth electrode, and to check the potentials between points of the installation which are liable to be touched simultaneously. Local improvements can then be effected at the points considered the most dangerous.

Appendix J
Grounding Methods for High-Voltage Stations With Grounded Neutrals

(Erdungsmassnahmen fur Höchstspannungsanlagen mit Geerdetem Sternpunkt)

Walter Koch.* Elektrotechnische Zeitschrift.
Vol. 71, No. 4, pp. 89–91, Feb. 1950

It is not economically feasible to provide grounding in high-voltage stations with grounded neutral, which will in toto limit contact potentials to ground electrodes and the connected apparatus to less than 125 volts. One has to deal with a multiplicity of potentials which may be established between the plant and the surroundings under short-circuit conditions.

Experiments with models show that by making the ground system in the form of a grid, areas within the system can be produced which will be safe. Means for safe entry into the grounding area will be given.

With a directly grounded neutral point there flows into the system at the fault point the so-called ground-fault current instead of the total single-phase short-circuit ground current (Ungrounded System). This ground-fault current depends upon the generating capacity of the power plants in the area and on the impedance of the ground circuit. The grounding systems of a solidly grounded network will carry a portion of the ground-fault current which may be a minimum for faults a great distance from the station and may be a maximum, namely the total ground-fault current, for a fault in the station.

While the grounding systems may be adjusted to eliminate dangerous contact potentials by suppression of ground short-circuit currents, this is not usually demanded of solidly grounded neutral systems because it does not appear to be practicable. For ground-fault currents above 1000 amperes, grounding systems of vast dimensions must be installed in order to meet the usual 125-volt contact-potential requirement. A numerical example will show this. The surface area of an outdoor substation may be 250 · 250 meters. Here one has the possibility of placing a ground plate of 62,500 square meters under the station. With an average

*English translation by T. W. Stringfield. Some portions on Petersen coil systems and German (VDE) regulations omitted.

ground resistivity of 100 ohm-meters and the equivalent circular plate diameter of 280 meters the ground resistance is

$$R = \frac{\rho}{2D} \text{ or } R = \frac{10{,}000}{2 \cdot 28{,}000} = 0.18 \text{ ohm}$$

With such a ground, a ground-fault current of 5000 amperes will produce a 900-volt potential above the more-distant surroundings which is many times the potential allowed by VDE. In spite of this, it has the indisputable advantage that the entire station on this metal plate will have no potentials between parts within itself that are worth mentioning. For persons inside the station there will not be the slightest danger from undue contact potentials at such a high current. There would be danger only if at the moment of fault one were to enter or leave the plant or touch it from the outside. It is not practical to construct such a ground plate. However, in order not to endanger the personnel of an electrical plant, ways must be sought to fulfill this requirement.

Besides the dangers to personnel, there will be some to the material of the control and communications equipment if it is not provided against. The sheaths of the control cables provide a connection between the controlled apparatus in the high-voltage bays and the control point. Thereby, a fault to ground in the station can cause a very large current to flow through the sheath and melt it. Communication cables which leave the plant will also conduct ground currents away since intentionally or unintentionally they come into contact with building construction parts. Thereby, the sheaths acquire the high potential of the station in their vicinity while the conductors approximate the potential of the more-distant surroundings, so that insulation failures may occur. So likewise the cables of the low-voltage plant, and the windings of control motors among others may be endangered by large potential differences. Indeed, for these reasons it is not permissible to rely only on a sufficient interconnection of all apparatus such as circuit breakers, transformer cases, frame parts, etc. To this all cable sheaths within the plant must also be connected; so likewise the control mechanisms in the switching station to which the control cables are connected. Basically the entire plant should be provided with a built-up ground mat for the ground-fault current, to which all equipment parts in the plant are connected. So likewise the existing neutral conductors of independent low-voltage systems should be tied to the ground mat. By this method there will be the least worry that significant potential differences will arise between the accessible metallic parts of the plant and the plant equipment so protected will be safe from failure.

Now, it is certain that considerable and therefore dangerous potentials can arise between the soil, the floors of buildings on one side and the metallic parts of the plant during the time of faults. Therefore one must also consider the safety of operating personnel who in the course of their work must touch such metal parts. For this purpose the operating position may be provided either with an insulated floor capable of withstanding the high potential or with a metallic grid in the floor and tied to the ground mat or provided with both. Such metallic foot grids have been previously used for protection in Siemens-Schuckert plants with ungrounded star neutral. They consisted of small meshed wire netting cemented into the floor and tied to the grounding system, and provided absolute protection to persons standing thereon and grasping operating controls in that a highly conducting shunt path was provided between hands and feet.

As mentioned in the introduction, a large metallic plate is a suitable protection against all step potentials and contact potentials within the plant. Since such a metallic plate installation is not realizable, the question arises on how far one can go in substituting a network of ground straps and the necessary mesh spacing in order to obtain tolerable potential differences.

The investigation of the potential distribution of complicated ground electrode arrangements, which such a ground mat is, is not possible by computation, since one can derive formulae only for simple electrode shapes and even simple combinations of these electrode shapes are not amenable to calculation. For mesh-type electrode arrangements with irregular depth of burial which is the way they are used for the purpose of potential control and other complicated grounding structures, one is led to the use of models. For this purpose such model measurements using an electrolytic tank were undertaken. A metal container filled with a conducting solution served as the semi-infinite space for the current diffusion. Figure 1 shows the circuit of the test arrangement. The potential distribution for model M can be obtained by a null method using the electrode S, the calibrated potentiometer P and telephone receiver T. In order to reduce the electrolytic effect of the chopped direct current supply on the model, a slowly rotating switch U was placed in the direct current supply leads.

The model of the ground mat consisted of a copper wire 0.2 millimeters in diameter arranged in a square with 120 millimeter sides and set on the surface. For the usual ground straps with a cross-section of 30 · 3 millimeters corresponding to an equivalent diameter of 23 millimeters this model represented a replica of a ground system with a length of

$$\frac{23}{0.2} \cdot 120 = 13800 \text{ millimeters}$$

or 13.8 meters on a side. After obtaining the potential distribution, the square was subdivided to contain four squares by the addition of a wire cross, the four subsquares were similarly subdivided until 64 subsquares were attained and in each case the potential inside the square was measured. As the mesh becomes finer the effect approaches a plate electrode. In Figures 2 through 5 the potential at the center point of each square is given in percent of the potential of the ground mat. The potential differences which characterize the step potentials and thereby the hazard are according to these figures for fine-mesh electrodes 11-20 percent of the total potential. The mesh spacing of the mat with 64 meshes is, according to the above-mentioned model scale, 13.8/8 = 1.7 meters. The potential distributions in cross-sections through the mats at A-B, C-D, E-F and G-H of Figures 2 through 5 are shown in Figure 6. In order to determine the effect of only a partial fine mesh inside the outer edge, the arrangement shown in Figure 7 was investigated and as shown in Figure 8 with further subdivision of a single mesh. From this it follows that in the area of a fine mesh the same relations (proportions) hold as in the complete meshing of the total grounding area. The still finer subdivision of a

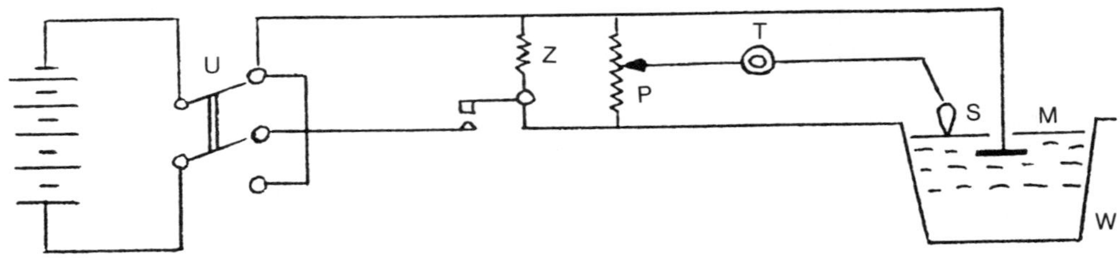

M = GROUND SYSTEM REPLICA
P = POTENTIOMETER
S = ELECTRODE
T = TELEPHONE RECEIVER
U = ROTATING SWITCH
W = ELECTROLYTIC TANK
A = INTERRUPTER (CHOPPER)

Figure 1.
Circuit for Obtaining Potential Distribution

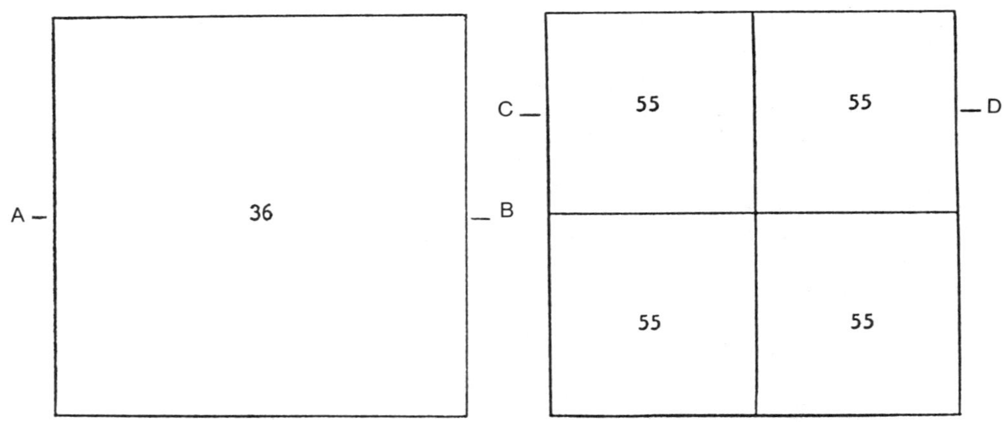

Figure 2. **Figure 3.**
Measured Potential Distribution for Various Ground Mats

single mesh results in a further raising of the potential inside the mesh, that is, a corresponding decrease of the potential differences and thereby the step potential.

The measurements show, as might be expected, that by using a fine mesh a considerable reduction in potential differences within the mat area can be obtained. Further, it is apparent that small protected areas can be produced by partial matting without completely matting the entire grounding area. Practical application of such finer meshing can be found principally in outdoor stations in the neighborhood of accessible equipment where the hazard is greatest.

A reduction of the effect, which will not completely eliminate potential differences, can be arrived at by a fill of course grit (gravel) to a depth of about

70	77	77	70
77	80	80	77
77	80	80	77
70	77	77	70

(E — F)

80	83	85	86	86	85	83	80
83	87	88	88	88	88	87	83
85	88	88	89	89	88	88	85
86	88	89	89	89	89	88	86
86	88	89	89	89	89	88	86
85	88	88	89	89	88	88	85
83	87	88	88	88	88	87	83
80	83	85	86	86	85	83	80

(G — H)

Figure 4. **Figure 5.**
Measured Potential Distribution for Various Ground Mats

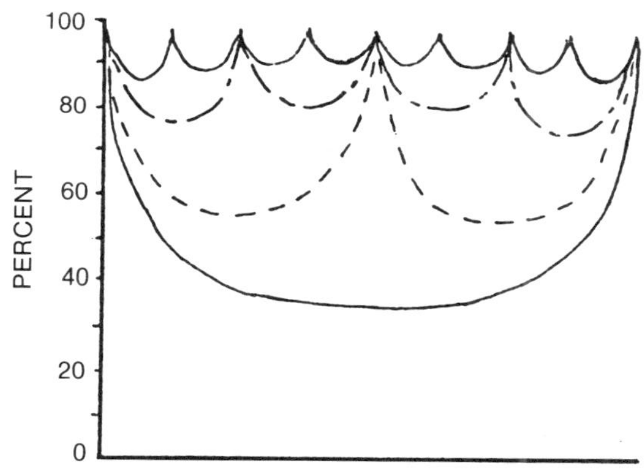

Figure 6.
Potential Distribution for a Ground Mat with Various Mesh Densities
Ground Mat Potential = *100 Percent*

1 centimeter over such a ground grid. With this everything practical has been done in order to minimize the hazard, if not to eliminate it entirely.

To be sure, there remain the locations of the passageways to the protected areas which remain a hazard when traveling over them during the time of a fault. Figure 6 shows the high potential drops at the edges of the wide meshed areas, where step potentials of about 45 percent of the total potential to the ground electrode can be encountered. If one must obtain absolute safety, then on the passageways one must resort to the so-called potential ramps in order to obtain a

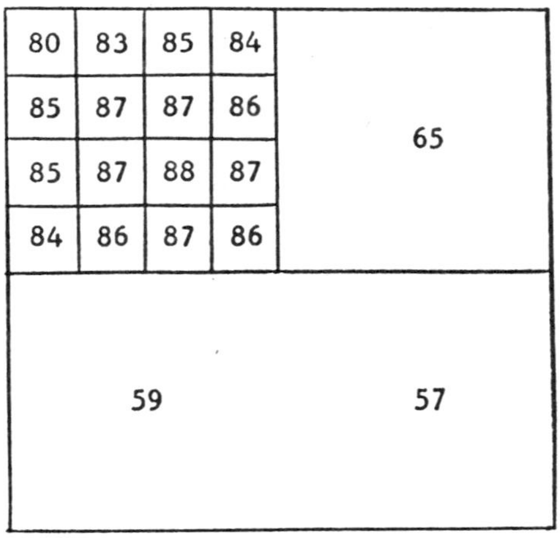

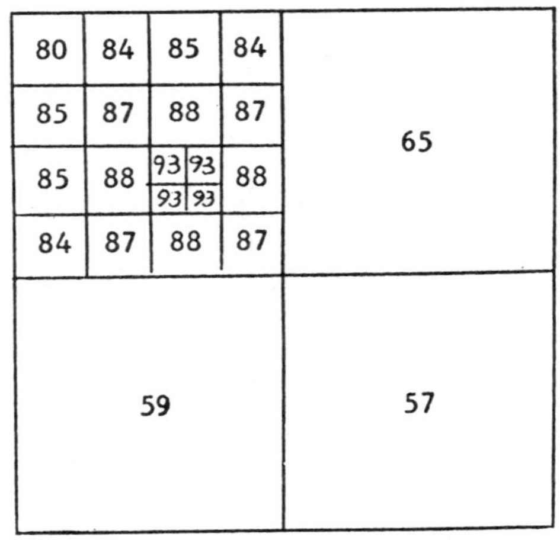

Figure 7. **Figure 8.**
Potential Distribution for Ground Mats with Fine Meshes in Portions

small, and as far as possible, uniform potential drop. Wooden passageways have likewise already been used in the Siemens-Schuckert works in 200-kv stations.

The means of potential control through grounding straps buried at progressively deeper depths is shown in principle in Figure 9, the effectiveness of which was proved by the leveling off of the potential surface in a model. Figure 10 shows the application of potential control around the footing of a tower when one does not desire to, or is not able to, employ a fence.

The magnitude of the expected step potential for a ground mat depends upon the ground resistance, the short-circuit current and the mesh density. If one takes the area of an outdoor substation 250 meters square, then a ground strap around the periphery will be 1000 meters long. Without regard to the cross connections and matted grounds, the ground resistance of this strap is $R = \frac{\rho}{\pi L} \ln \frac{2L}{d}$; where ρ is the ground resistivity (generally 100 ohm-meters) L is the length of the strap in centimeters and d is the equivalent diameter of the strap as a conductor with a semicircular cross-section (for the usual ground-strap d = 2.3 centimeters). With these figures R = 0.36 ohm. The resistance is thus only twice as great as for a solid plate 250 · 250 meters. The resistance will be reduced by the cross-connections which are required for tying in the apparatus to be grounded.

With a short-circuit current of, for example, 5000 amperes, the voltage to the ground system will be about 1800 volts. With a ground rating as shown in Figure 5 the greatest step potential to be expected will be about 11–12 percent of this value, or 200 volts, the effect of which on persons can be reduced effectively by using gravel fill. According to Figure 8, with a mesh spacing of 0.85 meters the potential inside the mesh is 7 percent of the ground mat potential and for a ground mat potential of 1800 volts the step potential can thereby be reduced below 125 volts if necessary.

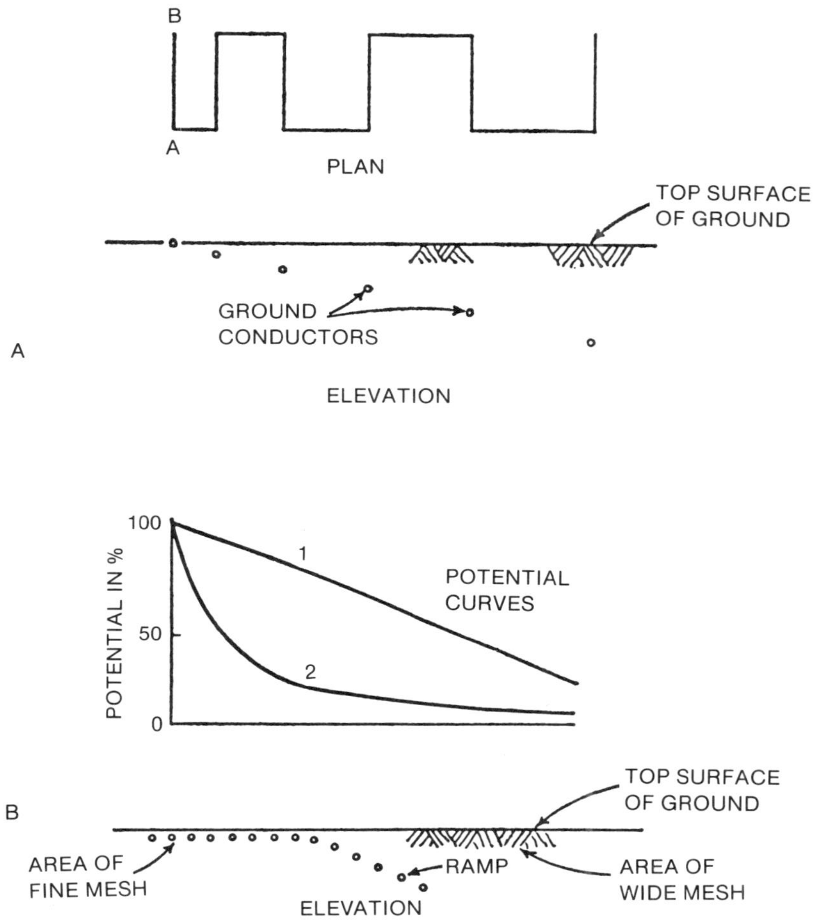

Figure 9.
Potential Distribution in a Ground Mat with Ramp (Curve 1) and without Ramp (Curve 2)

The systematic application of the protective measures described makes the separation of the operating ground from the protective ground superfluous. The separation of operating and protective grounds gives no protection for faults inside the station and from experience these must be considered. The installation of a separated star neutral ground system requires a tremendous amount of land outside the station. There is no advantage worth mentioning for this since a protective ground is still required inside the station. It therefore can only be recommended that the star-neutral point be connected to a suitable ground system as described in the foregoing or otherwise for a separate grounding system to employ the requisite materials for an ample development of the protective ground system.

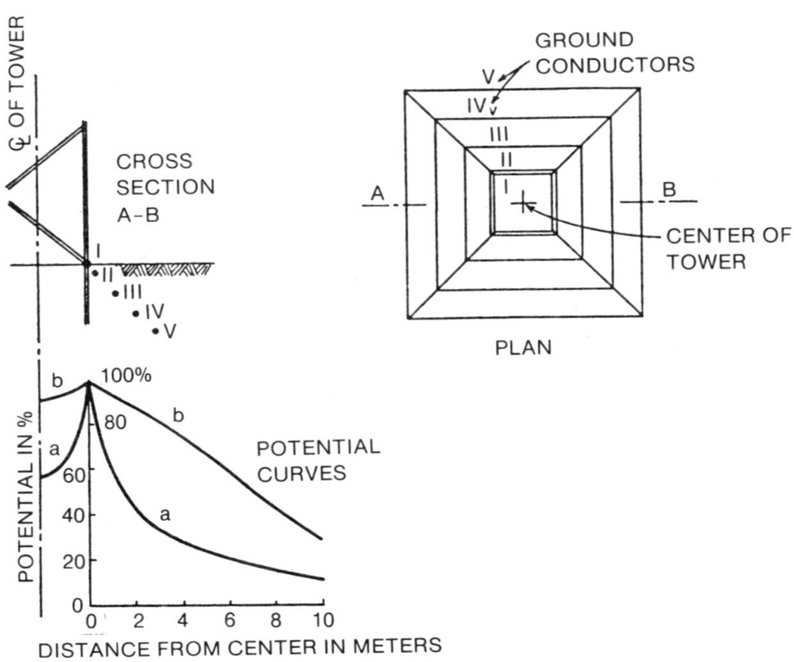

Figure 10.
Potential Distribution Around a Mast Footing in the Direction A-B for a Mast with Ramp (Curve b) and without Ramp (Curve a)

Tying together both systems (protective and operating) has the noteworthy advantage that for a ground fault within the station the ground-fault current component of the faulted station need not be carried by the ground mat but is conducted directly over the grounding conductors which are tied to the star neutral point. Also, one has only to reckon with the difference between the total ground-fault current and the station component, whereby there is a considerable reduction in ground mat potential and step potential.

The overhead ground wire of the outgoing station transmission lines may be advantageously connected to the station ground, and effectively reduce the total ground resistance; this is especially so where the ground wire which appears to be necessary for star neutral grounded systems with high ground-fault currents is of ample design.

SUMMARY

Large contact and step potentials under fault conditions must be considered in high-voltage stations using grounded star neutral point. Potential differences which may endanger cable insulation and low-voltage apparatus and facilities (for example, windings of control motors) may be eliminated by metallic interconnection of equipment housing, sheaths of control and service cables and their neutral conductors, and the construction parts in the control house. For protection of personnel at the danger points, narrow meshed ground mats with mesh spacing of about 1 meter will serve. The potential distribution of such ground mats may be investigated by means of electrolytic tanks. A separate operating ground for the star neutral point is not recommended, since connection of the latter to a general ground system, designed according to the viewpoint outlined herein, has advantages over separation. Approaches to parts of the ground system which have potential control can be made safe by the so-called potential ramps.

Index

D

E

F

G

H

I

J

K

L

M

N

O

P

R

S

T

V

W

X

Z